AF778976

Cyberology

This book provides an insightful examination of cybersecurity threats and their implications across various sectors, including healthcare, agriculture, and government. It covers fundamental concepts in cybersecurity, including cybercrime laws, cyberbullying, and emerging technologies. It discusses challenges in detecting and preventing cyber threats and analyzes trends in cyber warfare and resilience. This book serves as a resource for understanding the complex relationship between technology and society, equipping readers with the knowledge needed to navigate the evolving cyber landscape. It is for scholars, students, and professionals interested in cybersecurity and challenges.

This book

- introduces the fundamental principles and theories of cybersecurity, focusing on the relationship between humans and technology;
- details various cyber threats affecting the digital world, healthcare, and agriculture, emphasizing their real-world implications;
- discusses issues like cyberbullying, threats in cloud computing, and vulnerabilities in cyber-physical systems;
- examines the difficulties in detecting cybercrimes, addressing advanced persistent threats, and fostering cyber resilience;
- analyzes current trends in cyber warfare and terrorism, providing insights into new security challenges.

Cyberology

An Optimized Approach to the Cyber-World

Edited by

Mohd. Shahid Husain, Mohammad Faisal,
Halima Sadia, Tasneem Ahmed, Saurabh Shukla,
and Adil Kaleem

CRC Press
Taylor & Francis Group
Boca Raton London New York

CRC Press is an imprint of the
Taylor & Francis Group, an **informa** business
A CHAPMAN & HALL BOOK

Designed cover image: ShutterStock

First edition published 2025
by CRC Press
2385 NW Executive Center Drive, Suite 320, Boca Raton FL 33431

and by CRC Press
4 Park Square, Milton Park, Abingdon, Oxon, OX14 4RN

CRC Press is an imprint of Taylor & Francis Group, LLC

© 2025 selection and editorial matter, Mohd. Shahid Husain, Mohammad Faisal, Halima Sadia, Tasneem Ahmed, Saurabh Shukla, Adil Kaleem; individual chapters, the contributors

Reasonable efforts have been made to publish reliable data and information, but the author and publisher cannot assume responsibility for the validity of all materials or the consequences of their use. The authors and publishers have attempted to trace the copyright holders of all material reproduced in this publication and apologize to copyright holders if permission to publish in this form has not been obtained. If any copyright material has not been acknowledged please write and let us know so we may rectify in any future reprint.

Except as permitted under U.S. Copyright Law, no part of this book may be reprinted, reproduced, transmitted, or utilized in any form by any electronic, mechanical, or other means, now known or hereafter invented, including photocopying, microfilming, and recording, or in any information storage or retrieval system, without written permission from the publishers.

For permission to photocopy or use material electronically from this work, access www.copyright.com or contact the Copyright Clearance Center, Inc. (CCC), 222 Rosewood Drive, Danvers, MA 01923, 978-750-8400. For works that are not available on CCC please contact mpkbookspermissions@tandf.co.uk

Trademark notice: Product or corporate names may be trademarks or registered trademarks and are used only for identification and explanation without intent to infringe.

ISBN: 9781032526683 (hbk)
ISBN: 9781032529851 (pbk)
ISBN: 9781003409571 (ebk)

DOI: 10.1201/9781003409571

Typeset in Times
by KnowledgeWorks Global Ltd.

Contents

Preface

In today's rapidly evolving digital landscape, the interaction between humans and technology has grown increasingly intricate, becoming an indispensable aspect of our daily existence. We've experienced an unprecedented proliferation of interconnected systems that have deeply altered societies, economies, and even our own sense of self. As we navigate this dynamic and swiftly changing environment, it's becoming progressively imperative to grasp the intricate relationship between humans and technology.

Titled *Cyberology: An Optimized Approach to the Cyber-World*, this book delves into the exploration of the relationship between humans and technology, uncovering the myriad dimensions of this intricate bond at the fascinating crossroads of human psychology and technology. Esteemed experts in the realm of cyber science have had the privilege of observing the evolution of technology and its profound impact on human life over many years. From the advent of the Internet to the emergence of artificial intelligence, we've witnessed how these advancements have influenced the behavior, perceptions, and interactions of individuals.

Encompassing a broad array of subjects, ranging from cyber-world issues and challenges to the ethical and legal considerations stemming from novel technologies, this book endeavors to furnish readers with a comprehensive understanding of cyberology.

The book's chapters offer an extensive examination of various dimensions of cyberspace, meticulously analyzing its vulnerabilities and threats. It delves into the intricate landscape of vulnerabilities and cybersecurity issues, elucidating their ramifications on both individuals and organizations. Furthermore, the book provides valuable insights into a diverse array of tools and techniques aimed at mitigating these threats effectively.

Moreover, it navigates through the challenges posed by the perpetually shifting paradigm, driven by advancements in domains such as artificial intelligence, robotics, Internet of Things (IoT), and pervasive healthcare. These advancements necessitate a thorough exploration of strategies to address these challenges across sectors, including the food industry, healthcare, smart cities, agriculture, and sustainable development.

Additionally, the book prognosticates future directions, offering a glimpse into emerging trends like cyberbullying, cyber terrorism, and digital forensics. It advocates for the utilization of nature-inspired optimized approaches to mitigate risks in the cyber-world, thereby enhancing its security and productivity. Through this comprehensive exploration, the book endeavors to provide readers with a nuanced understanding of cyberspace complexities and equip them with the requisite knowledge and tools to navigate this dynamic terrain effectively.

EMERGING NEED

As we step into the twenty-first century, our reliance on technology escalates rapidly. Consequently, there's a growing demand for individuals well-versed in the intricacies of the digital realm. Cyberology emerges as the field tasked with bridging the gap between the tangible and virtual worlds, aiming to harness the potential of the digital age while comprehending the associated risks.

This book serves as an expedition into the realm of cyberology, furnishing readers with an accessible and thorough roadmap of discipline. Crafted by field experts, it elucidates fundamental concepts, theories, and practical applications within cyberology. It traverses through crucial subjects, including cybersecurity, artificial intelligence, data management, and digital ethics. Furthermore, it delves into the broader societal implications of this interdisciplinary domain and scrutinizes the challenges inherent in our journey through the digital epoch.

PURPOSE

This book aims to offer readers a thorough investigation into cyber science, presenting insightful analysis and practical perspectives on technology's significant influence on humanity. Adopting an interdisciplinary stance, each chapter draws upon diverse expertise and research from fields, including business, agriculture, healthcare, psychology, sociology, computer science, and economics. Through this multidisciplinary lens, the book delves into various facets of the intricate relationship between humans and technology, unraveling its complexities and scrutinizing its ramifications. Each chapter is dedicated to exploring distinct dimensions of this symbiotic interaction, shedding light on its multifaceted nature, and exploring the far-reaching implications it holds.

AUDIENCE

This book is essential reading for anyone interested in the possibilities of the digital future, and for those who want to gain a better understanding of the complexities of the cyber-world. It is a valuable resource for students, professionals, and anyone interested in the rapidly expanding field of cyberology.

We hope that this book will prove to be an illuminating and thought-provoking guide to the field for all readers.

EMPHASIS

This book serves as a thorough examination of the captivating and continually evolving realm of cyberology. Within its pages, readers will embark on a journey to grasp the fundamental principles of cyberology and its far-reaching implications, alongside its practical applications in real-world scenarios. Delving into cyberology's diverse applications across sectors like business, government, and education, the book offers detailed insights into its multifaceted role.

Moreover, ethical and legal dimensions of cyberology are meticulously explored, emphasizing responsible and ethical utilization of its tools and methodologies. Additionally, readers will benefit from the latest research findings and data within the field, along with practical tips and strategies for navigating the cyberological landscape successfully.

This book stands as an indispensable resource for those intrigued by cyberology, serving as both an introductory primer and an extensive manual in this rapidly expanding domain. Its wealth of knowledge makes it an invaluable asset for students, professionals, and researchers alike, promising to enrich their understanding and proficiency in cyberology.

WHAT YOU WILL LEARN

Alongside delving into the fundamental elements of cyberology, this book presents valuable perspectives on its diverse applications. Readers gain practical illustrations showcasing how cyberological principles manifest in daily life, fostering a comprehensive comprehension of the digital era's potential.

BOOK ORGANIZATION

The book explores the growing role of technology in shaping our interactions, encompassing communication, education, and employment. It comprises 20 chapters, each addressing various facets of cyberology.

Chapter 1 deals with ***Changing Paradigm in AI Application and Its Impact on Cybercrime: Review of Major Global Events.*** The convergence of artificial intelligence (AI) and cybersecurity represents a transformative leap forward in combating the ever-evolving landscape of cyber threats, enhancing the need to strengthen the overall defense systems of nations. This synergy automates analysis, identifies patterns, predicts potential vulnerabilities in large datasets, and employs real-time monitoring to mitigate risks. As a global concern, collaborative research becomes imperative to stay ahead, recognizing the potential havoc that AI can wreak if exploited by hackers. Drawing insights from industry, academia, and the foreign sector, this chapter explores challenges, opportunities, and potential ventures in the science and technology domain propelled by AI. Furthermore, this chapter delves into the measures, remedies, and strategies different countries have implemented encompassing technical solutions, policies, and best practices designed to detect, respond to, and prevent cyber threats. The surge in Internet-connected devices and escalating digital data transactions, coupled with increased reliance on cloud computing, has expanded potential risks. Interconnected activities of individuals, businesses, and governments underscore the critical need to safeguard digital systems. Cybersecurity assumes significance as it protects against hacking, unauthorized access, data breaches, and intricate cyberattacks. Cybersecurity approaches of some of the largest and most powerful countries such as the European Union, the United States of America, China, Israel, Russia, and India have been studied to mark the importance of 'Global Cooperation' in addressing cross-border cyber threats.

Chapter 2 deals with the ***Transformative Impact: Social Robots Redefining Industries and Services.*** When societies are changing rapidly owing to technological breakthroughs, the investigation of social robots as agents of positive change is becoming increasingly relevant. Acknowledging the encouraging results seen in the field of education and healthcare, there is a sincere desire to highlight the revolutionary possibilities of social robots, while also stressing the importance of ethical concerns and a human-centered approach. Socio-robots, often referred to as social or humanoid robots, are sophisticated robotic devices created to engage socially with people. Artificial intelligence, natural language processing, and other cutting-edge technologies have been built into them, allowing them to interact with humans through both verbal and nonverbal clues. Socio-robots are created to perform a variety of functions, including offering emotional support, engaging in social contact, and imparting knowledge. Service robots continue to permeate and automate the hospitality sector. The use of service robots in hotels has increased significantly in various countries. However, studies have examined how service robots detect and respond to human emotions, as one of their fundamental characteristics. The market size for service robots in the healthcare and hospitality sectors is projected to grow by 942 million USD during 2020–2024.

Chapter 3 deals with the ***Strategies for Cyber-Resilience: Combatting Advanced Persistent Threats in the Digital Age.*** The world of cybercrimes is a complicated place with ever-increasing obstacles to detection, prevention, and mitigation. Malicious actors have been using clever strategies to counter technological advancements, so it is important to have a greater awareness of critical concepts like cyber-resilience and advanced persistent threats (APTs). The enemies' sophistication is evident in the field of detection and prevention. These fraudsters easily get beyond conventional security measures by using cutting-edge tactics. Malware that can change its code in order to evade detection using signatures is an example of the difficult challenges that need to be overcome. Furthermore, encryption presents a special problem since, although protecting lawful activity, it also hides criminal intent. These worries are exacerbated by insider threats, which occur when authorized staff members use their access for improper purposes. The use of zero-day vulnerabilities to commit cybercrimes increases the need for efficient detection and quick action. A subcategory of cybercrimes known as APTs adds another layer of complication. These attacks continue to be persistent and subtle over extended periods of time. Their targeted nature necessitates specialized defenses above and beyond general security measures since they are frequently driven by political or competitive gains. Cyber-resilience emerges as a crucial concept in this context. It becomes essential to be prepared, which is demonstrated by having strong incident response plans, regular training, and updates. Strategic segmentation creates barriers against breaches by isolating vital systems. Strategies for backup and recovery, such as offline backups, lessen the damage caused by ransomware. Real-time insights for proactive defense are provided via cooperative threat intelligence sharing and continuous monitoring. A complete strategy that incorporates sophisticated detection systems, proactive threat intelligence, personnel education, and a watchful incident response plan is required to combat cybercrimes. Organizations must emphasize cyber resilience due to the constant growth of cyber threats, which creates an environment that can effectively withstand and recover from attacks.

Chapter 4 deals with the ***Unveiling the Digital Trails: A Survey of Contemporary Digital Forensics Tools and Techniques***. Although digital forensics techniques can be useful in civil and administrative matters, it is especially important in criminal investigations to make a strategic choice when deciding which tools to deploy. Human psychology is intrinsically linked to the field of digital forensics. The growing importance of digital evidence in both criminal and civil proceedings has prompted the creation of specialized techniques and approaches for its proper handling and interpretation. The importance of digital forensics today has been highlighted by its use in fields as varied as national security, business investigations, intellectual property protection, and civil litigation. The ever-evolving nature of technology presents various difficulties for digital forensics. Data capture and analysis are made more difficult by the prevalence of new devices, operating systems, and software. Digital evidence is difficult to access and analyze because of encryption and other data security techniques. Investigative needs must be balanced with the rights and interests of individuals and organizations in digital forensic investigations making privacy and data protection crucial ethical considerations. This chapter studies the rapidly evolving landscape of digital forensics, shaped by the continuous advancement in digital technologies and the increasing sophistication of cybercrimes. The aim is to provide a comprehensive and up-to-date survey of the field, highlighting not only the tools and techniques but also the critical role digital forensics plays in legal and cybersecurity contexts.

Chapter 5 deals with ***Issues and Challenges in Emerging Cyber Forensics and Traditional Cyber Forensics Methods.*** Cyber forensics plays an important role in identifying and reducing cybercrimes in the age of the digital world. This chapter discusses the fundamentals of gathering, storing, and evaluating digital evidence, and creates a baseline. The chapter identifies the challenges that emerge because of emerging technologies like blockchain, the Internet of Things (IoT), and encrypted communications channels. This chapter also provides insights into the complexity that arises because of emerging areas and how cyber forensics professionals are adapting to address them. This chapter also explores the legal and ethical dilemmas faced by cyber forensics experts as they strive to balance the imperative of evidence collection with the privacy rights of individuals. It shows the importance of adhering to legal frameworks and ethical guidelines while conducting any cyber forensics activity, emphasizing more on the need for constant awareness and adherence to evolving laws and regulations.

Chapter 6 deals with ***Optimized Approaches for Cyber Forensics Tools Applications Case Studies.*** This chapter presents innovative and optimized approaches for the application of cyber forensics tools through comprehensive case studies. The primary contribution lies in the development and validation of advanced methodologies that enhance the efficiency and effectiveness of cyber forensics investigations. Leveraging cutting-edge techniques, our proposed approaches address critical challenges in digital forensic analysis, such as data acquisition, preservation, and analysis, with a focus on minimizing processing time and maximizing accuracy.

Through meticulous case studies, we demonstrate the practical application of our optimized tools in real-world scenarios, showcasing significant improvements over traditional methods. Our results highlight the superior performance of the proposed cyber forensics approaches, including faster identification of digital evidence, reduced false positives, and enhanced overall investigation outcomes. Furthermore, the research explores the adaptability of these optimized tools across diverse cybercrime contexts, reinforcing their applicability and reliability in various scenarios. This study not only contributes to the advancement of cyber forensics methodologies but also provides valuable insights for practitioners and researchers in the ongoing battle against cyber threats.

Chapter 7 deals with the ***Cyber Threats and Security Frameworks in Cloud Computing.*** Cloud computing has transformed the way businesses store, manage, and process data, providing unparalleled scalability, flexibility, and cost-effectiveness. However, the growth of cloud technology creates new cybersecurity difficulties and vulnerabilities. This chapter provides a thorough examination of cyber risks and security frameworks in cloud computing settings. Various sorts of cyber risks, including data breaches, malware, DDoS attacks, insider threats, and misconfigurations, are recognized and examined in the context of cloud computing. Furthermore, this study investigates current security frameworks and best practices for mitigating these vulnerabilities, such as encryption, access restrictions, authentication systems, and monitoring solutions. This study, which synthesizes insights from academic research, industry reports, and real-world case studies, provides valuable insights into the evolving landscape of cyber threats in cloud computing and makes practical recommendations for implementing effective security measures to protect cloud environments.

Chapter 8 deals with ***Cyber Threats in Cloud Computing: Unveiling Vulnerabilities and How to Strengthen the Defense.*** In recent years, the rapid adoption of cloud computing has transformed the digital landscape, offering unparalleled opportunities for businesses and individuals to leverage scalable, economical, and flexible IT resources. However, as organizations migrate sensitive data and critical operations to the cloud, the accompanying surge in cyber threats has become an extreme concern. Numerous security concerns, such as data security and monitoring cloud usage by cloud computing providers, surround cloud computing. How can cloud computing users be sure that there are no security or availability problems with their information? This study attempts to pinpoint the most susceptible security risks in cloud computing so that vendors and end users can be informed about the main security risks related to cloud computing. Through our work, researchers and security experts will be able to understand the concerns of users and vendors and provide effective strategies to fortify the defenses. This chapter provides a comprehensive exploration of cloud computing security, from its foundational principles to the emerging threats and future considerations. The chapter also explores emerging threats and future considerations in cloud security, including challenges posed by serverless and edge computing, quantum computing, AI-driven attacks, etc. Organizations can confidently use the practical approaches and best practices discussed in this chapter to maintain the integrity and security of their assets stored in the cloud.

Chapter 9 deals with ***A Comprehensive Analysis of Vulnerabilities and Cybersecurity Issues for Digital Agriculture and Food Industry.*** The advancement of agriculture to smart farming has increased cyber threats. Smart farming ecosystems are, however, exposed to a variety of cybersecurity threats and vulnerabilities due to the employing of IoT and smart communication technologies. Ransomware is the cyberattack that affects the food industry the most frequently right now. Such hacks may cause economic instability in countries that heavily rely on agriculture. As per reports, food inflation has risen to 4.5% which was earlier 3% in the year 2023. Moreover, this year was declared 'El Nino' which can result in the growth of the hunger index in poor nations. So, modern cultures' food supply chains are essential components. It employs network connections and is fully digitalized, just like other facets of life. Nevertheless, the widespread practice of technical know-how comes with built-in security dangers and weaknesses, and both industries are now under more scrutiny than ever. The chapter presents a complete study on the protection and confidentiality related to food industries and agriculture sector data.

Chapter 10 deals with ***Unveiling Critical Success Factors of Zero-Trust Cybersecurity Implementation in Food Industry from a Stage of Change Perspective.*** The predominant goal of the current research is to ascertain the critical success factors (CSFs) for zero trust cybersecurity implementation (ZTCI) in the food and beverage industry by utilizing the Capability, Opportunity, Motivation, and Behaviour Model (COM-B) and the transtheoretical model (TTM). Following the TTM's five steps—pre-contemplation, contemplation, preparation, action, and maintenance—respondents are categorized. The statistical information used in this study is compiled over a five-year period from repeated cross-sectional samples of respondents from food and beverage enterprises about the ZTCI. The findings show that a model of organizational readiness (OR) and environmental readiness (ER) has the best fit and validity because all the suggested factors seem to have the capacity to change the organization's stage. When considered collectively, the findings point to the necessity for tailored policy interventions to encourage behavioral adoption based not just on the organizational and environmental context but also on the usage stage in question.

Chapter 11 deals with ***Understanding and Mitigating Cyber Threats in Smart Fruit Crops Safety: An Analysis of Vulnerability and Best Practices for Securing.*** This chapter explores how susceptible fruit crops are to cyberattacks and offers the best ways to secure agricultural systems. The first component of the study looks at the several cyber threats that fruit crops face in today's technology environment. Data breaches, ransomware attacks, malware infections, and vulnerabilities in the supply chain are a few of these dangers. Due to the operation disruption and data compromise caused by such attacks, the quality and safety of fruit products are seriously threatened. To comprehend the fundamental weaknesses of agricultural systems, the authors of this study looked into the crucial elements that make fruit crops vulnerable to online threats. The findings point to three primary issues: poor cybersecurity infrastructure, insufficient personnel training, and a lack of knowledge about cyber threats. Additionally, the attack increases with increased adoption of automated technologies, such as IoT devices and smart sensors, necessitating greater defenses. The chapter also lists suggested actions and potential fixes to strengthen the cybersecurity defenses of fruit production systems. Implementing multi-layered security procedures, conducting frequent security audits, and keeping an eye out for threats in real time are the first steps in protecting critical data and production processes. Educating farmers and agricultural stakeholders about cyber threats and promoting a cybersecurity culture within the industry are the components of effective risk reduction. To further develop consistent cybersecurity standards specifically for the agriculture industry, collaboration between researchers, decision-makers, and industry stakeholders is required. These rules should ensure that fruit crops are resilient to evolving cyber threats and account for the unique challenges they face. The study paves the way for a secure and resilient agricultural future in which the benefits of technology are used while lowering the associated risks.

Chapter 12 deals with ***Cyber Resilience in a Digital World: Securing Healthcare, Agriculture, and Food Industry against Emerging Threats.*** We are thrust into a future where cyberattacks can have devastating real-world consequences as we explore this unexplored area and realize that the digital revolution has created a complicated network of dependency. Electronic medical records, telemedicine consultations, and connected medical devices have changed the medical industry's traditional view, characterized by the privacy of doctor-patient interactions and the reliability of written records. While healthcare systems are gaining importance, they also offer new opportunities to hackers. These criminals are exploiting vulnerabilities in connected medical devices to steal personal health information and potentially compromise critical medical procedures. Successful cyberattacks in this area can lead to serious consequences, including loss of patient confidence in medical care and, in extreme cases, death. Similarly, in agriculture, the combination of precision farming and Internet of Things (IoT) technology has enabled data-driven agriculture, good resource allocation, and increased yields. However, due to its dependence on the Internet, agriculture is still vulnerable to communications that can be harmful to global production, consumption and supply chains, metals, and food. Protecting agriculture from cyber threats is not only a financial issue but also important in terms of supporting the masses. While the food industry plays an important role in the world of food, it also faces unique digital challenges. Digitized products and improved business processes, including electronic production processes, make them vulnerable to cyberattacks. Any damage to products, contamination of network equipment, or damage to consumer confidence as a result of an attack could have a significant impact on public health, financial security, and international economic relations. This chapter is dedicated to a comprehensive review of cyber threats affecting major businesses. The aim is to use qualitative research, case studies, and expert opinion to discover the weaknesses exploited by cybercriminals, identify vulnerabilities in each operation, and measure the outcome of a successful attack.

Chapter 13 deals with ***Cyberbullying: Understanding, Preventing, and Addressing Online Harassment.*** Cyberbullying is a pervasive and concerning issue that has permeated the fabric of our interconnected society. It encompasses a range of malicious behaviors where individuals exploit the anonymity and accessibility of the online world to intentionally harm, harass, or intimidate others. The anonymity offered by the digital veil allows perpetrators to engage in hurtful actions without revealing their identities, leading to a sense of detachment from the consequences of their actions. The screen becomes a mask behind which they can unleash vitriol and cruelty, perpetuating an environment of fear and insecurity for the victims. The chapter investigates the intricate digital relationships that enable cyberbullying and how it progresses. In extreme cases, cyberbullying can result in emotional suffering, worse grades, social isolation, and self-harm or suicide. To address this problem, you must first understand its expressions. Hurting communications, cyberstalking, doxxing, impersonation, and exclusionary practices are examples. It also discusses the reasons for cyberbullying, illuminating the complexities of digital human behavior. People become more aggressive online due to anonymity and the impact of disinhibition. Cyberbullying can easily traverse boundaries because of advances in technology and social media. Lack of empathy, peer pressure, and societal expectations all contribute to cyberbullying.

Chapter 14 deals with ***The Perils of Cyberbullying on Social Media: Detection and Prevention with Artificial Intelligence.*** This chapter explores the potential of artificial intelligence (AI) in tackling cyberbullying, a prevalent social issue plaguing online platforms like Snapchat, TikTok, and Facebook. In the Internet age, cyberbullying is a pervasive social problem that has an enormous effect on people of all ages. Cyberbullying detection is a challenging and complicated problem. Cyberbullying can be hidden or disguised using a variety of methods, making it difficult for computer programs or human reviewers to detect. Machine learning and AI may open up many possibilities to prevent cyberbullying. AI-based systems can be trained to identify patterns and trends in cyberbullying behavior, even when it is hidden or disguised. AI can also be used to develop preventive measures, such as identifying and intervening with potential bullies before they begin cyberbullying others. Detecting bullying, harassment, and hate-related texts using supervised machine learning algorithms, and natural language processing techniques, AI-enabled cyberbullying, online social networks, human-AI collaborative approaches are some of the promising methods for using AI to detect and prevent cyberbullying. These strategies could all be successful in identifying and preventing cyberbullying. At the same time, AI-based systems are susceptible to bias and can be tricked by cunning cyberbullies. The chapter advocates for employing AI in conjunction with other strategies, such as social media responsibility campaigns and digital literacy education. This chapter also covers the hazards of cyberbullying on social media and how AI may be used to recognize and prevent it.

Chapter 15 deals with ***Cybersecurity Challenges and Countermeasures in Smart Cities for Sustainable Environment Development: A Comprehensive Analysis of Threats and Resilience Strategies.*** Rapid urbanization is occurring in urban areas of developing nations, which has led to problems with urban planning such as an inconsistency between supply and demand for infrastructure services and utilities, traffic jams, pollution, a loss of natural green cover, the typology of peri-urban settlements harming the natural and built environments and unreliable transportation, a lack of access to medical care, a lack of adequate water and electricity supplies, and other issues are all brought on by the rapid development in urbanization. Urbanization is a significant urban development process that is linked to resource consumption, economic activity, and population concentration in metropolitan areas and is always growing in complexity. The development of sensor technologies and the Internet of Things (IoT) has accelerated, making it easier to construct IoT-based smart cities. However, due to their close connection with people's everyday lives, the security issues associated with smart cities have become the top concern of the people amid the rapid market expansion of the city. Modern technology is utilized in the planning and implementation of thoughtful strategies for sustainable development, safe and secure communities, and equitable growth. Intelligent solutions that can accelerate economic growth by enhancing infrastructure and services, creating jobs, and raising the standard of living in smart cities are provided by digital technology and their interactions with physical systems.

Chapter 16 deals with ***Cyber Risk Management in IoT Security Technologies.*** More and more cyberattacks are occurring on the Internet of Things (IoT). Corporate operations, compliance, reputation, and the economy have been negatively impacted as a result. The surge in cyberattacks is a result of the widespread use of IoT sensors in industries such as smart relationships, medical monitoring systems, automated production, and logistics. By protecting IoT assets and user privacy, cybersecurity in IoT aims to condense cybersecurity threats that are present for both businesses and consumers. The development of new cybersecurity technologies is leading to both potential and issues in IoT cybersecurity management. The complex cybersecurity challenges associated with IoT systems cannot be addressed by a comprehensive risk management structure alone. Information on IoT privacy technology and security risk management strategies is included in this chapter. A four-layer structure for IoT risk administration for cybersecurity will also be discussed, as will the shortcomings of IoT risk administration for cybersecurity. Linear programming in this chapter discusses the optimal way to distribute funds among multiple IoT vulnerability projects. The demonstration of IoT cyber threat management for smart hotel accommodation is explained to support risk assessment through proof-of-concept validation.

Chapter 17 deals with the ***Landscape of Cyber Terrorism on Internet and AI-Powered Countermeasures.*** The combined power of the Internet and cyberspace connects the complete world on a single communication platform. Almost all physical communication turns into digital communication. Terrorist organizations exploit the potential power of the Internet and cyberspace to augment their activities in the form of cyber terrorism. In order to achieve the objective of terrorist organizations, cyber terrorism is employed to perform all their activities, such as planning, training, execution, communication, and fundraising. The purpose of this study is to create awareness about cyber terrorism so that differentiation can be provided between cyber terrorists and common people because nobody knows who is accessing the Internet. This study explores the awareness, threats, and roots of cyber terrorism on the Internet. This study explores the mechanism of fundraising for terrorist organizations, i.e., the fuel of cyber terrorism. This study critically examines secure communication, i.e., end-to-end encryption using social media by the terrorist organizations, and anonymous communication using the dark web, including sell and purchase of illegal things by the terrorist organizations. Finally, this study critically explores the solutions of artificial intelligence for the countermeasures of central investigation agencies to combat cyber terrorism.

Chapter 18 deals with ***Hybrid Secure Clustered Nature-Inspired Decision Support System for Secure Data Communication in Cyberspace.*** Industry Revolution 4.0 involves the integration of a wireless sensor network (WSN). WSNs are prone to attack due to their energy-constrained sensor nodes which minimizes the potential of these cyber networks. Protecting the WSN is a hotly contested study area for researchers. Therefore, implementing a powerful security algorithm is necessary to protect it from various cyberattacks. This study suggests a hybridized Secured Clustered Nature-inspired Decision Support System (SCNIDSS) to protect cyber networks from various attacks where a decision support system based on a nature-inspired approach has been implemented. The hybrid trust model is the main contribution to the proposed work. Further simulation has been carried out to validate the performance and the results of the proposed work are compared with the existing models in terms of various parameters and scenarios of attacks and densities. In Attack and Density Variations Scenario, the improved result shows less energy consumption 0.0583 (J), minimum delay 0.2212 (seconds), maximum throughput 37.76 (%), higher packet delivery ratio (PDR) 88.96 (%), minimum overhead 5.409 (%), and minimum energy consumption 0.0259 (J), less delay 0.2661 (seconds), increased throughput 70.225 (%), high PDR 79.675 (%), and less overhead 6.639 (%), respectively.

Chapter 19 deals with ***Transforming Network Protection via Open Innovation Strategies.*** Open innovation has become a crucial strategy for many businesses, especially in the technology and pharmaceutical sectors, to maintain their competitiveness and promote ongoing growth. Companies can access a worldwide network of talent and ideas through open innovation, promoting a culture of cooperation and accelerating invention at a faster rate than would be possible with old, closed models. Agile and adaptable security solutions have also emerged because of this collaborative environment, improving the capacity to identify and thwart sophisticated cyberattacks. Open innovation has promoted information exchange and skill growth, fostering the creation of a workforce capable of handling challenging security issues in the quickly changing technology environment.

Chapter 20 deals with the ***Emerging Role of Digital Library in Evaluating Copyright Regulations.*** The landscape of copyright regulation has undergone a dramatic change as a result of the rise of digital libraries. The influence of digital libraries on copyright regulation has increased more than ever in the current digital era, as knowledge and information are more readily available online. Digital libraries are now a potent force in the fight for a fair handling of copyright. They act as knowledge repositories, providing a substantial selection of digitized books, research papers, old documents, and numerous multimedia materials. This abundance of digital content has spurred crucial discussions about copyright reform and the necessity for intellectual property rights. Digital libraries must maintain a balance between the protection of intellectual property rights and achieving universal access to knowledge. Copyright laws are very important to writers and might impede the free flow of ideas. More easygoing copyright regulations are supported by digital libraries because they facilitate the digitization and distribution of works, which benefits both the public and writers.

In summary, *Cyberology: An Optimized Approach to the Cyber-World* beckons you to embark on a captivating exploration of the cyberological realm, where the lines between humanity and technology become indistinct, unveiling fresh opportunities. Let us venture together into the complex web of the human-technology dynamic, embracing both its challenges and its potential.

Acknowledgments

We are immensely grateful for the enriching journey this work has provided us in the realm of cyberology. Numerous individuals, both directly and indirectly, have contributed to the realization of this book. We extend our heartfelt appreciation to each person involved in the inception and culmination of this endeavor.

Our sincere gratitude goes to the CRC publication team, Taylor & Francis group with special mention to Isha Singh, whose unwavering encouragement and support were indispensable in bringing this book to fruition. Additionally, we express our thanks to esteemed technical experts: Dr. T J Siddiqui (Professor, University of Allahabad), Dr. Rashid Ali (Professor, AMU Aligarh), Dr. M Z Khan (Assistant Professor, Bisha University, KSA), and Dr. Nawazish Naveed (Assistant Professor, UTAS, Oman), for their invaluable contributions in timely, impartial reviews, proof-reading, and constructive feedback.

We are appreciative of all experts who generously dedicated their time to review submissions and provide critical insights. Gratitude also extends to our friends and colleagues for their support whenever needed.

Our acknowledgment would be incomplete without expressing deep gratitude and reverence to our parents and family members, whose unwavering support has been a constant source of strength throughout our endeavors. This book is dedicated to them, for it is their confidence in us that has shaped us into who we are today.

We extend our thanks to the contributors whose expertise and insights have enriched the content of this book. Likewise, we appreciate the readers whose curiosity and thirst for knowledge drive the pursuit of understanding.

Finally, we acknowledge the organizations and individuals who granted permission to utilize research materials and information essential for this book's completion.

Last but not least, we are profoundly thankful to the Almighty for showering blessings upon us throughout our journey.

Editors

About the Editors

Dr. Mohd. Shahid Husain is a Research Professional and Faculty member with 14 years of teaching and research experience. He is currently working as an Assistant Professor in the College of Computing and Information Sciences, University of Technology and Applied Sciences, Oman.

He obtained his M.Tech. in Information Technology (spl: Intelligent System) from the Indian Institute of Information Technology, Allahabad (www.iiita.ac.in), India. He received his Ph.D. in Computer Science & Engineering from Integral University, Lucknow (www.iul.ac.in), India. His areas of interest include Artificial Intelligence, Information Retrieval, Natural Language Processing, Data Mining, Web mining, Data Analytics, Sentiment Analysis, and Computer Networks & Security.

He has published 6 books, 15 book chapters, and more than 30 research papers in journals/conferences of international repute. He was involved with many sponsored projects as PI/Co-PI. Currently, he is working on two funded projects submitted to TRC (The Research Council), Oman. He is also contributing his knowledge and experience as an invited speaker, doctoral committee member, session chair, member of the editorial board/advisory committee, and TPC in various international journals/conferences of repute. Dr. Mohd Shahid Husain is an active member of different professional bodies, including IEEE Young Professionals, IEEE-TCII, ISTE, CSTA, and IACSIT.

He qualified UGC-NET (National Eligibility Test) in June 2014 and GATE (Graduate Aptitude Test in Engineering) in 2008.

He obtained two specialization certificates in Python and Applied Data Science from the University of Michigan in 2020.

Dr. Mohammad Faisal is currently working as a Professor and Head of the Department of Computer Application, Integral University, Lucknow, Uttar Pradesh, India. He has more than 17 years of teaching and research experience teaching courses in both undergraduate and postgraduate levels. His areas of interest are Software Engineering and Cyber Security. He has one patent, two authored and six edited books on contemporary trends in computer application. He has published quality research papers in journals and national and international conferences of repute. He has also been contributing his knowledge and experience as a member of the editorial board/advisory committee and TPC in various international journals/conferences of repute. He is also an active member of various professional bodies such as IAENG, CSTA, ISOC-USA, EASST, HPC, ISTE, IAENG, and UACEE.

Dr. Halima Sadia is an Associate Professor in the Department of Computer Science & Engineering, at Integral University, Lucknow. She has over 17 years of teaching and research experience. She has completed her Ph.D. in Computer Science & Engineering from Abdul Kalam Technical University, Lucknow, India. Her research interests include requirement engineering, cybersecurity, and blockchain in the healthcare sector. She has published multiple research articles in Scopus-indexed journals and national and international conferences of repute. Dr. Halima Sadia has also authored three books in the field of Computer Science & Engineering. She is passionate about mentoring the next generation of computer scientists and engineers and is contributing her knowledge and experience as a member of the editorial board, advisory committee, and technical program committee. She is an active member of different professional bodies, including ISTE, IAENG, UACEE, CSTA, ISOC-USA, EASST, and HPC-Houston.

Dr. Tasneem Ahmed received his Ph.D. degree in 2016 in image processing from Indian Institute of Technology (IIT) Roorkee, India. Currently, he is working as an Associate Professor in the Department of Computer Application at Integral University Lucknow, India. His research interests include digital image processing and computer vision, optical and microwave satellite image processing, image classification, data fusion, time series analysis, and SAR data analysis for land cover classification. He has edited 3 books, published 23 research papers in various SCIE and Scopus indexed journals and 19 conference papers in various national and international conferences/seminars/symposiums and 16 book chapters.

He is actively involved in R&D projects, Ph.D. supervision, organizing international conferences and events as a TPC member, editorial board member, and reviewer of various international journals, and involved in other key administrative tasks in the university.

Dr. Saurabh Shukla is working as an Assistant Professor, and Head of the Department, Computer Science at the Indian Institute of Information Technology Lucknow (IIITL). He has received his Postdoc from the University of Galway, Ireland in the Unit of Semantic Web, Data Science Institute (DSI), Insight SFI Centre of Data Analytics. He obtained his Ph.D. from Universiti Teknologi PETRONAS (UTP), Malaysia in the research area of Healthcare Internet of Things (IoT). He was awarded M.Tech. from the Indian Institute of Information Technology (IIIT), Allahabad, India in 2010 in the research area of intelligent systems. He completed B.Tech. from Dr. A.P.J. Abdul Kalam Technical University, Uttar Pradesh (U.P.), Lucknow, India in 2008.

His research interests are Healthcare Internet of Things, fog computing (FC), cloud computing, machine learning, and the blockchain mechanism. He is currently working on Cooperative Energy Trading System (CENTS) project for an efficient peer-to-peer energy trading system in a Smart Grid (SG) network. He has academic experience of more than seven years and published 1 book, 6 book chapters, and around 20 papers in various international journals and conferences. In recognition of his articles in Q1/Q2 journals and quality publications, he has received many awards and certificates.

He received research grants like UTP-Hitachi Collaborative Research, Long Term Research Grant Scheme 2018, UTP-PAF KIET International Collaborative Research Funding (ICRF) 2019, and Collaborative Research Funding (CRF) 2019.

Adil Kaleem is an experienced Industry Consultant on Cloud, Big Data, ETL, and Data Warehousing, in varying sectors like Telco, retail, banking, not-for-profit, and education. Currently he is working as Cloud Delivery Technical Architect in NCS, Australia. His experience constitutes across broad work areas from DevOps to Design & Architecting Solutions. Having lived through the Hadoop era and now experiencing Cloud in market, he worked through some of the most innovative companies and teams like TCS, Uniting, Teradata Australia Pty Ltd, Teradata India Pvt. LTD, IBM India Pvt Ltd paving their way in incorporating new techs in their businesses.

Contributors

Syed Adnan Afaq
Department of Computer Application, Integral University
Lucknow, India

Afsaruddin Khan
GLA University
Mathura, India

Faheem Ahmad
Department of IT, University of Technology and Applied Science
Al-Musannah, Sultanate of Oman

Muhammad Kalamuddin Ahmad
Department of Computer Application, Integral University
Lucknow, India

Shish Ahmad
Department of Computer Science & Engineering, Integral University
Lucknow, India

Tasneem Ahmed
Department of Computer Application, Integral University
Lucknow, India

Sarfaraz Alam
Department of Computer Application, Integral University
Lucknow, India

Anis Ali
Department of Management, College of Business Administration, Prince Sattam Bin Abdulaziz University
Al Kharj, Saudi Arabia

Kashif Asad
Department of Computer Application, Integral University
Lucknow, India

Danish Ather
Department of IT and Engineering, Amity University
Tashkent, Uzbekistan

Priyanka Bajpai
Department of Commerce & Business Management, Integral University
Lucknow, India

Neha Bhushan
Amity School of Communication, Amity University
Noida, India

Mohammad Faisal
Department of Computer Application, Integral University
Lucknow, India

Mohd. Faizan Farooqui
Department of Computer Application, Integral University
Lucknow, India

Praveen Gujjar J
Faculty of Management Studies, JAIN (Deemed-to-be University)
Bengaluru, India

Mohd Haroon
Department of Computer Science & Engineering, Integral University
Lucknow, India

Firoz Husain
Department of Commerce & Business Management, Integral University
Lucknow, India

Mohd Husain
Islamic University of Madinah
Madinah, Saudi Arabia

Pham Quang Huy
University of Economics
Ho Chi Minh City (UEH), Vietnam

Mohammad Ishrat
Department of CSE, Koneru Lakshmaiah Education Foundation
Vaddeswaram, India

Sumbul Javed
Lal Bahadur Shastri Group of Institutions
Lucknow, India

Farhina Sardar Khan
Department of Commerce & Business Management, Integral University
Lucknow, India

Wasim Khan
Department of CSE, Koneru Lakshmaiah Education Foundation
Vaddeswaram, India

Vipin Khattri
Faculty of Computer Science and Engineering, Poornima University
Jaipur, India

Vu KienPhuc
University of Economics
Ho Chi Minh City (UEH), Vietnam

Naveen Kumar H N
Department of Electronics and Communication Engineering, Vidyavardhaka College of Engineering
Mysore, India

Rishi Kumar
Department of Computer Science and Engineering, Graphic Era (Deemed to be University
Dehradun, India
Universiti Teknologi Petronas
Seri Iskandar, Malaysia

Shubham Kumar
Department of Computer Application, Integral University
Lucknow, India

Neyha Malik
Department of Commerce & Business Management, Integral University
Lucknow, India

Mohd. Noman Mansoori
Integral University
Lucknow, India

Syed Shahid Mazhar
Department of Commerce & Business Management, Integral University
Lucknow, Inida

Monia Mohammed Al Farsi
Department of Information Technology, University of Technology & Applied Science
Al Musannah, Oman

Sasmita Padhy
School of Computing Science and Engineering, VIT Bhopal University
Bhopal, India

Harold Andrew Patrick
Faculty of Management Studies, JAIN (Deemed-to-be University)
Bengaluru, India

Guru Prasad M S
Department of Computer Science and Engineering, Graphic Era (Deemed to be University)
Dehradun, India

Shahana Gajala Qureshi
School of Computing Science and Engineering, VIT Bhopal University
Bhopal, India

Mohammad Shiblee
Taif University
Taif, Saudi Arabia

Mohammad Siddique
Faculty of Information Technology, Majan University College
Muscat, Sultanate of Oman

Azra Siddiqui
BIS Department, WITS University
Johannesburg, South Africa

Deepak Kumar Singh
Indian Institute of Information Technology
Lucknow, India

Soni Singh
Department of Computer Science and Engineering, Lovely Professional University
Punjab, India

Saurabh Srivastava
Department of Computer Science & Engineering (Data Science), Moradabad Institute of Technology
Moradabad, India

Amrit Suman
Sharda School of Engineering & Technology, Sharda University
Greater Noida, India

Preetam Suman
School of Computing Science and Engineering, VIT Bhopal University
Bhopal, India

Manish Madhava Tripathi
Department of Computer Science & Engineering, Integral University
Lucknow, India

Preetika Tripathi
Integral University
Lucknow, India

Priyanka Tripathi
Integral University
Lucknow, India

Nayyar Ali Usmani
German Multinational Bank
Berlin, Germany

Saman Uzma
Cubeight Solutions
Sydney, Australia

Harish Chandra Verma
ICAR-Central Institute of Subtropical Horticulture
Lucknow, India

Gausiya Yasmeen
Department of Computer Application, Integral University
Lucknow, India

1 Changing Paradigm in AI Application and Its Impact on Cybercrime

Review of Major Global Events

Neyha Malik, Firoz Husain, Anis Ali, and Sumbul Javed

1.1 INTRODUCTION

Artificial intelligence (AI) and cybersecurity mark a transformative leap forward in the detection and prevention of ever-evolving cyber threats and strengthening the overall defense system of a country. It helps in automating analysis, identifying patterns, and predicting potential vulnerabilities in large datasets and henceforth mitigating risks with enhanced real-time monitoring strategies. It is not a local issue but rather a global concern; therefore, research and collaboration are imminent to stay ahead in this because AI can wreak havoc if exploited by hackers. The insights incorporated in this chapter have been taken from the experiences and happenings of industry, academia, and foreign sectors in order to highlight the relevant challenges and opportunities as well as the potential ventures in the science and technology domain by the emergence of AI. This technology has vast potential for application which may have far-reaching impacts on the future industry and society at large. The contribution of this study is far-reaching in making the reader aware of the various strategies and infrastructure erected at the global level to mitigate a global threat. This will help in benchmarking the best practices. Technology is deeply rooted in every aspect of our lives in this digital age. Henceforth, cybersecurity has gained unprecedented importance with the burgeoning internet-connected devices; the potential risks have expanded significantly related to the rising digital data transactions and increasing reliance on cloud computing. Since the activities of individuals, businesses, and governments have become increasingly interconnected, the need to safeguard and maintain the integrity of digital systems has become a paramount priority. Cyberattacks can lead to severe consequences, such as financial losses, compromised sensitive data and personal information, breaches in critical infrastructure, and even threats to national security.

Cybersecurity protects vulnerable computer data networks from cyber threats of hacking, unauthorized access, data breaches, and complicated cyberattacks. It

DOI: 10.1201/9781003409571-1

includes measures like a combination of technical solutions, policies, and best practices designed to detect, respond, and in the future prevent cyber threats, ensuring the availability, reliability, and confidentiality of digital data.

1.2 LITERATURE REVIEW

AI offers a robust solution to the intricate cyberattacks increasingly happening in the complicated digital landscape through efficiently harnessing advanced algorithms and machine learning powers. This synergetic real-time relationship between AI and cybersecurity empowers organizations with unprecedented capabilities to predict, detect, and counteract cyber threats, mitigating risks and fortifying digital defenses. From bolstering threat detection to automating incident response, the fusion of AI and cybersecurity ushers in a new revolutionized era of proactive protection in the digital realm. Various researchers have given their insights into the future of AI and its role as a boon or bane for cybersecurity. "Artificial intelligence (AI) is increasingly reshaping service by performing various tasks, constituting a major source of innovation, yet threatening human jobs." This theory mentions four types of intelligence required for accomplishing varied jobs, which can be known as empathetic, mechanical, intuitive, and analytical and firms have to decide between human and machine to complete them. AI has a double-edged impact which has revolutionized multidisciplinary research attention but has its own threats to the job market [1]. AI can be defined as a "system's ability to correctly interpret external data, to learn from such data, and to use those learnings to achieve specific goals and tasks through flexible adaptation." The authors presented a framework in which they showed the potential risk of AI that "helps organizations think about the internal and external implications of AI, which we label the Three C Model of Confidence, Change, and Control." With the advent of the third-generation AI known as artificial super intelligence (ASI), there is a need for legislation to set guidelines for controlling procedures to avoid AI becoming ungovernable. The author asserted that "just as humans can never truly understand how chimpanzees think, despite the fact that they share 99% of our DNA, we will not be able to understand how an ASI system thinks. This limits our ability to control such systems, which again makes them appear more dangerous than useful" [2].

According to renowned English theoretical physicist and cosmologist, "the development of full artificial intelligence could spell the end of the human race"; on the other hand, according to American businessman and philanthropist Bill Gates, "humans should be worried about the threat posed by Artificial Intelligence" [3]. "AI has become more popular today due to Big Data, advanced algorithms, and improved computing power and storage," "With the rapid advancement of Big Data technologies, e.g. improved computing storage capability and super-fast speed of data processing machines, AI is being revitalized with the availability and power of Big Data" [4]. Dwivedi et al. have asserted that "the era of AI systems has progressed to levels where autonomous vehicles, chatbots, autonomous planning and scheduling, gaming, translation, medical diagnosis, and even spam fighting can be performed via machine intelligence" [5]. However, all industries ranging from manufacturing, finance, healthcare management, logistics, retail, and supply chain to name a few now face the significant impact of AI which may potentially be disrupted by the emergence of AI capabilities if not checked in real time. Some authors "suggest that AI will be more effective if it is deployed in ways that augment (rather than replace) human managers" [6]. Raisch and Krakowski "explore the automation and augmentation concepts in the management domain. Whereas automation implies that machines take over a human task, augmentation means that humans collaborate closely with machines to perform a task" [7].

On the other hand, authors in positive studies that favor AI suggest that there is a need to design policies "favorable to the development of a good AI society" [8]. Ben Shneiderman has proposed human-centered AI (HCAI), a high-definition technological framework that creates "Reliable, Safe & Trustworthy" (RST) products containing outstanding computer automation coupled with high-level human control [9]. According to an article by Deepak Maheshwari, "the (Indian) President last week gave her approval to the Digital Personal Data Protection Act (DPDPA), almost six years after the government had constituted an expert committee chaired by Justice B. N. Srikrishna for this purpose and the Supreme Court's landmark 2017 judgment affirming privacy as a fundamental right." According to the author, "By 2030, the Indian economy would have grown way beyond $5 trillion and 6G networks would be rolling out." It is further envisioned that "if we, the people can discretely and securely control and access our personal data—stored across trusted federated goods and services and determine; if, what and how much information we would like to share with whom and when; with assurance that they would use such information only for the purpose that we have consented to and authorized, that should be evidence enough about the robustness, relevance and resilience of the DPDPA" or "any cyber legislation" [10].

Cyberspace "is a virtual digital space through which interconnection is established between computers and smartphones connected within the Internet of Things environment." It is difficult for companies to detect cyber challenges and obstacles in cybersecurity performance. In this chapter, the most relevant practices of "data transfers between electronic devices safely and without the presence of malicious software" were discussed to suggest policies to contain cybercrime and implement significant strategies to make a safe digital environment [11]. Farid et al. focused on the applications of digital information security management (DISM) practices and implementation "to ensure the privacy and protection of data and resources in an electronic environment such as academic libraries due to their strict privacy and data security policies" [12]. There are other cybersecurity standards, viz. ISO/IEC 27000 (International Organization for Standardization/International Electro-Technical Commission), that are present; however, there is a need for customized cybersecurity standards that are unique to a firm's cybersecurity risk profile [13]. Another study "impresses the need for the TMT to attend to high breach costs as part of firm responsibility towards stakeholders" [14]. Large databases, blockchain technology, hybrid working, numerous digital platforms, electronic commerce, etc. need national-level information infrastructure. However, this gives rise to IT security, "including the absence of a comprehensive information policy of states, information leakage, use of unlicensed software, data loss due to spyware, cybercrime, cyber-attacks, cyber wars, cyber terrorism." Henceforth there is a need of "critical infrastructure development from the standpoint of information security in the system of national security of states and restructuring of national economies" [15].

Another form of cybersecurity is "sharing information." This word-of-mouth sort of technique prevents the compromising an organizational data by the cyber attacker and also helps reduce cyber threats by preventing them from exploiting any vulnerability that may be used to launch multiple organization attacks [16]. "Cyber security is guarding computer systems, data, network, and other resources against unauthorized access and malicious users." Cyberspace has a plethora of cyber criminals who target systems either handled by unaware employees or have major vulnerabilities. On top of it, a proper cyber protection mechanism for measuring the strength of existing cybersecurity is not present which can find undetected threats that possibly become security incidents. The authors have proposed a taxonomy for cybersecurity that acts as "Metrics" that can be used for measuring the strength of cyber protection and evaluation of an organization. The suggested set of metrics for this taxonomy is "protection mechanism, the interaction of users, vulnerabilities, defense system, threats, and situation encounter outcomes" [17].

1.3 AI CAPABILITY FOR THREAT DETECTION AND INCIDENCE REPORT ANALYSIS AND PREVENTION

Cybersecurity with AI has a synergetic effect on processing mammoth data, recognizing patterns, and making informed decisions in real time. AI plays an important role in threat and incidence report "detection, analysis, and prevention" within the realm of cybersecurity, significantly enhancing an organization's ability to safeguard its digital assets.

1.3.1 Benefits of AI in Threat Detection

The benefits of AI in threat detection are discussed as follows:

- *Data Protection and Encryption*: AI aids in identifying sensitive data and implementing encryption measures to safeguard information at rest or in transit.
- *Pattern Recognition*: Machine learning models excel at identifying complex patterns in vast amounts of big data leading to earlier and more accurate threat detection that might go unnoticed by human analysts in uncovering subtle indicators of attacks.
- *Threat Intelligence (Threat Detection and Prevention)*: AI enables organizations to be aware and updated about the latest techniques used by cyber criminals. AI-driven systems examine "system logs, network traffic & user actions" to detect unusual patterns that may signal potential cyber threats and reduce false positives. Models generated by machine learning continuously update themselves with new data in order to adapt to evolving threats.
- *Behavioral Analysis (User Authentication and Access Control)*: AI monitors and analyzes the behavior of users to establish baseline unusual patterns, detect anomalies to enable real-time risk assessment, and enhance authentication processes. This helps identify compromised accounts, unauthorized access, or insider threats that might otherwise go unnoticed and provide safe access control.

- *Malware detection and mitigation*: AI algorithms aid in identifying malware and classifying it based on its traits, behavior, and code analysis to contain malicious software.
- *Anomaly Identification*: AI algorithms create a standard interface across various networks, systems, and user bases. Deviations from this baseline signal potential threats or breaches, enabling rapid response.
- *Phishing and Fraud Prevention*: AI detects phishing and fraud attempts by assessing email content and sender behavior and filters out malicious messages with the help of social engineering tactics.
- *Automated Incident Response*: AI-driven incident response platforms can autonomously assess and initiate automated actions to contain and mitigate threats. This accelerates response times and reduces manual intervention.
- *Predictive Analytics*: By analyzing information from historical data and trends, AI forecasts potential cyber threats and identifies emerging trends, allowing organizations to proactively strengthen their firewall rules defenses.
- *Network Security Optimization*: AI optimizes network security by segmenting traffic, identifying unusual patterns, and dynamically adjusting AI-driven firewalls and intrusion detection system rules that automatically adapt and optimize security measures based on real-time analysis of network traffic to prevent unauthorized access.

In essence, the integration of AI empowers cybersecurity professionals to stay one step ahead of cyber adversaries. It complements human expertise by automating routine tasks, enabling rapid responses, and providing invaluable insights into emerging threats. The synergy between AI and cybersecurity represents a formidable defense mechanism in the dynamic and ever-evolving digital landscape. Once the threat detection is initiated, AI helps in threat analysis by bringing into light the potential risks through data processing, i.e. efficiently sifting through massive datasets, extracting relevant information for analysis, and identifying correlations that could indicate sophisticated attacks. Another AI threat analysis is malware analysis wherein AI-powered tools analyze the characteristics and behavior of malware, categorizing and identifying novel threats to aid in their containment and removal. In vulnerability assessment, AI identifies and prioritizes potential vulnerabilities within systems, applications, and networks, helping security teams focus on critical issues. This enables organizations to focus on critical areas for patching and updates and help in critical infrastructure.

It is a time-honored adage that "prevention is better than cure" and the same applies to the cyber world also. Threat prevention requires an iron-clad system and religious monitoring of the same. The threat prevention measures include the following important steps among many such as real-time monitoring of system logs, network traffic, and cyber users' activities; adaptive access controls reducing the risk of unauthorized access; phishing detection by analyzing emails and identifying suspicious content, URLs, or sender behavior to prevent employees from falling victim to scams. Through the automation of mundane tasks, human decision-making is enhanced, and handling extensive datasets is made possible by AI-empowered cybersecurity teams to detect, analyze, and prevent threats more effectively. Its ability to evolve and learn from new data ensures that it stays current in identifying and countering emerging threats, making it an indispensable weapon to win the battle against cybercrime.

1.3.2 Benefits of AI in Improving Incident Response

AI brings substantial benefits to incident response processes henceforth enhancing efficiency and minimizing the potential for human error. It not only accelerates response times but also enhances the overall quality of incident handling. By leveraging AI's capabilities, organizations can effectively address cyber threats, reduce the risk of human error (human emotions, fatigue, or distractions), and maintain a robust cybersecurity posture. AI can help improve incident response which has the benefits of speed and automation where AI can automatically detect and analyze threats in real time, accelerating the entire incident response lifecycle. This reduces the time between detection and containment, limiting the potential damage caused by cyberattacks. A 24/7 network traffic and system log monitoring ensures that incidents are identified promptly, even outside regular working hours. It can handle a larger volume of alerts and incidents without proportionally increasing the workload on human analysts. Real-time decision-making based on predefined rules and machine learning models can be ensured through AI use. This agility allows for immediate response actions, mitigating threats before they escalate. AI enriches incident data (data enrichment) with contextual information from various sources, helping analysts understand the scope and impact of incidents more comprehensively. This aids in making informed decisions. AI can automate workflow optimization of routine jobs, allowing expert analysts to concentrate on more intricate and strategic facets of incident response. This optimization leads to improved overall efficiency. One of the interesting capabilities of AI is adaptive learning wherein the systems learn from historical incident data and continuously improve their capabilities. They proactively do threat hunting by analyzing historical data, identifying trends, and uncovering potential vulnerabilities or threats that might otherwise remain hidden consequently adapting to new attack techniques and evolving alongside emerging threats, enhancing the effectiveness of incident response over time.

1.4 GROWING COMPLEXITY OF ATTACKS AND THEIR POTENTIAL IMPACT ON BUSINESSES AND INDIVIDUALS

The growing complexity of cyberattacks presents significant challenges for both businesses and individuals, as attackers employ intricate techniques to exploit vulnerabilities and achieve their malicious objectives. The potential impacts of these sophisticated attacks are far-reaching and can result in the following:

For Businesses

- *Financial Losses*: Cyberattacks may result in financial setbacks, involving the pilfering of funds, unnecessary expenses of ransoms, and data restoration.
- *Data Breaches*: Compromised customer data can lead to breaches of trust, and reputational damage, and additionally, there is a risk of legal liabilities arising from failure to comply with data protection regulations.
- *Operational Disruption*: Attacks like ransomware cause undesirable disruptions in business operations, reduced productivity, and possible financial consequences due to service disruptions.
- *Intellectual Property Theft*: Businesses with valuable intellectual property are at risk of theft, potentially leading to lost competitive advantage or revenue.

- *Supply Chain Disruptions*: Attackers targeting suppliers and partners can disrupt the supply chain, impacting production schedules and delivery of goods and services.
- *Regulatory Penalties*: Failure to adhere to data protection regulations can lead to significant fines and legal consequences.
- *Prestige Loss*: Publicized cyber incidents damage a company's reputation, eroding customer trust and loyalty.

For Individuals

- *Identity Theft*: Personal information stolen in data breaches can be used for identity theft, leading to financial losses and long-term consequences.
- *Financial Fraud*: Stolen financial data can lead to unauthorized transactions, credit card fraud, and other financial scams.
- *Privacy Violations*: Breached personal data can lead to an invasion of privacy, including exposure to sensitive conversations, medical records, and private photos.
- *Social Deception*: Cyber adversaries employ advanced methods to manipulate individuals into revealing sensitive personal information or threatening them into engaging in detrimental consequences.
- *Loss of Digital Assets*: Attacks like ransomware can lead to the loss of personal files, photos, and other digital assets if victims refuse to pay the ransom.
- *Emotional Distress*: Victims of cyberattacks often experience emotional distress and anxiety due to the violation of their personal space and security.
- *Impact on Employment*: Cyberattacks that compromise personal and professional information can impact an individual's career and job prospects.

The growing complexity of cyberattacks underscores the need for comprehensive cybersecurity measures encompassing routine software updates, robust authentication protocols, employee training, adoption of intrusion detection mechanisms, and regular preparation as well as monitoring of incident response plans. Both businesses and individuals must remain vigilant and proactive in safeguarding their digital assets against an ever-evolving threat landscape.

1.5 CASE STUDY: SOLARWINDS SUPPLY CHAIN ATTACK

Here is a brief discussion of one out of many important case studies that highlight cybersecurity incidents:

***Date of Incident*:** Discovered in December 2020.

- *Overview*: "SolarWinds is a major software company based in Tulsa, Okla., which provides system management tools for network and infrastructure monitoring, and other technical services to hundreds of thousands of organizations around the world. Among the company's products is an IT performance monitoring system called Orion" (Saheed Oladimeji, Sean Michael Kerner)

- *Published*: 03 Nov 2023. The SolarWinds (Oklahoma, U.S.) supply chain attack is one of the noteworthy recent cybersecurity events. Attackers exploited a loophole in the SolarWinds Orion IT monitoring software, enabling them to inject harmful code into their software updates which then dispersed into SolarWinds' consumer base containing individuals, government agencies, critical infrastructure providers, and private sector organizations.
- *Attack Vector*: The cyber attackers infiltrated SolarWinds' software environment and implanted malware into the authentic software updates. Entities that downloaded these compromised updates unknowingly introduced the malicious code into their networks.
- *Impact*: The incident affected large number of organizations globally. These attackers now have access to sensitive organizational data and systems which included even the highly protected U.S. government agencies, rich Fortune 500 companies, and sophisticated cybersecurity firms. The incident highlighted the vulnerability of software supply chains, where attackers target trusted vendors to infiltrate multiple downstream organizations. While attribution remains a subject of investigation, the attack bore similarities to the work of nation-state actors, potentially elevating the scale of espionage.
- *Response*: Multiple cybersecurity firms and government agencies conducted extensive investigations to uncover the extent of the attack and identify the attackers. Stakeholders identified the malicious code, took steps to remove it, and took measures to strengthen their security. SolarWinds publicly acknowledged the incident, shared the technical details, and released patches to address the vulnerability.
- *Lessons Learned*: The incident emphasized the importance of continuous vigilance in software supply chains to detect and respond to potential breaches. Organizations reevaluated their security strategies to implement zero trust principles, assuming that perimeter defenses can be breached. Another important lesson is the sharing of the highlights of the incident as threat intelligence among organizations and the cybersecurity community.
- *Conclusion*: The SolarWinds supply chain attack underscores the evolving sophistication of cyber threats and the critical importance of securing software supply chains. It serves as a stark reminder that even trusted vendors can become vulnerable to cybercrime, and organizations must adopt a premeditated comprehensive approach to ensure cybersecurity.

1.6 CYBERSECURITY MEASURES IMPLEMENTED BY VARIOUS COUNTRIES

International collaboration in cybersecurity is crucial in today's interconnected digital landscape. Cyber threats transcend national borders, and no single country can address them effectively in isolation. Collaborative efforts enable the exchange of threat intelligence and resources, leading to benchmarking and improved cybersecurity strategies and responses. International collaboration in cybersecurity requires building trust, overcoming geopolitical challenges, and respecting the sovereignty of individual nations. Platforms like international conferences, information-sharing forums, and collaborative initiatives provide the foundation for countries to work together in countering cyber threats and creating a safer digital environment for everyone. Different countries have implemented various remedies and strategies to leverage AI in cybersecurity. Here are some examples of the approaches taken by several countries.

1.6.1 European Union

The General Data Protection Regulation (GDPR), enacted in 2018, enforces exhaustive data protection and privacy regulations across EU member states, compelling organizations to implement stringent cybersecurity measures and implement data privacy laws. The regulation has 99 articles in 11 chapters. GDPR monitors organizations handling the personal data of those residing in the EU, irrespective of their geographic location. Its impact extends beyond data privacy to cybersecurity practices as well. Here's an overview of GDPR and its impact:

Key principles of GDPR are discussed as follows:

- *Lawful and Transparent Collection and Dissemination of Data*: Entities must have a legitimate, specified, and explicit reason to "Responsibly" collect and process "Accurate" personal data only that is necessary for the intended purpose to ensure "Data Minimization" and should be "Retained" only for the necessary time limit.
- *Integrity, Accountability, and Confidentiality*: Organizations must demonstrate compliance with GDPR principles to prove that they are adhering to data protection requirements and are required to enact appropriate security measures to protect personal data from cyber breaches, forbidden access, and potential data losses.

The impact of GDPR cybersecurity measures and data privacy are discussed as follows:

- *Enhanced Data Security*: GDPR's data protection requirements encourage organizations to implement robust cybersecurity mechanisms to safeguard personal data against leaks, breaches, and unauthorized access.
- *Breach Notification*: GDPR warrants prompt notification of cyberattacks to affected parties, i.e. the organization and individuals, encouraging swift responses to cybersecurity incidents.
- *Default Protected Designs*: Organizations are required to incorporate data protection measures from the initial production stage of products, services, and systems, ensuring that privacy and security are integral to their offerings.
- *Data Protection Officer (DPO)*: Organizations are advised to appoint a DPO responsible for monitoring data protection strategies, which may include cybersecurity measures.

- *Consent Requirements*: Obtaining explicit and informed consent for data processing activities highlights the importance of transparency, giving individuals more control over their personal data.
- *Cross-Border Data Transfers*: GDPR sets strict criteria for transferring personal data outside the EU, requiring organizations to ensure that data protection standards are upheld even in foreign jurisdictions.
- *Financial Penalties*: Failure to comply with GDPR may lead to substantial fines to set benchmarking to prioritize data privacy and cybersecurity as the prime motives of an organization.

GDPR's impact on cybersecurity and data privacy lies in its ability to drive organizations toward responsible data handling practices, improved cybersecurity measures, and heightened awareness of individuals' rights. By fostering a culture of data protection, GDPR enhances trust between organizations and their customers while strengthening cybersecurity practices to defend against cyber threats and data breaches.

Another strategic agency of the European Union to support innovation in sketching industrial policy that resonates with its cybersecurity efforts is The European Cybersecurity Competence Center (ECCC) (2021), working under the Network of National Coordination Centers (NCCs). The ECCC initiative reflects its recognition of the critical importance of cybersecurity in the contemporary digital age. It helms the commitment to strengthening cybersecurity capacities, fostering innovation, and addressing the ever-evolving cyber threats across member states. This initiative aims to consolidate resources, expertise, and collaboration to enhance the EU's overall cybersecurity resilience. ECCC aims to fortify its member states against evolving cyber threats while positioning Europe as a hub of cybersecurity excellence. Following is the overview of ECCC's efforts in this direction, which are as follows:

- *International Collaboration*: ECCC brings together a wide range of stakeholders, including cybersecurity experts, policymakers, public and private sector industry players, academic and research institutions, centers, organizations, and international partners to share their insights, best practices, knowledge, and expertise on a global scale driving innovation and collectively address Europe's collective cybersecurity resilience contributing to a more secure digital world.
- *Research and Innovation*: At the heart of the ECCC's mission is to advance R&D in cybersecurity technologies, tools, and methodologies. It supports the creation of cutting-edge technology that can effectively counter emerging cyber threats and produce tangible solutions. Testbeds and experimentation provide platforms to develop, validate, and test new cybersecurity technologies and practical solutions to be later implemented in real-world scenarios.
- *Human Capital Development*: The center emphasizes developing a skilled and capable cybersecurity workforce by offering training, workshops, education, and professional development opportunities to defend against cyber threats effectively.

- *Incident Response Reporting*: It aims to improve rapid incident response capabilities by fostering collaboration and information sharing, enabling real-time, quick, and effective responses to cyber incidents.
- *Global Competitiveness*: The center nurtures a thriving cybersecurity industry and fosters an innovation ecosystem to boost Europe's global competitiveness in the field of cybersecurity.
- *Technology Transfer*: It supports the research outcomes transfer onto practical applications by facilitating collaboration between researchers, industry partners, and end-users.
- *Policy Support*: The center's activities play a significant role in informing and contributing to EU policy developments, and shaping regulatory frameworks and strategies related to cybersecurity. By aligning its initiatives with policy objectives, the center contributes to shaping Europe's cybersecurity landscape.
- *Holistic Approach*: By combining research, innovation, education, and industry collaboration, the center takes a holistic approach to addressing cybersecurity challenges. Its initiatives span across various sectors such as finance, energy, healthcare, and transportation developing their critical infrastructure to emerging technologies.

The establishment of the EU's Cybersecurity Competence Center underscores the union's commitment to safeguarding its digital future. By unifying expertise, driving innovation, and fostering collaboration, the ECCC empowers Europe to effectively navigate the complexities of the digital age while addressing emerging cyber threats with confidence.

1.6.2 United States

U.S. Department of Homeland Security (DHS) is a central force in driving cybersecurity best practices through its various agencies and initiatives for "keeping America safe." By fostering collaboration, providing resources, and promoting awareness, DHS contributes to a more secure and resilient cyber landscape, ensuring critical infrastructure and digital systems are better protected against cyber threats. The DHS focuses on enhancing critical infrastructure cybersecurity and improving incident response capabilities. U.S. Cyber concerns have emphasized public-private partnerships for sharing threat intelligence and promoting collaboration.

Highlighting the role of the U.S. DHS in promoting cybersecurity best practices: DHS plays the following significant roles in promoting cybersecurity best practices across various sectors to enhance the nation's overall cybersecurity resilience:

- *Cybersecurity and Infrastructure Security Agency (CISA)*: CISA is the operational lead of DHS responsible for safeguarding both public and private sector organizations' critical infrastructure (industrial control systems, ICS security), including energy, transportation, and water, against cyber and physical threats that could impact industrial control systems by providing guidance, tools, and resources.

- *Information Sharing and Collaboration*: DHS facilitates information dissemination among public entities (viz. federal, state, local, tribal, and territorial governments) as well as private sector partners. This collaborative approach helps stakeholders stay informed about emerging threats and vulnerabilities.
- *National Cyber Awareness System*: CISA operates the National Cyber Awareness System, which provides regular alerts, tips, and information to individuals, businesses, and government agencies about current cybersecurity threats and vulnerabilities.
- *Election Security*: In collaboration with state and local governments, DHS plays a role in securing the nation's election systems, offering cybersecurity support and threat intelligence to safeguard election processes.
- *Cyber Exercises and Training*: DHS conducts cyber exercises and simulations to help organizations and government agencies practice responding to cyber incidents, ensuring readiness in the face of real-world threats.
- *Federal Cybersecurity Initiatives*: DHS coordinates federal cybersecurity initiatives and programs, ensuring alignment and collaboration across government agencies to enhance the nation's overall cybersecurity posture.
- *National Initiative for Cybersecurity Education (NICE)*: DHS supports NICE, an initiative that focuses on establishing a skilled cybersecurity workforce through education, training, and professional development programs.
- *Cybersecurity Grant Programs*: DHS administers grant programs that provide funding to government entities (viz. federal, state, local, tribal, and territorial) to enhance their cybersecurity capabilities and preparedness.
- *Promoting Cybersecurity Research*: DHS supports research efforts aimed at developing innovative cybersecurity solutions, technologies, and strategies to address emerging threats.

The National Institute of Standards and Technology (NIST) Cybersecurity Framework (2014) is a widely recognized and utilized resource. Their motto is "*Helping organizations to better understand and improve their management of cybersecurity risk.*" Initiatives like the NIST Cybersecurity Framework provide managing and reducing cybersecurity risk guidelines to organizations. The U.S. National Institute of Standards and Technology (NIST) Cybersecurity Framework is an elaborate set of guidelines, standards, and benchmarking practices developed in response to the increasing frequency as well as sophistication of cyber threats, providing a flexible framework that can be adapted to various industries and sectors.

The NIST Cybersecurity Framework aims to bolster enhancing cybersecurity stance through the adoption of a structured risk-based roadmap to strengthen defenses, respond effectively to incidents, and ensure the resilience of critical systems and data.

Key components of the NIST Cybersecurity Framework are discussed as follows:

- *Framework Core*: This central component consists of five functions that represent key cybersecurity activities: Identify, Protect, Detect, Respond, and Recover to guide entities in developing their cyberspace protection strategies.

- *Categories*: Each of the above-mentioned "function" is broken down into categories that define cybersecurity outcomes. For instance, the "Identify" function contains categories viz. "Risk Assessment, Asset Management, and Business Environment."
- *Subcategories*: Further breaking down categories, subcategories provide more detailed actions and outcomes that organizations can implement to achieve their cybersecurity goals.
- *Informative References*: The framework also provides references to existing standards, guidelines, and practices from various sources, such as NIST publications and industry-recognized cybersecurity documents.

The benefits of the NIST Cybersecurity Framework are discussed as follows:

- *Risk Management*: The framework highlights a risk-centric approach to cybersecurity, helping entities identify and prioritize risks based on their business context and critical assets.
- *Customization*: It's flexible and can be tailored to fit organizations of different sizes, industries, and levels of cybersecurity maturity.
- *Common Language and Collaboration*: The framework provides a common language for discussing and addressing cybersecurity risks, facilitating communication and collaboration between internal and external stakeholders, and fostering a holistic approach to cybersecurity.
- *Continuous Improvement*: The iterative nature of the framework encourages continuous improvement as organizations assess, adapt, and enhance their cybersecurity practices over time.
- *Alignment with Regulations*: The framework can help organizational entities confer with structured rules, regulations, processes, and standards.

1.6.3 China

Chinese Cybersecurity Law (2017) emphasizes data localization, security assessments of network products, and stricter control over data leaving their country. It is an exhaustive and rather complex legal framework that aims to regulate various aspects of cybersecurity within the country. The law is multifaceted and covers data protection, critical information infrastructure, and cross-border data transfers. The government encourages collaboration between state-owned enterprises, private companies, and foreign companies operating in China to develop cybersecurity technologies and solutions. Foreign companies must implement data localization, navigate complex compliance requirements, and assist Chinese authorities when requested.

Below are the key provisions of China's Cybersecurity Law showing its approach to regulating cyberspace, protecting critical infrastructure, and its significant implications for domestic and foreign companies operating within China:

- *Data Localization*: The law requires critical information infrastructure operators to store important and personal data within mainland China. However, this provision affects industries like finance, energy, transportation, and telecommunications. Companies are required to invest in data storage infrastructure within China, potentially incurring additional costs.

- *Cross-Border Data Transfers*: The law stipulates that transferring important and personal data outside of China requires a strict mandatory security assessment. This provision impacts foreign companies operating in China when they are transferring data abroad. Further, companies must invest in cybersecurity measures for protecting user data, establish breach reporting mechanisms, and ensure compliance with data protection obligations.
- *Technology Transfer and Security Reviews*: Critical network products and services must undergo National Security Reviews before being offered or sold in China to comply with local regulations in order to evaluate potential risks and threats to national security, aligning with China's focus on protecting critical infrastructure.
- *Real-Name Registration*: When signing up, online users are required to provide real-name identification to ensure accountability for their cyber behavior.
- *User Data Protection and Privacy*: Companies are obligated to comply with robust data protection measures in place to secure user data, and they must report data breaches on time to authorities and affected individuals.
- *Surveillance and Inspection*: Government agencies have the authority to inspect and supervise network operators' online activities and content, security measures, and compliance with local regulations. "It requires network operators to store select data within China and allows Chinese authorities to conduct spot-checks on a company's network operations" (Wiki).
- *Emergency Response Planning and Reporting*: Companies must establish cybersecurity incident response plans and report significant incidents to concerned authorities.
- *Critical Information Infrastructure (CII) Upgrades*: Critical information infrastructure operators are subject to stricter regulations and security reviews, required to ensure maximum security of their networking systems, and facilities to meet the law's requirements. They are required to cooperate with law enforcement agencies in cybersecurity and criminal investigations.
- *Digital Sovereignty*: Aligned with China's perspective to maintain control over information and communication technologies, Digital Sovereignty is exercised where the government asserts authority over its cyberspace and data flows.
- *Technological Advancement*: China is investing heavily in cybersecurity R&D. It seeks to develop homegrown technologies to reduce dependence on foreign solutions and protect against potential vulnerabilities.

In summary, China's commitment to strengthening cybersecurity and safeguarding critical information infrastructure has far-reaching implications for both domestic and foreign companies operating within the country. The country places·significant emphasis on user data protection, and data sovereignty by maintaining a firm grip on the digital realm. China has the vision of leveraging digital technologies for economic growth while safeguarding national security and maintaining social stability. Though some raise concerns about freedom of expression and potential barriers to international cooperation, some view these measures as necessary for cybersecurity.

1.6.4 Israel

Israel is known for its strong cybersecurity capabilities and has established cybersecurity innovation hubs and research centers to foster expertise in the field. Collaboration between the military and private sector has contributed to Israel's reputation as a cybersecurity powerhouse. The country's prowess in this area can be attributed to a combination of factors, including its unique security challenges, strong technological ecosystem, and emphasis on collaboration between the military, academia, and industry.

Israel's military intelligence, particularly *Unit 8200*, has played a pivotal role in fostering cybersecurity expertise and innovation in the country. The unit's unique combination of intelligence operations, technological prowess, and emphasis on continuous learning has contributed significantly to Israel's reputation as a global cybersecurity powerhouse. Unit 8200 recruits young talented individuals with exceptional skills in mathematics, computer science, engineering, and related fields. These recruits undergo rigorous training that equips them with advanced technical skills. Their hands-on experience in offensive and defensive cyber operations enables them to understand the tactics, techniques, and procedures of cyber adversaries. Unit 8200 emphasizes teamwork and collaboration. Unit 8200's operational excellence extends to cyber operations and intelligence analysis. Members develop a deep understanding of threat actors, vulnerabilities, and attack vectors, and therefore, they act as Startup Founders who bring cutting-edge globally recognized products and services.

Israel's cybersecurity reputation and its innovative approaches are discussed as follows:

- *Unique Security Landscape*: Israel's geopolitical situation and history of conflicts have driven the country to develop advanced cybersecurity capabilities. With the constant threat of cyberattacks, Israel has had a practical incentive to invest heavily in securing its digital infrastructure and critical systems.
- *Strong Technological Ecosystem*: Israel boasts a thriving technology ecosystem characterized by a high density of tech startups, research institutions, and skilled professionals. This ecosystem provides a fertile ground for cybersecurity innovation to flourish.
- *Military Involvement*: The Israeli Defense Forces (IDF) has an imperative role in restructuring the country's cyberspace security expertise. The IDF's Unit 8200 (similar to the U.S. National Security Agency—NSA) focuses on signals intelligence, cyber warfare, and other advanced technological operations. Many of its members go on to contribute to the country's cybersecurity industry.
- *Collaboration Model*: Israel encourages collaboration between the governments, military, academia, and private sector. This collaborative approach helps to pool resources, share insights, and create a holistic response to cybersecurity challenges.

- *Startups and Innovation*: Israel has produced numerous cybersecurity startups that have gained international recognition as they develop cutting-edge technologies, products, and solutions addressing a wide range of cybersecurity issues.
- *Investment in Research and Development*: The Israeli government invests in research and development (R&D) in cybersecurity, fostering a culture of innovation. This investment has led to the creation of advanced tools and technologies to defend against cyber threats.
- *Incubators and Accelerators*: Israel hosts numerous incubators and accelerators focused on cybersecurity, which provide startups with mentorship, funding, and resources to develop their solutions and bring them to market.
- *Public-Private Partnerships*: Israel's cybersecurity strategy leverages public-private partnerships, encouraging collaboration between government agencies, academia, and industry. This approach ensures a coordinated response to emerging threats.
- *Global Reputation and Exports*: Israel's cybersecurity solutions are highly regarded globally. The country exports its cybersecurity products, technologies, and expertise to international markets, contributing to its reputation as a cybersecurity leader.
- *Education and Training*: Israel places a strong emphasis on education and training in cybersecurity. This includes specialized programs and courses at universities, as well as training opportunities within the military.

Overall, Israel's reputation as a cybersecurity powerhouse is a result of its proactive and innovative approach to addressing the challenges posed by the digital age. Its combination of technological expertise, collaboration, and strategic investments has propelled it to the forefront of cybersecurity innovation and defense.

1.6.5 Russia

Russia has developed a national cybersecurity strategy aimed at strengthening national resilience to cyber threats and enhancing international cooperation on cybersecurity issues. Initiatives like Russia's "Sovereign Internet" seek to establish more control over domestic internet traffic and protect critical infrastructure. Russia's national cybersecurity efforts are aligned with its broader national security objectives. The country places a strong emphasis on protecting critical infrastructure, national interests, and state secrets in the digital realm. Russia invests in the development of cybersecurity technologies, tools, and solutions to enhance its cyber capabilities. This includes both defensive and offensive capabilities.

Following is an overview of Russia's national cybersecurity strategy:

- *State-Sponsored Cyber Activities*: Russia has been associated with state-sponsored cyber activities that have gained international attention. These activities include cyber espionage, disinformation campaigns, and attacks targeting other nations' governmental and critical infrastructure systems.

- *Legal and Regulatory Framework*: Russia has developed cybersecurity laws and regulations that provide a legal foundation for addressing cyber threats and ensuring information security. These laws define the tasks and obligations of government agencies, organizations, and individuals.
- *Critical Information Infrastructure Protection*: Russia recognizes the significance of safeguarding critical information infrastructure, such as energy, finance, telecommunications, and transportation systems.
- *International Engagement*: Russia engages in international discussions and initiatives related to cybersecurity. The country participates in international dialogues on cybersecurity norms, cooperation, and the development of legal frameworks governing cyberspace.
- *Public-Private Collaboration*: Collaboration between the government, private sector, academia, and research institutions is important for Russia's cybersecurity efforts.
- *Capacity Building and Education*: Russia focuses on developing a skilled cybersecurity workforce through educational programs, training initiatives, and research activities. This helps nurture expertise in cybersecurity and fosters the growth of a capable workforce.
- *Incident Response*: Effective incident response mechanisms are essential for addressing and mitigating cyber incidents. Russia likely maintains incident response teams and frameworks to handle cyberattacks and breaches.
- *Digital Sovereignty*: Russia emphasizes its right to exercise control over cyberspace, including the ability to regulate online content and ensure that digital platforms comply with national laws and regulations.
- *Bilateral Agreements*: Russia has engaged in bilateral discussions with certain countries on cybersecurity issues. These agreements may include sharing threat intelligence, exchanging best practices, and coordinating efforts to address cyber threats.
- *Information Sharing*: In some instances, Russia has shared information with other countries about cyber threats and attacks targeting shared interests. Such information sharing contributes to a collective understanding of the evolving cyber landscape.
- *Cybersecurity Drills*: Russia has participated in joint cybersecurity exercises and drills with other nations. These exercises simulate cyberattacks and responses, fostering collaboration and testing incident response mechanisms.

1.6.6 India

India's National Cyber Security Policy outlines strategies for strengthening cybersecurity capabilities, including setting up cybersecurity operations centers and promoting cybersecurity education. Government initiatives such as the Cyber Swachhta Kendra (Botnet Cleaning and Malware Analysis Centre) provide cybersecurity support to citizens and organizations. The National Cyber Security Policy (2013) provides a comprehensive framework for safeguarding the nation's digital infrastructure, promoting secure digital transactions, and responding to cyber threats. The policy focuses on five strategic pillars:

- *Secure*: Creating a secure cyber ecosystem by building and promoting capabilities to protect information and critical infrastructure.
- *Cyber Resilient*: Strengthening the capacity to endure and capability to bounce back to normal from the harm inflicted by cyberattacks.
- *Inclusive*: Ensuring a balanced and inclusive growth of the digital economy.
- *Responsive*: Developing mechanisms for real-time information sharing and responding to cyber incidents.
- *Assured*: Establishing a comprehensive regulatory framework to ensure cybersecurity.

Initiatives and Programs (Malik, N., et al. 2023):

- *National Critical Information Infrastructure Protection Centre (NCIIPC)*: NCIIPC identifies critical information infrastructure sectors, assesses vulnerabilities, and develops strategies to protect critical systems and networks.
- *National Cyber Coordination Centre (NCCC)*: Monitors internet traffic and serves as a hub for cyber threat intelligence sharing and analysis.
- *CERT-In (Indian Computer Emergency Response Team)*: CERT-In is the central authority for acknowledging and answering cybersecurity incidents. It issues advisories and alerts and coordinates incident response.
- *Cyber Awareness Programs*: Various government agencies, such as the Ministry of Electronics and Information Technology (MeitY), organize cybersecurity awareness campaigns, workshops, seminars, and webinars to educate users about online risks, safe online behavior, and data protection.
- *National Cyber Security Coordinator*: The National Cyber Security Coordinator oversees cybersecurity initiatives and facilitates coordination between government bodies, industry, and academia.
- *Skill Development and Training*: Initiatives such as the Cyber Swachhta Kendra (Botnet Cleaning and Malware Analysis Centre) promote cybersecurity awareness, education, and skill development.
- *Cyber Exercises and Drills*: India conducts cybersecurity exercises and drills to enhance preparedness, coordination, and incident response capabilities.
- *Collaboration with the Private Sector*: India encourages public-private partnerships to strengthen cybersecurity across industries and sectors.
- *International Cooperation*: India engages in international dialogues and cooperation to address global cybersecurity challenges and enhance information sharing.
- *National Cyber Security Awareness Month*: India observes October as National Cyber Security Awareness Month, during which a series of events, workshops, and awareness campaigns are conducted to educate users about cybersecurity best practices.
- *Online Resources and Portals*: Government agencies provide online resources, guidelines, and resources to help individuals, students, and businesses understand cybersecurity risks and adopt preventive measures.

While India's cybersecurity initiatives have made significant strides, challenges such as the evolving threat landscape, data protection, and rapid digitalization remain. Future directions may include strengthening legislative frameworks, promoting research and innovation, and further enhancing public-private collaboration. India has been actively working to enhance cybersecurity awareness and collaboration through various initiatives aimed at educating the public, businesses, and government agencies about cyber threats and best practices. Additionally, efforts are being made to foster collaboration between different stakeholders to collectively address cybersecurity challenges [18].

1.7 DISCUSSION

- Importance of Global Cooperation in Addressing Cross-Border Cyber Threats

  AI holds significant potential to facilitate information sharing and threat intelligence among nations in the field of cybersecurity. Its capabilities in processing and examining extensive volumes of data, detecting patterns, and identifying oddities can greatly enhance the effectiveness of international collaboration in countering cyber threats. Here's how AI can contribute to this effort:
  - *Enhanced Data Analysis*: AI can examine mammoth cyber threat data accessed from multiple sources, including malware samples, network logs, and security incident reports. The expeditious identification of patterns and correlations through AI can help nations understand the global threat landscape more comprehensively.
  - *Early Threat Detection*: AI-powered systems can detect emerging threats in real time by monitoring global cyber activities. This early detection enables timely responses and prevents the rapid spread of threats across borders.
  - *Contextual Insights*: AI can provide contextual insights by correlating threat intelligence with geopolitical events, such as international conflicts or political tensions. This helps nations understand the motivations behind cyberattacks and predict potential targets.
  - *Automating Everyday Tasks*: Tasks like data collection, categorization, and initial analysis are redundant and can be automated through AI in order to consequently free up cybersecurity experts to direct their attention to more advanced tasks that require expert judgment and decision-making.
  - *Improved Attribution*: AI can aid in attributing specific cyberattacks which will help in analyzing attack patterns and techniques; AI systems can help identify commonalities and distinguish between various threats.
  - *Data Sharing and Privacy*: AI can enable secure and privacy-preserving data sharing. Techniques like federated learning allow organizations to collaborate on threat analysis without directly sharing sensitive data.

- *Global Threat Heat Maps*: AI can create real-time threat heat maps that visualize the intensity and distribution of cyberattacks globally. These visualizations can aid in identifying trends and hotspots of malicious activities.
- *Predictive Analysis*: AI algorithms can predict potential cyber threats based on historical data, helping nations prepare and take preventive measures.
- *Rapid Response Coordination*: AI-powered platforms can automate communication and coordination between nations during cyber threat incidents. This speeds up the exchange of information and decision-making on implementing mutual strategies for mitigating threats.
- *Continuous Learning*: AI models can continually learn from new threat data, adapting their detection capabilities to evolving attack methods.

While AI offers substantial global benefits, challenges exist, such as the need for international consensus on data-sharing protocols, data privacy concerns, and potential discrepancies in AI algorithms among others. Collaborative efforts must also consider legal, political, social, and cultural factors that may impact information sharing. Creating a secure digital future presents both significant challenges and opportunities as technology continues to evolve and shape various aspects of our lives. Balancing innovation and convenience with cybersecurity is crucial to ensure a safe and resilient digital environment. Here's a discussion of some key challenges and opportunities in this endeavor:

- Need for advanced tools and strategies to combat cyber threats

  To combat cyber threats in the rapidly evolving digital landscape, there is a dire need for advanced tools and strategies. Traditional cybersecurity measures have become obsolete for against these sophisticated persistent attacks. To effectively address these challenges, organizations and individuals must adopt advanced approaches, as follows:

  - *Threat Intelligence Platforms*: Advanced threat intelligence platforms aggregate data from various sources to provide real-time insights into emerging threats, enabling proactive defense strategies.
  - *Machine Learning and AI*: Machine learning and AI algorithms can examine and analyze extensive quantities of data to identify patterns and oddities subsequently strengthening threat detection mechanisms and response capabilities.
  - *Behavioral Analytics*: Advanced analytics tools monitor user and system behavior to identify deviations from the norm, helping to detect insider threats and unknown attack vectors.
  - *Zero Trust Architecture*: This strategy presupposes that no user, network, or device is inherently trustworthy and demands thorough verification at every stage of access, reducing the attack surface.
  - *Multi-Factor Authentication (MFA)*: MFA enhances security by necessitating users to furnish multiple forms of verification prior to access authorization.

- *Security Orchestration and Automation*: These tools automate incident response workflows, improving efficiency and reducing the time it takes to contain and mitigate cyber threats.
- *Endpoint Detection and Response (EDR)*: EDR solutions observe and address activities on endpoints in real time, allowing organizations to identify and contain threats at the device level.
- *Deception Technology*: Deception tools create decoy environments to lure attackers away from valuable assets, providing more time to identify and counter cyber threats.
- *Cloud Security Solutions*: Cloud-native security tools provide continuous monitoring and protection for cloud-based assets, ensuring data security in dynamic environments.
- *Threat Hunting*: It involves proactively mitigating compromises within networks and enabling early detection of hidden threats.
- *Collaborative Threat Sharing*: Sharing threat intelligence within industry groups and communities helps organizations collectively defend against common threats.
- *Security Awareness Training*: Ongoing training for employees helps them recognize and respond to phishing attempts, social engineering, and other common attack vectors.
- *Incident Response Plans*: Having standard incident response plans ensures that organizations can effectively and efficiently mitigate the impact of cyber incidents.
- *Regulatory Compliance*: Following industry regulations and standards can help guide security practices and ensure a baseline level of protection.

The evolving nature of cyber threats necessitates a proactive and multi-layered approach to cyberspace security. By implementing advanced tools and strategies, organizations and individuals can stay ahead of attackers and effectively mitigate the risks posed by increasingly sophisticated cyber threats.

1.8 CONCLUSION

There is an ongoing need for innovative approaches to cybersecurity through AI and international cooperation. AI's ability to process and analyze cyber threats and thereby share them across global intelligence platforms can significantly enhance global cybersecurity. As nations face increasingly sophisticated cross-border cyber threats, leveraging AI for international collaboration has the potential to strengthen defenses, respond more effectively to incidents, and foster a safer digital environment for everyone. In conclusion, the challenges and opportunities in creating a secure digital future require a multifaceted approach that involves technological innovation, collaboration, education, regulation, and individual responsibility. Addressing these challenges while capitalizing on opportunities will be essential to safeguarding our digital lives and promoting a secure and resilient digital ecosystem.

REFERENCES

1. Huang, Ming-Hui & Rust, R (2018). Artificial Intelligence in Service. *Journal of Service Research*, 21 (2), 155–172.8.
2. Kaplan, A. & Haenlein, M. (2019). Siri, Siri, in My Hand: Who's the Fairest in the Land? On the Interpretations, Illustrations, and Implications of Artificial Intelligence. *Business Horizons*, 62 (1), 15–25. https://www.researchgate.net/publication/328761767_Siri_Siri_in_my_hand_Who's_the_fairest_in_the_land_On_the_interpretations_illustrations_and_implications_of_artificial_intelligence
3. https://www.bbc.com/news/technology-30290540
4. Duan, Y., Edwards, J.S., & Dwivedi, Y. (2019). Artificial Intelligence for Decision Making in the Era of Big Data – Evolution, Challenges and Research Agenda. *International Journal of Information Management*, 48, 63–71. DOI 10.1016/j.ijinfomgt.2019.01.021
5. Dwivedi, Yogesh K., et al (2021). Artificial Intelligence (AI): Multidisciplinary Perspectives on Emerging Challenges, Opportunities, and Agenda for Research, Practice and Policy. *International Journal of Information Management*, 57, art. no. 101994, https://doi.org/10.1016/j.ijinfomgt.2019.08.002
6. Davenport, T., Guha, A., Grewal, D., & Bressgott, T. (2020). How Artificial Intelligence Will Change the Future of Marketing. *Journal of the Academy of Marketing Science*, 48 (1), 24–42. https://doi.org/10.1007/s11747-019-00696-0
7. Raisch, S. & Krakowski, S. (2021). Artificial Intelligence and Management: The Automation–Augmentation Paradox. *Academy of Management Review*, 46 (1), 192–210. https://doi.org/10.5465/2018.0072
8. Cath, C., Wachter, S., Mittelstadt, B., Taddeo, M., & Floridi, L. (2018). Artificial Intelligence and the 'Good Society': the US, EU, and UK Approach. *Science and Engineering Ethics*, 24 (2), 505–528. https://doi.org/10.1007/s11948-017-9901-7
9. Shneiderman, Ben (2020). Human-Centered Artificial Intelligence: Reliable, Safe & Trustworthy. *International Journal of Human-Computer Interaction*, 36 (6), 495–504. https://doi.org/10.1080/10447318.2020.1741118
10. Maheshwari, Deepak (2023) https://www.thehindubusinessline.com/opinion/after-25-years-a-data-protection-law/article67206414.ece
11. Mijwil, M., Mohammad, Aljanabi, & ChatGPT. (2023). Towards Artificial Intelligence-Based Cybersecurity: The Practices and ChatGPT Generated Ways to Combat Cybercrime. *Iraqi Journal for Computer Science and Mathematics*, 4 (1), 65–70. https://doi.org/10.52866/ijcsm.2023.01.01.0019
12. Farid, G., Warraich, N. F., & Iftikhar, S. (2023). Digital Information Security Management Policy in Academic Libraries: A Systematic Review (2010–2022). *Journal of Information Science*, 0 (0). https://doi.org/10.1177/01655515231160026
13. Goel, S. & Shawky, H.A (2009). Estimating the Market Impact of Security Breach Announcements on Firm Values. *Information & Management*, 46, 404–410.
14. Shaikh, F. A. & Siponen, M. (2023). Information Security Risk Assessments Following Cybersecurity Breaches: The Mediating Role of Top Management Attention to Cybersecurity. *Computers & Security*, 124, 102974, ISSN 0167-4048, https://doi.org/10.1016/j.cose.2022.102974.
15. Khaustova, Viktoria, ŢÎRLEA, Mariana Rodica, Dandara, Livia, Trushkina, Nataliia, & Bîrca, Iulita. (2023). DezvoltareaInfrastructuriiCritice Din Punct De Vedere Al SecurităţiiInformaţiilor, Development of Critical Infrastructure from The Point of View of Information Security. (2023). *RevistaUnivers Strategic*, 53, 170–188. *Central and Eastern European Online Library*. https://www.ceeol.com/search/article-detail?id=1107664

16. Amini, Mahyar & Bozorgasl, Zavareh. (2023). A Game Theory Method to Cyber-Threat Information Sharing in Cloud Computing Technology (January 2023*). International Journal of Computer Science and Engineering Research*, 11 (4). Available at SSRN: https://ssrn.com/abstract=4370033
17. Bhol, S.G., Mohanty, J.R. & Pattnaik, P.K. (2023). Taxonomy of Cyber Security Metrics to Measure Strength of Cyber Security, *Materials Today: Proceedings*, 80 (3), 2274–2279. ISSN 2214-7853. https://doi.org/10.1016/j.matpr.2021.06.228.
18. Malik, N., Husain, F., Ali, A. & Elahi, Y. A. (2023). Recent Advances in Cyber Security Laws and Practices in India-Implementation and Awareness. *Advances in Cyberology and the Advent of the Next-Gen Information Revolution*. IGI Global Publication, https://doi.org/10.4018/978-1-6684-8133-2.ch012.

2 Transformative Impact *Social Robots Redefining Industries and Services*

Farhina Sardar Khan, Priyanka Bajpai, and Syed Shahid Mazhar

2.1 INTRODUCTION

Social (or sociable) robots are designed to interact with people in a natural, interpersonal manner, often to achieve positive outcomes in diverse applications, such as education, health, quality of life, entertainment, communication, and tasks requiring collaborative teamwork [1]. What makes it difficult to create human-like robots [2]? Socio-robots, often referred to as social robots [3] or humanoid robots, are sophisticated robotic devices created to engage socially with people. Artificial intelligence, natural language processing, and other cutting-edge technologies have been built into them, allowing them to interact with humans through both verbal and nonverbal clues [4]. Research on social robots primarily comprises research on algorithmic problems to expand a robot's capabilities to improve its communication with humans [5]. Socio-robots are created to perform a variety of functions, including offering emotional support, engaging in social contact, and imparting knowledge [6]. They are frequently used in educational contexts to help students with their studies, in hospital settings to help patients with their daily routines, and in customer services to offer individualized assistance [7]. The capacity of social robots to detect and respond to human emotions is a fundamental characteristic. Human-robot intertwining aura is shown in Figure 2.1

FIGURE 2.1 Human-robot intertwining aura. (Source: website https://builtin.com/robotics/social-robot.)

DOI: 10.1201/9781003409571-2

2.2 THEORETICAL BACKGROUND

When societies are changing rapidly owing to technological breakthroughs, the investigation of social robots as agents of positive change is becoming increasingly relevant [8]. The motivation was to learn how these smart tools can be used to address important problems in social connectivity, healthcare, and education. Although few abilities trump keen observation and quality in a bedside manner, technological skills and experience are definitely on the rise [9]. With online education, technology has become a bridge between schools, instructors, and students. In addition to virtual lectures and online exams, you have patient simulations, augmented reality, and other training mechanisms that prepare students to enter increasingly tech-heavy careers upon graduation [9]. Acknowledging the encouraging results seen in these fields, there is a sincere desire to highlight the revolutionary possibilities of social robots, while also stressing the importance of ethical concerns and a human-centered approach. Social media provides new ways to connect directly with subscribers, including a feature that allows users to submit questions. Studying public health, d'Almeida and Villavecer discuss another social media channel where forums allow for the creation of individual channels to connect with smaller communities [10]. This investigation is driven by a desire to promote a nuanced understanding of how technology can improve the human experience and imagine a time when social robot benefits are seamlessly incorporated into society while upholding our moral underpinnings and fundamental beliefs [11].

Socio-robots, often referred to as social or humanoid robots, are sophisticated robotic devices created to engage socially with people [12]. Artificial intelligence, natural language processing, and other cutting-edge technologies have been built into them, allowing them to interact with humans through both verbal and nonverbal clues. Socio-robots are created to perform a variety of functions, including offering emotional support, engaging in social contact, and imparting knowledge. Several studies have adopted multimodal approaches by combining different modalities based on information captured by multiple robot sensors. For instance, some studies have used Kinect cameras to recognize emotions based on human facial expressions and gait, as presented in a recent study. Other studies combined facial and speech, body gestures, and voice to detect human emotions and improve human-robot interaction (HRI) or navigation. Additionally, some researchers converted speech to text and applied natural language processing (NLP) to recognize emotions from text and speech [13]. They are frequently used in educational contexts to help students with their studies, in hospital settings to help patients with their daily routines, and in customer service to offer individualized assistance [14]. Service robots continue to permeate and automate the hospitality sector [15]. The use of service robots in hotels has increased significantly in various countries. Consequently, there has been an increase in research in the hotel industry that focuses on customer-robot interactions and service robots [16]. However, studies have examined how service robots detect and respond to human emotions, as one of their fundamental characteristics. The market size for service robots in the healthcare and hospitality sectors is projected to grow by 942 million USD during 2020–2024 [16]. A study conducted by researchers at the Audencia Business School in France found that most hotel customers experience positive emotions, such as joy, love, awe/surprise, interest, and excitement when served by robots [17]. According to an article published in the International Hospitality Review, designing effective autonomous service robots requires an understanding of HRI, a relatively young discipline dedicated to understanding, designing, and evaluating robotic systems for use by or with humans. HRI has not yet received sufficient attention in hospitality robotic design, much like human-computer interaction (HCI) in property management system design in the 1980s [18].

They have sensors that can recognize human facial expressions and body language and change how they react as a result. Therefore, they are beneficial for offering emotional support to those who are under stress or anxiety [19]. Another crucial feature of social robots is their capacity to learn from their interactions with people. They are programmable to change their behavior in response to user feedback, which helps them to perform better over time [20, 21].

Socio-robots have the potential to be very helpful; however, there are concerns about how they will impact society as a whole. Some scientists worry that as people become increasingly reliant on technology for social interaction, they will become more isolated as a result of the widespread use of socio-robots [22]. Overall, the technology sector known as socio-robots is interesting and expanding quickly and has the potential to drastically change many facets of people's lives [22]. Robotics that demonstrates abilities beyond human capabilities is mesmerizing, including superior lifting, carrying, and running in all-terrain landscapes. [23]. It will be crucial, as with any new technology, to carefully assess the ethical implications of its use and ensure that it is implemented constructively and responsibly. These sophisticated machines can perceive, comprehend, and react to human input, enabling meaningful interactions and conversations [24]. Social robots have become a promising technology with the development of artificial intelligence, robotics, and HCI [25, 26]. They have the potential to revolutionize many facets of society. Social robots are intended to work in dynamic and unstructured situations, such as homes, hospitals, schools, and public areas, in contrast to standard industrial robots, which are primarily built for repetitive activities under controlled conditions [27]. They have sensors, actuators, and cutting-edge algorithms that allow them to perceive their surroundings, comprehend how people behave, and adjust their behavior accordingly [22]. There are numerous potential advantages and uses of social robots in various industries. They can help healthcare personnel with patient care, rehabilitation, and monitoring while also giving people in need of companionship and emotional support [28]. As per the International Federation of Robotics, the automation of production is accelerating around the world: 74 robot units per 10,000 employees is the new average of global robot density in the manufacturing industries [29]. Furthermore, an article in "The Lancet" suggests that frontline healthcare workers have a higher risk of testing positive for COVID-19 than the general public [30, 31]. Social robots can personalize learning and increase student engagement in the classroom by serving as instructors, mentors, and study partners. Additionally, they can act as social companions for children with autism or other developmental disabilities, promoting their social and emotional growth. In addition, social robots can solve loneliness and social isolation, which are serious problems in contemporary societies.

This study aims to discuss the multifaceted dimensions of social robot influence on industries and services. By exploring the evolving landscape shaped by these robotic entities, this study seeks to provide valuable insights from their integration. This motivation is grounded in the need to comprehend the broader societal, economic, and technological shifts triggered by the widespread adoption of social robots, ultimately contributing to a comprehensive understanding of their transformative impact.

2.3 METHODOLOGY

As different studies have explained that social robots can improve human experiences and solve societal issues, their integration into a variety of areas has attracted increasing attention. Social robots have been utilized in healthcare to offer companionship and support to patients as well as in education to aid in teaching and learning [32]. The purpose of this review is to provide a broad overview of research on the applications of social robots in different sectors. The various uses of social robots in healthcare, education, and social interactions are highlighted in this literature survey. Even though these studies show their potential advantages, it is crucial to approach their integration with a thoughtful awareness of the difficulties, challenges, and moral issues that may arise [33]. Future studies should concentrate on improving social robots' capacities while preserving human values to provide conditions for the peaceful cohabitation of intelligent machines and people.

2.3.1 Healthcare Sector

Patient care and professional support are about to undergo a paradigm shift with the introduction of social robots into the healthcare industry. Focusing on the effects of social robots on patient outcomes and the improvement of medical procedures, this literature review offers a comprehensive analysis of research on the application of social robots in diverse healthcare environments. Social robots have proven useful in the healthcare industry for assisting medical personnel and improving patient care. According to Belpaeme et al., social robots have a favorable effect on patients' well-being because of their capacity to keep an eye on vital signs, help with everyday chores, and offer companionship [34]. Furthermore, Winkle et al. [45] investigated how social robots reduce the workload of healthcare workers, especially in settings where human resources are scarce.

2.3.2 Education Sector

Educational research highlights the revolutionary potential of social robots in customized learning environments. An important study demonstrating the effectiveness of social robots as tutors and mentors in delivering individualized learning experiences was conducted by Belpaeme et al. (2018). This study highlights how social robots may modify lesson plans to meet the specific requirements of every student, increase engagement, and facilitate a more efficient learning environment. Johnson and Smith (2019) add to the body of literature by examining how social robots affect kids with special needs. Studies have emphasized the importance of social robots in offering a comforting and accepting atmosphere so that children who are experiencing extraordinary challenges can improve their social and communication abilities.

Studies conducted in 2019 by Dautenhahn et al. demonstrated how social robots improve student participation and engagement. This study investigated how these robots can foster an engaging and dynamic learning environment that motivates students to participate actively and work together. The growing use of social robots in education raises ethical questions and issues that require further discussion.

2.3.3 Social Interactions

One possible approach to address social isolation and loneliness is the development of social robots. According to a study by Lee and Kim (2021), social robots can be beneficial in helping with everyday chores, meaningfully conversing with the elderly, and offering companionship. These results demonstrate that social robots may enhance the mental health of people who do not have close social relationships. There are beneficial effects of social robots in reducing social isolation among seniors. The study highlights the ways in which these conversational robots converse with their elderly counterparts, offer companionship, and help with everyday tasks—all of which have a positive impact on their mental health and general quality of life. According to Broadbent et al. (2017), social robots can help with mental health issues by offering emotional support. This study demonstrates how these robots provide emotional support and companionship, which has a favorable impact on the mental health of people who do not have close social connections.

2.3.4 Challenges and Ethical Considerations

Even though the literature shows encouraging results, it is important to recognize the difficulties and moral dilemmas that arise when integrating social robots. In a study published in 2012, Sharkey and Sharkey expressed concerns regarding the possible dehumanization of healthcare and stressed the importance of carefully weighing ethical implications. To guarantee the responsible and moral deployment of social robots across multiple industries, future research ought to give precedence to tackling these issues. Addressing privacy problems and the need for informed permission in HRIs [35] adds to the literature. This study highlights the necessity of explicit consent procedures and open lines of communication, particularly in situations where social robots may gather and use personal data.

Winkle et al. [45] addressed technical obstacles and dependability problems related to social robots. This study emphasizes the importance of improving social robots' capabilities to guarantee their reliable and efficient operation, especially in vital industries such as healthcare. The socioeconomic effects of automation were examined in a study by Brynjolfsson and McAfee [7], along with the use of social robots and possible employment displacement. The study highlights the necessity of careful policy deliberations and community readiness to lessen detrimental effects on employment. As artificial intelligence is incorporated into social robots, questions regarding justice and bias have surfaced. In his discussion of the moral issues surrounding algorithmic decision-making, Diakopoulos (2016) emphasized the importance of dealing with biases to guarantee fair results and stop societal injustices from continuing.

2.4 APPLICATION OF SOCIAL ROBOTS IN DIVERSE INSTITUTIONAL LOCATIONS

The way we connect, work, and live has been revolutionized by the use of social robots in numerous fields. Some instances of social robots being utilized across different industries are as follows.

2.4.1 Healthcare

Social robots are becoming increasingly popular in healthcare settings owing to their ability to perform a multitude of tasks. These robots can assist medical personnel in providing care to patients and contribute to enhancing their overall wellness [36]. One of the major advantages of social robots is their ability to monitor vital signs and remind patients to take medication. They can also offer emotional support to those who need it, which is particularly useful in situations where patients may feel isolated or lonely [37]. Social robots can provide cognitive, physical, and social support, as well as coaching activities, such as exercise training, education, and monitoring [38]. Furthermore, Figure 2.2 presents a clear glimpse of social robots that can aid physical therapy by guiding patients through exercises and tracking their progress. This can be particularly beneficial for patients who require ongoing rehabilitation following an injury or surgery [39]. Overall, the use of social robots in healthcare is a promising development with the potential to improve patient outcomes and enhance the quality of care provided by medical professionals [40].

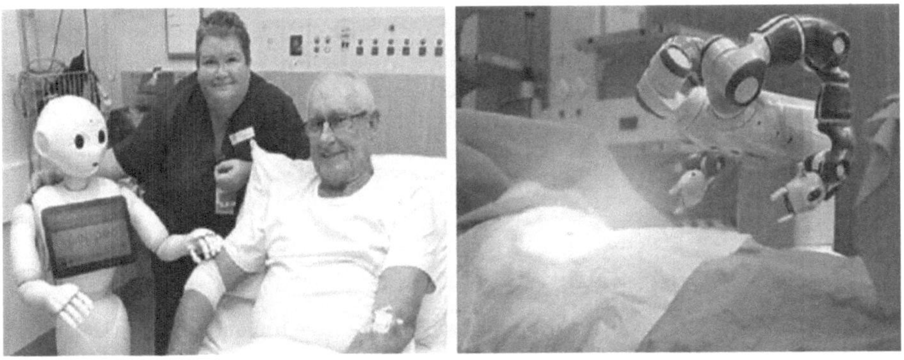

FIGURE 2.2 Social robots in medical hospitals. (Source: website https://www.bing.com/images/search?q=social+robots+in+healthcare&form=HDRSC3&first=1.)

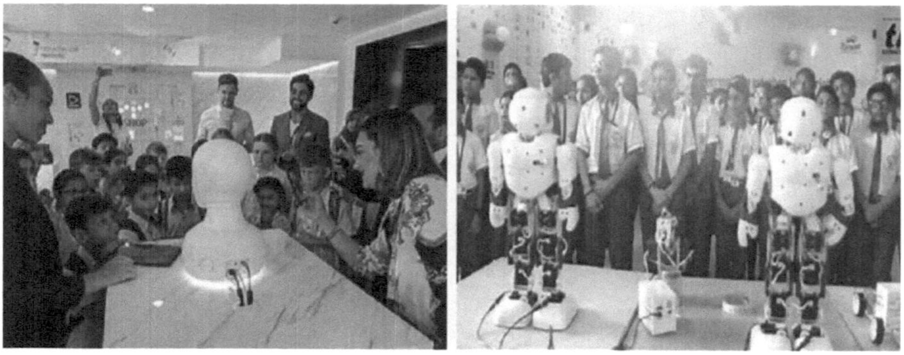

FIGURE 2.3 Social robots in educational sector. (Source: website https://miar.ub.edu/issn/0258-2724.)

2.4.2 Education

Studies have shown that the use of social robots in education has several benefits. Figure 2.3 shows the integration of social robots in an education sector. In a study conducted by researchers at Yale University, it was found that children who interacted with a social robot during a learning activity showed higher levels of engagement and motivation than those who did not [41, 42]. Additionally, a study by the University of California, San Diego found that social robots can be effective in teaching vocabulary to young children with autism. These findings highlight the potential of social robots to revolutionize the way we approach education and provide personalized learning experiences for students [43].

2.4.3 Customer Service

Social robots are widely used in various industries, such as retail, hospitality, and banking to assist customers [44]. Technology can also be used to develop innovative products, suitable service models, structured expansion, and appropriate regulatory norms [45]. These machines can guide users through different processes, answer frequently asked questions, and provide relevant information [46]. By doing so, they can significantly reduce wait times, enhance customer satisfaction, and offer a consistent and personalized experience [47].

2.4.4 Social Assistance

Social robots have been designed to provide companionship and emotional support, making them valuable tools for reducing social isolation [48]. They are equipped with the ability to engage in conversations, play games, and provide entertainment, which can be helpful in improving their mental well-being [48]. Elderly individuals, those with disabilities, and people who live alone can benefit greatly from

FIGURE 2.4 Social robots for social assistance. (Source: https://techcultak.pages.dev/posts/living-with-social-robots/.)

having a robot companion. By providing a friendly and interactive presence even for children shown in Figure 2.4, these robots can help alleviate feelings of loneliness and isolation [49].

2.4.5 Manufacturing and Industrial Settings

In recent years, social robots have become increasingly common in manufacturing and industrial settings. These robots are designed to work alongside human employees, helping them to perform tasks that require a high level of precision or pose a safety risk to human workers [50]. One type of social robot that is particularly popular in these settings is the cobot or collaborative robot [51]. Cobots are designed to work closely with human workers, enhance productivity and efficiency, and improve safety. With their advanced capabilities and sophisticated sensors [52], cobots can help streamline a wide range of industrial processes, from inventory management to quality control to assembly line operations [51]. As an increasing number of companies turn to social robots such as cobots to help them achieve their goals, it seems likely that we will continue to see these advanced machines playing an increasingly important role in the manufacturing and industrial sectors [53].

2.4.6 Public Spaces

Social robots are being used and becoming popular in public settings such as airports, museums, and shopping centers because of their ability to provide informative and interactive experiences for visitors. They are highly versatile and capable of guiding visitors through directions, explaining exhibitions, and participating in interactive displays to enhance their overall visits [54].

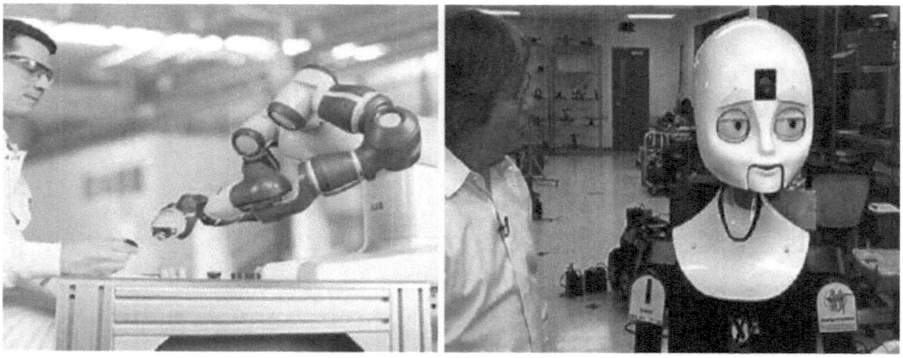

FIGURE 2.5 Social robot in research and development. (Source: Social robots in research and development – Bing images.)

2.4.7 Research and Development

Social robots play a crucial role in the advancement of robotics and artificial intelligence (AI) research and development. They provide an excellent platform for testing new algorithms and technologies, modeling social behaviors, and examining the interactions between humans and robots [27]. Furthermore, social robots have been used by researchers to experiment with theories, explore various applications as shown in Figure 2.5, and gain a deeper understanding of the complexities of social dynamics and human behavior [55]. As such, they are invaluable tools for the development of intelligent and empathetic robots that can effectively interact with humans in various settings.

2.5 INTEGRATING SOCIAL ROBOTS WITHIN THE HEALTHCARE SECTOR

Social robots have shown considerable promise in revolutionizing the healthcare industry by assisting patients, supporting medical personnel, and enhancing overall healthcare results. The following are some crucial uses of social robots in the medical field.

2.5.1 Patient Monitoring and Assistance

Social robots can assist with patient monitoring, data gathering, and upcoming medical procedures or medication reminder reminders [56, 57]. Additionally, they can support you as you go about your everyday business, such as getting out of bed, moving about, or exercising for recovery [58].

2.5.2 Elderly Care

With care for the elderly as the population ages, social robots can provide them with priceless care. They can keep a company, engage in conversation, play games,

FIGURE 2.6 Social robots entertaining elderly people. (Source: https://www.scpr.org/programs/airtalk/2020/09/16/66026/should-you-get-a-robot-for-your-grandparents-we-ex/.)

or even work on their brains by presenting mental challenges or memory drills [59]. Figure 2.6 exhibits how social robots can accompany aging citizens and also monitor senior folks' health and alert authorities or medical personnel to any emergencies [56].

2.5.3 Mental Health Support

People with mental health conditions can benefit from social robots both emotionally and practically. They can offer relaxation techniques, participate in talk therapy, and act as unbiased listeners [60]. Social robots can offer a confidential and safe space where people can express their emotions and receive assistance [60, 61].

2.5.4 Autism Spectrum Disorder (ASD) Therapy

Treatment for Autism Spectrum Disorder (ASD): ASD treatment: Social robots have been useful as therapy tools for children with ASD [62]. By interacting in controlled and predictable ways with these robots, children can develop their social and communication abilities [63].

2.5.5 Rehabilitation and Physical Therapy

Social robots can guide patients through exercises, monitor their progress, and provide feedback during physical therapy and rehabilitation sessions [64]. These robots might offer positive reinforcement, watch over movements, and ensure a perfect form, boosting the effectiveness of therapy and improving patient outcomes [65].

2.5.6 Hospital Assistance

By moving test samples, delivering medication and supplies to various departments, and traversing the facility on their own, social robots can assist hospitals. As a result, healthcare workers' workload is reduced, productivity is increased, and the likelihood of human error is decreased [66, 67].

2.5.7 Health Education and Information

Social robots have the potential to provide education to patients regarding their health conditions, available treatments, and lifestyle options. They can provide tips on maintaining a healthy lifestyle, address frequently asked medical questions, and assist in keeping track of an eye on one's health [58]. Social robots can also remind users to take their meds, schedule appointments, and follow up with their healthcare professionals [56].

2.6 THE IMPLEMENTATION OF SOCIAL ROBOTS IN THE EDUCATION SECTOR

The use of social robots in education has revolutionized the learning process, providing both students and teachers with numerous benefits. Some of the significant uses of social robots in education include the following.

2.6.1 Personalized Learning

Social robots can customize teaching strategies to match students' individual learning requirements and preferences. They offer interactive courses, content delivery at a pace that suits each student, and quick feedback, all of which contribute to personalized and engaging learning experiences [34].

2.6.2 Special Education Support

Social robots can help students with unique educational requirements such as ASDs or learning difficulties. By acting as social companions, they can assist students in developing their social and communication skills, creating an accepting and secure learning environment that promotes independence and self-assurance [41, 68].

2.6.3 Language Learning

Social robots can aid language learning by involving students in dialogue, pronunciation drills, and vocabulary exercises [69]. They enhance language acquisition by offering opportunities for interactive practice, modeling proper pronunciation, and providing quick feedback [70, 71].

2.6.4 STEM EDUCATION

Social robots can teach science, technology, engineering, and math (STEM) concepts interactively and engagingly. They can assist with practical experiments, demonstrate how science works, and guide students through problem-solving exercises to encourage critical thinking and creativity [34, 72]. Two types of robots have been used in education. The first was used to teach and encourage children in STEM subjects.

2.6.5 VIRTUAL TEACHING ASSISTANTS

Social robots can serve as virtual teaching assistants, assist teachers in running classrooms, address student inquiries, and present lesson materials. By handling administrative tasks, providing specialized assistance to students, and coordinating group activities, they can free up teachers' time [34, 68, 73].

2.6.6 COLLABORATIVE LEARNING

Social robots that promote group conversations, teamwork, and peer relationships can assist collaborative learning environments. They can delegate responsibilities, encourage communication, and promote cooperation among students, thereby creating a positive and engaging learning experience [74]. There are already many examples of social robots being used in educational settings. Two such examples are the Robocamp and Robotour programs, which are innovative co-learning concepts that utilize social robots as collaborative learning tools.

2.6.7 EMOTIONAL SUPPORT

Social robots can provide emotional support and companionship to students, particularly in challenging or stressful situations. Although social robots could be used as complementary tools, working in tune with human teachers to create effective and student-centered learning environments, it is essential to consider ethical concerns, such as privacy, data security, and the role of human educators in maintaining a balanced approach to technology integration. By doing so, social robots can improve learning outcomes, provide personalized instruction, and increase access to education [34, 68].

2.7 USE OF SOCIAL ROBOTS IN CUSTOMER SERVICES

The way businesses connect with their consumers is changing as an increasing number of social robots are being used in customer support positions across numerous industries. The following are some crucial ways in which social robots are used in customer service.

2.7.1 INFORMATION AND ASSISTANCE

Social robots can serve as interactive information kiosks to route users to products, answer queries, and provide product information. Compared to conventional static displays or signage, these robots can provide a more engaging and interactive experience, ensuring that clients obtain accurate and current information [75, 76].

2.7.2 Order Placement and Checkout

Social robots can help customers place orders and complete transactions. They can assist with secure payment alternatives, recommend products based on taste, and walk customers through the ordering process. These machines streamline customer service and shorten lines at checkout stations [77, 78].

2.7.3 Product Demonstrations

Social robots can be developed to highlight the major advantages, illustrate functionality, and highlight product features. They can offer compelling and dynamic product demonstrations that draw clients and help them comprehend the value the product offers [1, 27].

2.7.4 Customer Feedback and Surveys

Social robots can conduct surveys and collect customer feedback on behalf of a company. They could converse with customers, ask specific questions, and record their responses. Using these robots, companies can gather valuable data, evaluate customer satisfaction, and identify areas for improvement.

2.7.5 Queue Management

Social robots can help with queue management and direct consumers to the appropriate service area. Customers can receive real-time wait time updates [79], be informed when it is their turn to speak with a representative, and be directed to those who are available. Lowering annoyance and increasing the effectiveness of the service improves the entire customer experience [80].

2.7.6 Multilingual Support

Social robots can provide multilingual help, bridge linguistic divides, and enhance accessibility for users from various linguistic backgrounds. They can offer language translation services, facilitating clear communication and comprehension between clients and companies [81].

2.7.7 Brand Engagement and Marketing

Social robots can serve as brand ambassadors by promoting products, services, and brand messaging. They can interact with customers, deliver personalized marketing content, and create memorable experiences that leave a lasting impression [82]. These robots can contribute to brand awareness, customer loyalty, and positive brand associations [83].

2.8 EMPLOYING SOCIAL ROBOTS FOR AID IN SOCIAL SERVICES

These robots are designed to interact with people in a friendly and helpful manner, providing a range of services and assistance that can be tailored to meet everyone's needs [84]. The use of social robots in social assistance has the potential to revolutionize the way we provide care and support to those who need it the most [85]. In a variety of settings, social robots have proven effective in offering people social support and assistance. The following are some important uses of social robots for social assistance [86].

2.8.1 Companionship

People who are socially isolated or lonely may find companies with social robots [87]. They can converse, play games, and entertain, fostering a sense of community and lessening feelings of loneliness [31, 88].

2.8.2 Emotional Support

Social robots can provide empathy and emotional support to people experiencing difficult circumstances or battling mental health problems [89]. They can offer individualized coping mechanisms, reassuring words, and attentive listening, all of which contribute to greater emotional well-being [60].

2.8.3 Cognitive Stimulation

Social robots can help people engage in cognitive workouts and activities to improve their mental acuity. They can offer puzzles, memory exercises, and other mental exercises that encourage mental activity and support the maintenance of cognitive functions [90].

2.8.4 Assistance for the Elderly

Social robots can help the elderly in several ways. They can assist with daily chores, including scheduling appointments, reminding people to take medications, and organizing duties [91]. Social robots can also keep an eye on elderly people's health, monitor for falls or accidents, and notify caregivers or emergency services [92].

2.8.5 Support for Individuals with Impairments

People with impairments can receive support from social robots adapted to their requirements [93]. They can assist in personal care duties, mobility, and environmental navigation. Social robots can help people with cognitive limitations by reminding them of appointments and helping them with everyday duties [94].

2.8.6 Rehabilitation and Therapy

Social robots can be employed in therapeutic environments to assist in therapeutic and healing procedures. They can direct people through workouts, offer commentary on their actions, and track their development [58]. Social robots provide engagement and motivation, which increase rehabilitation program outcomes [95].

2.8.7 Development of Social Skills

Social robots can help people, especially young children, improve their social and communication skills [63]. They can offer a non-judgmental and secure setting for practicing social interactions, assisting people in navigating social situations, and enhancing their social skills [96].

2.9 USE OF SOCIAL ROBOTS IN MANUFACTURING AND INDUSTRIAL SETTINGS

Social robots are increasingly being employed in manufacturing and industrial settings to boost productivity, efficiency, and safety [97]. The following is a list of notable applications of social robots in the manufacturing and industrial sectors.

2.9.1 Task Requiring Collaboration

Social robots can cooperate with people to perform tasks that require cooperation [98]. They can help with laborious, repetitive tasks, including material handling, assembly line work, and packaging. Cobots, or collaborative robots, are built to work safely next to people, allowing for productive human-robot collaboration [99].

2.9.2 Quality Control

Social robots can be designed to inspect products for quality control [100]. They can perform measurements with extreme precision and accuracy using sensors and cameras to evaluate product quality, spot flaws, and identify errors [101]. By streamlining the quality control procedure, social robots can minimize human mistakes and guarantee a constant level of product quality [102].

2.9.3 Inventory Management

Social robots can help with inventory management duties, such as stocktaking, item locators, and supply chain management [103]. They can move around warehouse settings, monitor inventory levels, and provide current information on stock availability [104].

2.9.4 Maintenance and Repairs

In an industrial context, social robots can be used for maintenance and repair. They can perform standard maintenance procedures, inspect equipment, and identify spot anomalies [105]. Social robots can help with preventative maintenance procedures, equipment uptime, and downtime reduction [106].

2.9.5 Dangerous Environments

They can be outfitted with sensors and cameras to monitor the environment, perform inspections, and carry out jobs that could be dangerous [107] for human workers.

Social robots can reduce human exposure to hazardous conditions, thereby enhancing safety.

2.9.6 Training and Skill Development

Social robots can be used for training by providing employees [108] with a controlled and secure environment to practice tasks and learn new abilities. They can provide instructions and real-time feedback while simulating various situations. Social robots increase the value of training, lower the cost of training [109], and assist employees in mastering their jobs.

2.9.7 Data Gathering and Analytics

Social robots can gather information on a range of manufacturing and industrial operations, including output rates, energy usage, and machinery performance [110]. They can analyze data and offer insightful recommendations for decision-making, proactive maintenance, and process improvement [111].

2.10 INTEGRATION OF SOCIAL ROBOTS IN PUBLIC SPACES

In recent years, there has been a substantial increase in interest and attention surrounding the deployment of social robots in public settings. Social robots are created to engage and communicate with people in a way similar to how people naturally interact with one another. The following are some typical uses and advantages of social robots in public settings.

2.10.1 Customer Service

Social robots can assist and advise customers in malls, hotels, airports, and other public areas [112]. They can give advice, and instructions, respond to frequently asked inquiries, and improve the entire client experience [1].

2.10.2 Education

Social robots can also be used in libraries, museums, and schools [113]. All ages will learn more effectively and enjoy it thanks to their interactive courses, informative tours, and information on a variety of topics.

2.10.3 Healthcare

Hospitals and rehabilitation facilities are two places where social robots have been used [35]. They can help with patient monitoring, remind people to take their medications [27], and spend time with patients, especially the elderly or those who are struggling with mental health issues.

2.10.4 Entertainment

People can be entertained by social robots in public places such as theme parks, exhibits, and events. They can entertain audiences through dances, interactive games, and other activities to create a distinctive and interesting experience [114].

2.10.5 Socialization and Companionship

By offering companies and conversing with people in public places, social robots can help to fight social isolation and loneliness [95]. They can provide a sense of connection with human Figure 2.7, listen to people's tales, and provide emotional support.

2.10.6 Public Safety

Social robots can be used to improve public area security protocols. They can monitor the environment, detect suspicious activities, and alert the necessary authorities when needed. Additionally, they can alert the public to safety alerts and emergency information during a crisis.

2.10.7 Accessibility

Social robots can aid in the navigation of people with disabilities in public spaces. They can promote inclusion and accessibility by offering guidance, providing information in accessible formats, and acting as communication links between people and their surroundings.

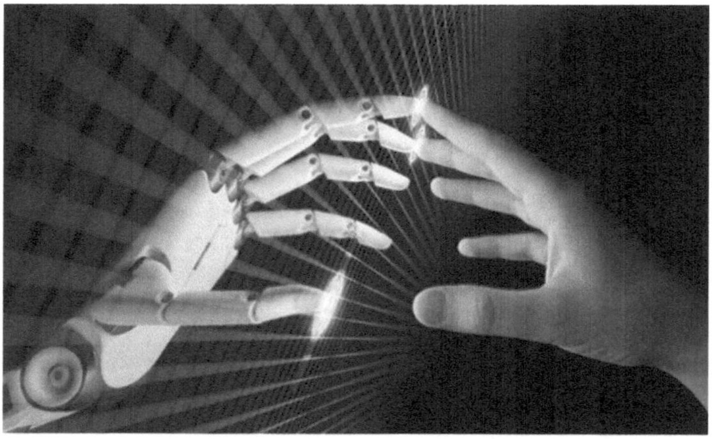

FIGURE 2.7 Human-robot interface. (Source:https://www.mediapost.com/publications/article/361244/iab-artificial-intelligence-group-to-build-standar.html.)

2.11 INCORPORATING SOCIAL ROBOTS IN THE REALM OF RESEARCH AND DEVELOPMENT

Research and development (R&D) with social robots provides numerous opportunities to investigate HRIs, research social behaviors, and create novel technologies. The following are some applications of social robots in research and development.

2.11.1 Human-Robot Interaction (HRI) Studies

In the controlled experiments, social robots were used to examine how people perceive and interact with them [115]. To learn how individuals react to social cues, convey emotions [88], and build rapport with robots, researchers look at things such as robot design, communication methods, and gestures [116].

2.11.2 Psychology and Social Sciences

Social robots are useful research tools for examining social behavior and human psychology [64]. They have been employed by researchers to conduct experiments on social impact, empathy, trust, and group dynamics [117].

2.11.3 Cognitive Development and Learning

Studies of human learning and cognitive development have used social robots. They can help observe and examine how people pick up new abilities, information, and social behaviors [118]. Together with social robots, researchers can study topics such as language learning, imitation, and problem-solving [119].

2.11.4 Autism and Special Needs

Social robots have shown promise for people with ASD and other specific needs [120]. They can also enhance social abilities, communication, and emotional comprehension.

2.11.5 Robotic Assistants and Collaborators

Social robots are employed as research platforms to create robotic helpers that can collaborate with people in a variety of settings [27]. For instance, researchers have examined how social robots may help with manufacturing, healthcare, and home chores [121].

2.11.6 User Experience and Acceptance

Researchers have investigated the manner in which social robots are received by users in various contexts. They looked at variables affecting users' opinions, attitudes, and trust in robots [122]. Users' feedback is used to improve the interaction capabilities and design of robots to ensure that they meet user wants and expectations [123].

2.11.7 Future Technology Development

The creation of intelligent robots, natural language processing, computer vision, and machine learning algorithms is influenced by the results of HRI studies and user feedback [124]. Researchers can learn more about human behavior, develop new technologies, and come up with original solutions to problems [125] that exist in the real world by utilizing social robots in R&D [126].

2.12 PROSPECTS FOR THE FUTURE DEVELOPMENT OF SOCIAL ROBOTS

The transition of robots into human has been witnessed by the countries around the world in almost every sector of economy as exhibits in Figure 2.8 that clearly conveys the glimpse of humanoid robots. Social robots may develop into sophisticated personal assistants who can plan, manage everyday duties, and offer tailored recommendations [127]. They can serve as observant and opportunistic companions, assisting with housework, organizing events, and shopping. Social robots could have a significant impact on older people's care and companionship as a result of the aging population [128, 129]. They may help with medical reminders, keep track of vital signs, and provide emotional support, which would ease the strain on healthcare systems and enhance seniors' general well-being [41, 130]. Social robots can revolutionize education by acting as individualized tutors and study partners. They can adapt to individual learning styles, deliver tailored lessons, and provide feedback, making learning more engaging and effective across various subjects and age groups [68, 131].

Social robots may help with emergency reactions in urgent circumstances [132]. They might offer real-time information, direct people to safety, and help first responders to identify and reduce threats. With AI and sensor capabilities, they may help

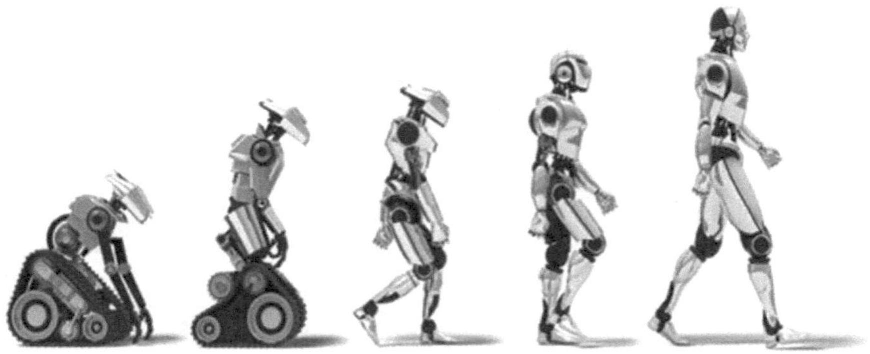

FIGURE 2.8 The transition of robots into humans. (Source: https://robohub.org/envisioning-the-future-of-robotics/.)

with sustainability and environmental monitoring projects, compile information on pollution levels, carry out ecological research [133], and aid in spreading awareness of environmental problems. Social robots will continue to engage and amuse people in diverse settings for amusement and relaxation purposes. They may engage with audiences, perform in theme parks, museums [54], and theaters, and offer immersive and engaging experiences [134].

As technology develops and our understanding of how humans and robots interact evolves, the potential applications of social robots are constantly growing [135]. With further research, development, and integration, social robots have the potential to become indispensable companions, assistants, and collaborators in numerous areas [136], positively impacting our daily lives and society.

2.13 DISCUSSION AND CONCLUSION

The diverse range of uses of these smart devices provides information about the advantages and disadvantages they may have in different industries [137]. The study's conclusions shed light on several important issues that influence conversations regarding the incorporation of social robots into many fields. These robots' capacity to keep an eye on vital indicators, help with everyday chores, and be a friend is viable option for increasing patient satisfaction and reducing medical staff workload. This result emphasizes how social robots can be used to solve important problems in healthcare settings, particularly when resources are scarce. Social robots can completely transform conventional teaching approaches by meeting each student's specific needs and improving learning outcomes as a whole [138].

The use of social robots is a significant technological achievement with broad ramifications for many facets of society. These interactive devices have the potential to revolutionize social integration, healthcare, education, and customer services. Social robots present special possibilities in public areas by offering help, knowledge, and entertainment to improve the overall user experience. They have the power to alter how we interact with the world, improving accessibility, inclusivity, and engagement in public settings. A deeper understanding of social dynamics, human behavior, and cognitive processes is also made possible by employing social robots in research and development. These studies encourage the development of intelligent technologies and innovations in fields such as psychology, education, and healthcare [139]. However, it is essential to address ethical issues and ensure that the social robot deployment is performed responsibly. To encourage the adoption and long-term success of these technologies, privacy protection, user trust, and upholding acceptable boundaries are crucial. Future social robot applications include individualized help, senior care, therapy, and collaborative workspaces. The capabilities and potential uses of social robots will continue to grow with the development of artificial intelligence and robotics. The use of social robots represents a paradigm shift in HRI and opens up new possibilities for support, learning, and engagement. Social robots have the potential to help create a society that is more connected, diverse, and technologically advanced, with careful ethical considerations and ongoing research [140].

REFERENCES

1. Cynthia Breazeal, Kerstin Dautenhahn and Takayuki Kanda, Social Robotics. In: Siciliano, B., Khatib, O. (eds.) *Springer Handbook of Robotics*. Cham: Springer Handbooks, Springer, 2016. https://doi.org/10.1007/978-3-319-32552-1_72
2. Stefan Schaal, "Is Imitation Learning the Route to Humanoid Robots?" Trends in Cognitive Sciences, 3(6), pp. 233 -242, 1999.
3. T. Belpaeme, J. Kennedy, A. Ramachandran, B. Scassellati and F. Tanaka, "Social Robots for Education: A Review," *Science Robotics*, 3(21), p. eaat5954, 2018. https://doi.org/10.1126/scirobotics.aat5954.
4. Edith Law, Dana Kulic, Pierre-Yves Oudeyer, Didier Roy and Jessy Ceha, "Identifying Functions and Behaviours of Social Robots during Learning Activities: Teachers' Perspective," *Cornell University*, 2021.
5. Claudia Muhl, Britta Wrede, Martina Hielscher-Fastabend, Gerhard Sagerer and F. Hegel, "Understanding Social Robots," *2009 Second International Conferences on Advances in Computer-Human Interactions*, pp. 169–174, 2009.
6. Jacob Biba, "https://builtin.com/robotics/social-robot," 10 March 2023. [Online]. Available: https://builtin.com/robotics/social-robot.
7. Erik Brynjolfsson and Andrew McAfee, The Second Machine Age: Work, Progress, and Prosperity in a Time of Brilliant Technologies. W. W. Norton & Company.
8. Jennifer Connolly, Leanne Hides, David J Kavanagh and Nicole L. Robinson, "Social Robots as Treatment Agents: Pilot Randomized Controlled Trial to Deliver a Behavior Change Intervention," *Internet Interventions*, 20, pp. 1–11, 2020.
9. Timon Kaple, https://www.edumed.org/resources/technology-in-healthcare-education/, August 2022. [Online]. Available: https://www.edumed.org/resources/technologyin-healthcare-education/.
10. D. Villavecer, "https://www.himss.org/resources/accelerating-use-social-media-health-education," May 2021. [Online]. Available: https://www.himss.org/resources/accelerating-use-social-media-health-education
11. Luís Proença, "TeleHealth and Digital Healthcare," 2024. [Online]. Available: https://www.mdpi.com/journal/healthcare/topical_collections/E-learning_Digital_Training_Healthcare_Education
12. https://en.wikipedia.org/wiki/Social_robot, "https://en.wikipedia.org/wiki/Social_robot," June 2022. [Online]. Available: https://en.wikipedia.org/wiki/Social_robot.
13. Raquel Patiño, José Diaz-Amado, Yudith Cardinale and Marco Quiroz, "Group Emotion Detection Based on Social Robot Perception," *MDPI*, 2022.
14. Rohit Singla and Christopher Nguan, https://www.intechopen.com/chapters/81719, 11 May 2022. [Online].
15. Iis P. Tussyadiah, Jason Stienmetz and Aarni Tuomi, "Applications and Implications of Service Robots in Hospitality," *Cornell Hospitality*, 12(1), pp. 232–247, 2021.
16. LisaC.WanandSungwooChoi,TheRiseofServiceRobotsintheHospitalityIndustry:Some Actionable Insights" Oct 2021. [Online]. Available: https://www.bu.edu/bhr/2021/10/04/the-rise-of-service-robots-in-the-hospitality-industry-some-actionable-insights/
17. Times, "Hotel customers experience positive emotions when they interact with physical robots," 2023. [Online]. Available: https://travel.economictimes.indiatimes.com/news/research-and-statistics/research/hotel-customers-experience-positive-emotionswhen-they-interact-with-physical-robot-reveals-new-research/101816689
18. Galen R. Collins, "Improving Human–Robot Interactions in Hospitality Settings," *International Hospitality Review, Emerald Insights*, Vol. 34 No. 1, pp. 61–79. https://doi.org/10.1108/IHR-09-2019-0019.
19. Marc Becker, Emir Efendić and Gaby Odekerken-Schröder, "Emotional Communication by Service Robots: a Research Agenda," *Journal of Service Management*, 33(5), pp. 675–687, 2022.

20. Evan Ackerman, "Meet Moxie, a Social Robot That Helps Kids With Social-Emotional Learning," IEEE Spectrum, 57(12), pp. 34–35, 2020.
21. Robin Fearon, "https://www.discovery.com/science/emotional-robotscmachines-thatrecognize-human-feelings,"02July2020.[Online].Available:https://www.discovery.com/science/emotional-robotscmachines-that-recognize-human-feelings.
22. Troy Miner, https://www.zivarobotics.com/robotics-social-impact-robots-changing/, 10 Feb 2023. [Online]. Available: https://www.zivarobotics.com/robotics-socialimpact-robots-changing/.
23. Katina Michael, "Robots and SocioEthical Implications," Technology and Society, ieee.org, 2018.
24. Yuliia Kniazieva, https://labelyourdata.com/articles/machine-perception-in-artificial intelligence, Oct 2022. [Online]. Available: https://labelyourdata.com/articles/machine-perception-in-artificial-intelligence.
25. Begoña Román Maestre, Carme Torras and Júlia Pareto Boada, "Ethics for social robotics: A critical analysis," *TRAITS-HRI,* 2022.
26. Jessica K. Barfield, "Discrimination Against Robots: Discussing the Ethics of Social Interactions and Who Is Harmed," Paladyn, Journal of Behavioral Robotics, 12(3), pp. 245–260, 2023. https://doi.org/10.1515/pjbr-2023-0012.
27. Anna Henschel, Guy Laban and Emily S. Cross, "What Makes a Robot Social? A Review of Social Robots from Science Fiction to a Home or Hospital Near You," *Current Robotics Reports*, 9(1), pp. 9–19, 2021.
28. Sara L. Lopes, Aristides I. Ferreira and Rui Prada, "The Use of Robots in the Workplace: Conclusions from a Health Promoting Intervention Using Social Robots," *International Journal of Social Robotics*, 15(4), pp. 893–905, 2023.
29. H. Erikson and M. Salzmann-Erikson, "https://www.delveinsight.com/blog/robotics-in-healthcare," DelveInsight Blog, 2021.
30. PAL Robotics, "https://blog.pal-robotics.com/exploring-the-potential-and-challenges-ofsocial-robots-in-healthcare/," Aug 2023. [Online]. Available: https://blog.pal-robotics.com/exploring-the-potential-and-challenges-of-social-robots-in-healthcare/.
31. S. D. Wood, "How can social robots relieve healthcare providers under pressure?" Oct 2020. [Online]. Available: https://furhatrobotics.com/blog/social-robotsrelieve-healthcare-providers/.
32. Oliver Korn, Social Robots: Technological, Societal and Ethical Aspects of Human-Robot Interaction, SpringerLink, 2019.
33. Expert Panel, Forbes Technology Council, "https://www.forbes.com/sites/forbes techcouncil/2019/09/03/15-socialchallenges-ai-could-help-solve/?sh=19ba4a0e3533," 03 September 2019. [Online].
34. Tony Belpaeme, James Kennedy, Aditi Ramachandran, Brian Scassellati and Fumihide Tanaka, "Social Robots for Education: A Review," *Science Robotics | Review*, 3(21), pp. 1–9, 2018.
35. Nikolaos Mavridis, Thirimachos Bourlai and Dimitri Ognibene, The Human-Robot Cloud: Situated Collective Intelligence on Demand. In *2012 IEEE International Conference on Cyber Technology in Automation, Control, and Intelligent Systems (CYBER)* (pp. 360 -365). IEEE, May, 2012.
36. Laura Aymerich-Franch and Iliana Ferrer, "Socially Assistive Robots' Deployment in Healthcare Settings: A Global Perspective," *Employment Social Robots Healthcare*, 20, pp. 1–28, 2021.
37. Heba Soffar, "Educational robotics, robot teachers, social robots review, features advantages & drawbacks," July 2023. [Online]. Available: https://www.online-sciences.com/robotics/educational-robotics-robot-teachers-social-robots-review-features-advantages/.
38. Len Calderone, "https://www.roboticstomorrow.com/article/2019/11/robots-assistingwith-physical-therapy/14439," Nov 2019. [Online]. Available: https://www.roboticstomorrow.com/article/2019/11/robots-assisting-with-physical-therapy/14439.

39. Carlos A. Cifuentes, Maria J. Pinto, Nathalia Céspedes and Marcela Múnera, "Social Robots in Therapy and Care," *Current Robotics Reports*, 1, pp. 59–74, 2020.
40. Min Hun Lee, Daniel P. Siewiorek, Asim Smailagic, Alexandre Bernardino and Sergi Bermúdez i Badia, "Enabling AI and Robotic Coaches for Physical Rehabilitation Therapy: Iterative Design and Evaluation With Therapists and Post-Stroke Survivors," *International Journal of Social Robotics*, 16, pp. 1–22, 2022.
41. William Weir, "Robots help children with autism improve social skills," YaleNews, Yale, 2018.
42. Karen Hao, "Robots that teach autistic kids social skills could help them develop," Feb 2020. [Online]. Available: https://www.technologyreview.com/2020/02/26/916719/ai-robots-teach-autistic-kids-social-skills-development/.
43. Amrita Bhatty and Farhina Sardar Khan, "Public Sector Banks' Role in Promoting Financial Inclusion: An Empirical Analysis of the Perspective of Financial Inclusion Schemes' Beneficiaries," *International Management Review*, 19, pp. 143 -149, 2023.
44. RBI, "FinTech: The Force of Creative Disruption," RBI Bulletin, 2020.
45. K. Winkle, F. Althaus and B. Nakisa. The Impact of Social Robots on Patient Care: A Scoping Review. *Frontiers in Robotics and AI*, 7, 45, 2020.
46. Aaron Hall, "Robots Revolutionizing Customer Service: Benefits, Drawbacks, and Examples," 12 Aug 2023.
47. Moojan Ghafurian, Colin Ellard and Kerstin Dautenhahn, "Social Companion Robots to Reduce Isolation: A Perception Change Due to COVID-19," In *Human-Computer Interaction–INTERACT 2021: 18th IFIP TC 13 International Conference, Bari, Italy, August 30–September 3, 2021, Proceedings, Part II 18* (pp. 43–63). Springer International Publishing.
48. E. Team, "Robots found to improve mental health and loneliness in older people," 20 Sept 2020. [Online]. Available: https://eandt.theiet.org/2020/09/08/robots-found-improve-mental-health-and-loneliness-older-people.
49. A. Sauppé and M. Mutlu, "The Social Impact of a Robot Co-Worker in Industrial Settings," in *Conference: the 33rd Annual ACM Conference*, USA, 2015.
50. WiredWorkers, "https://www.wiredworkers.io/cobot/," 2020. [Online].
51. M. Urwin, "https://builtin.com/robotics/cobot," 6 DEC 2023. [Online].
52. F. Insights, "Cobots: The Future of Industrial Robotics in 2023 & Beyond," 3 DEC 2023. [Online]. Available: https://www.futurebridge.com/industry/perspectives-industrial-manufacturing/cobots-the-future-of-industrial-robotics-in-2023-beyond/.
53. Mehdi Hellou, JongYoon Lim, Norina Gasteiger, Minsu Jang and Ho Seok Ahn, "Technical Methods for Social Robots in Museum Settings: An Overview of the Literature," *International Journal of Social Robotics*, 14(8), pp. 1767–1786, 2022.
54. Alaina Orr, Elin A. Björling, Maya Cakmak and Patrícia Alves-Oliveira, "Connecting the Dots of Social Robot Design From Interviews With Robot Creators," 2 June 2022. [Online]. Available: https://www.frontiersin.org/articles/10.3389/frobt.2022.720799/full.
55. D. O. Korn, "Social Robots onA New Perspective in Healthcare," *Engineering & Technology*, 9(12), 2020. https://researchoutreach.org/wp-content/uploads/2020/05/Oliver-Korn.pdf.
56. A. University, "Robotics in Healthcare: Past, Present, and Future," January 2022.
57. J. Biba (2023). What Is a Social Robot? https://builtin.com/robotics/social-robot.
58. Adina M. Panchea, François Ferland and Pouyan Asgharian, "A Review on the Use of Mobile Service Robots in Elderly Care," *Robotics*, 11(6), 2022.
59. Tony J. Prescott, Julie M. Robillard and Katarzyna Kabacińska, "Socially Assistive Robots as Mental Health Interventions for Children: A Scoping Review," *International Journal of Social Robotics*, pp. 919–935, 2021.

60. Xeven solutions, "https://www.xevensolutions.com/blog/the-role-of-social-robots-for-mental-health/," 16 Sept 2022. [Online]. Available: https://www.xevensolutions.com/blog/the-role-of-social-robots-for-mental-health/.
61. T Caprì, L Pignolo, R. Minutoli, F. Marino, C. Failla. A. Andrea Arnao, G. Tartarisco, A. Cerasa, G. Pioggia. Alfio Puglisi, "Social Humanoid Robots for Children With Autism Spectrum Disorders: A Review of Modalities, Indications, and Pitfalls," *Children*, 9(7), 2022.
62. Keith R. Laws, Karen Irvine, Silvana E. Mengoni, Alister Baird, Shivani Sharma and Athanasia Kouroupa, "The Use of Social Robots With Children and Young People on the Autism spectrum: A Systematic Review and Meta-Analysis," *PLOS ONE*, 17(6), e0269800. 2022.
63. Shelly Levy-Tzedek and Allison Langer, "Emerging Roles for Social Robots in Rehabilitation: Current Directions," *ACM Transactions on Human-Robot Interaction*, 10(4), pp. 1–4, 2021.
64. Jaelson Castro, Cinthya Lins, Judith Kelner, Maria Lencastre, Óscar Pastor and Larissa Rodrigues da Costa, "On the Use of Social Robots for Rehabilitation: The Case of NAO Physio," *SpringerLink,* pp. 507–517, 2023.
65. C. S. González-González, "Social Robots in Hospitals: A Systematic Review," *Applied Sciences*, 11(13), p. 5976, 2021. https://doi.org/10.3390/app11135976.
66. R. Matheson, "Study: Social robots can benefit hospitalized children," Massachusetts Institute of Technology, MIT Weekly Newsletters, USA, 2019.
67. J. Guggemos, S. Seufert, S. Sonderegger and M. Burkhard (2022), Social Robots in Education: Conceptual Overview and Case Study of Use. In: Ifenthaler, D., Isaías, P., and Sampson, D.G. (eds.), *Orchestration of Learning Environments in the Digital World. Cognition and Exploratory Learning in the Digital Age.* Cham: Springer. https://doi.org/10.1007/978-3-030-90944-4_10.
68. Vasfiye Geçkin, Cansu Oranç, Ezgi Mamus, Aylin C. Küntay, Tilbe Göksun and Junko Kanero, "Social Robots for Early Language Learning: Current," *Child Development Perspectives*, 12(3), pp. 146–151, 2018.
69. J. V. Rianne van den Berghe, "Social Robots for Language Learning: A Review," *Review of Educational Research*, 89(2), pp. 259–295, 2019.
70. Brechtje Jansen, Victoria Mondaca Bustos, Veerle L. N. F. Hobbelink, Daniel Preciado Vanegas and Elly A. Konijn, "Social Robots for (Second) Language Learning in (Migrant) Primary School Children," *International Journal of Social Robotics*, 17(6), e0269800. pp. 827–843, 2022.
71. J. D Redcay, Redcay's STEM-oscope Model: Connecting STEM Education, Social Robots, and Metacognition, Millersville University, Lancaster, PA, USA: IntechOpen Journals, 2019.
72. O. E. Outlook, "Pushing the Frontiers with Artificial Intelligence, Blockchain and Robots https://www.oecd-ilibrary.org/contact," OECD iLibrary News, 2023.
73. Y. Koren, G. Gordon and Rinat B. Rosenberg-Kima, "Robot-Supported Collaborative Learning (RSCL): Social Robots as Teaching Assistants for Higher Education Small Group Facilitation," *Front Robot AI*, 6, p. 148, 2019.
74. B. Turner, "Brian Turner last updated September 26, 2023," https://www.techradar.com/best/best-interactive-kiosk-providers, 2023.
75. N. Sakovich, "What Is an Information Kiosk? Definition, Functioning, Types, Benefits," https://www.sam-solutions.com/blog/information-kiosk/, 2023.
76. Alicia A. Grandey and K. Morris, "Robots Are Changing the Face of Customer Service," *Harvard Business Review*, 2023.
77. Matthias Schulte-Althoff, David Dillmann, Emmanuelle Maurer, Yannic Niedenzu, Philipp Schließer, Daniel Fürstenau and Gabriella Volpe, "Humanoid Social Robots and the Reconfiguration of Customer Service," In *The Future of Digital Work: The*

Challenge of Inequality: IFIP WG 8.2, 9.1, 9.4 Joint Working Conference, IFIPJWC 2020, Hyderabad, India, 2020, Proceedings (pp. 310–325). Springer International Publishing. December 10–11, 2020.

78. Christine Mathies, Paul G. Patterson and Amelia Amelia, "Customer Acceptance of Frontline Service Robots in Retail Banking: A Qualitative Approach," *Journal of Service Management*, *33*(2), pp. 321–341. 2021.
79. A. Gupta, A. Pal, B. Sriswathi, G. Mathur, S. Kumar Alaria and Aditi Vijay, "IoT Social Distancing & Monitoring Robot For Queue," *International Journal of Engineering Trends and Applications (IJETA)*, 8(4), pp. 20–25, 2021.
80. https://referralrock.com/blog/social-media-brand-ambassadors/, "Social Media Brand Ambassadors: Start a Social Media Ambassador Program," 23 August 2022. [Online]. Available: https://referralrock.com/blog/social-media-brand-ambassadors/.
81. R. Tyagi, "Robots as Brand Ambassadors to Empower your Team: Why Augment your Staff With a Robot as a Brand Ambassador?" https://www.aldebaran.com/en/blog/news-trends/robots-brand-ambassadors-empower-your-team-why-augment-your-staff-robot-brand, 2021.
82. Meia Chita-Tegmark and M. Scheutz, "Assistive Robots for the Social Management of Health: A Framework for Robot Design and Human–Robot Interaction Research," *International Journal of Social Robotics*, 13(2), pp. 197–217, 2020.
83. Yale University, "Socially Assistive Robotics," Yale University, 2024.
84. Julie Bouchard, Kevin Bouchard, G Sébastien and Adam Robaczewski, "Socially Assistive Robots: The Specific Case of the NAO," *International Journal of Social Robotics*, 13, pp. 795–831, 2020.
85. F. E. Angelika Penner, "Germ-Free Robotic Friends: Loneliness During the COVID-19 Pandemic Enhanced the Willingness to Self-Disclose Towards Robots," *Robotics*, 2022.
86. University of Glasgow Report, "Social Robots Could Be an Effective Tool To Combat Loneliness," University of Glasgow, 2023.
87. Denis Trayman, Chen Levy, Adi Manor, Mario Mikulincer, Oren Zuckerman and Hadas Erel, "Enhancing Emotional Support: The Effect of a Robotic Object on Human–Human Support Quality," *International Journal of Social Robotics*, 14(1), pp. 257–276, 2022.
88. S. L. Timothy and J. Legg, "https://www.healthline.com/health/mental-health/brain-exercises," 17 February 2023. [Online]. Available: https://www.healthline.com/health/mental-health/brain-exercises.
89. Baiju, "https://roboticsbiz.com/top-seven-companion-and-social-robots-for-elderly-people/," 9 May 2020. [Online]. Available: https://roboticsbiz.com/top-seven-companion-and-social-robots-for-elderly-people/.
90. Alessio Antonini, Enrico Motta and Gianluca Bardaro, "Robots for Elderly Care in the Home: A Landscape Analysis and Co-Design Toolkit," *International Journal of Social Robotics*, 14(3), pp. 657–681, 2022.
91. F. Escalona and Ester Martinez-Martin, "Socially Assistive Robots for Older Adults and People With Autism: An Overview," *Electronics*, 9(2), p. 367. 2020.
92. G. Olbring and S. Yumakulov, "Social Robots: Views of Staff of a Disability Service Organization," *International Journal of Social Robotics*, 6, pp. 457–468, 2014.
93. D. H. García, P. G. Esteban and H. R. Lee, "Social Robots in Therapy and Care," *Human-Robot Interaction (HRI)*. In *2019 14th ACM/IEEE International Conference on Human-Robot Interaction (HRI)* (pp. 669–670). IEEE. 2019.
94. B. Eren, "The Use of Music Interventions to Improve Social Skills in Adolescents With Autism Spectrum Disorders in Integrated Group Music Therapy Sessions," *Procedia - Social and Behavioral Sciences*, 197, pp. 207–213, 2015.
95. A. Haleem, R. Pratap Singh, S. Rab, R. Suman and Mohd Javaid, "Significant Applications of Cobots in the Field of Manufacturing," *Cognitive Robotics*, 2, pp. 222–233, 2022.

96. J. Biba, "What Is a Social Robot?" https://builtin.com/robotics/social-robot, 2023.
97. Ami Eidels, Michael Arthur-Kellyand and Despoina Damianidou, "The Use of Robots in Social Communications and Interactions for Individuals With ASD: a Systematic Review," *Advances in Neurodevelopmental Disorders*, 2, pp. 357–388, 2020.
98. Daniel J. Rea, James E. Young, Ehud Sharlin and S. Banh, "Inspector Baxter: The Social Aspects of Integrating a Robot as a Quality Inspector in an Assembly Line," in *HAI '15: Proceedings of the 3rd International Conference on Human-Agent Interaction*, New YorkNYUnited States, 2015.
99. L. Ruiz-Garcia, G.P. Moreda, Renfu Lu, N. Hernandez-Sanchez, E.C. Correa, B. Diezma, B. Nicolaï, J. García-Ramos and M. Ruiz-Altisent, "Sensors for Product Characterization and Quality of Specialty Crops—A Review," *Computers and Electronics in Agriculture*, 74(2), pp. 176–194, 2010.
100. G. Kovács, L. Dudás and Rabab Benotsmane, "Economic, Social Impacts and Operation of Smart Factories in Industry 4.0 Focusing on Simulation and Artificial Intelligence of Collaborating Robots," *Social Sciences MDPI*, 8(5), p. 143, 2019.
101. R. Katoch, "IoT Research in Supply Chain Management and Logistics: A Bibliometric Analysis Using Vosviewer Software," *Materials Today: Proceedings*, 56(5), pp. 2505–2515, 2022.
102. P. Mauricio Ribera, Jennifer A. Farris, Randall G. Kirk and Scott J. Mason, "Integrating the Warehousing and Transportation Functions of the Supply Chain," *Transportation Research Part E: Logistics and Transportation Review*, 39(2), pp. 141–159, 2003.
103. Hokeun Kim, Charles C. Kemp and Daehyung Park, "Multimodal Anomaly Detection for Assistive Robots," *Autonomous Robots*, 43, pp. 611–629, 2019.
104. L. S. Jasiulewicz-Kaczmarek Małgorzata, "Maintenance 4.0 Technologies – New Opportunities for Sustainability Driven Maintenance," *Management and Production Engineering Review*, 11, pp. 74–87, 2020.
105. Micah H Clark and David J. Atkinson, "Methodology for Study of Human-Robot Social Interaction in Dangerous Situations," in *HAI '14: Proceedings of the Second International Conference on Human-Agent Interaction*, 2014.
106. M. Donnermann, F. Rossin and B. Lugrin, "Pepper on the Job: Applying Social Robots in Employee Training," in *2023 32nd IEEE International Conference on Robot and Human Interactive Communication (RO-MAN)*, 2023.
107. J. Wirtz, "Organizational Ambidexterity: Cost-Effective Service Excellence, Service Robots, and Artificial Intelligence," *Organizational Dynamics (Science Direct)*, 49(3), pp. 1–9, 2020.
108. Yiming Miao, Gaoxiang Wu, Mohammad Mehedi Hassan, Iztok Humar and Long Hu, "iRobot-Factory: An Intelligent Robot Factory Based on Cognitive Manufacturing and Edge Computing," *Future Generation Computer Systems*, 90, pp. 569–577, 2019.
109. Y. Kwon, J. Lu, Z. Wu, X. Ma and Zhenhui Peng, "Design and Evaluation of Service Robot's Proactivity in Decision-Making Support Process," in *CHI '19: Proceedings of the 2019 CHI Conference on Human Factors in Computing Systems*, 2019.
110. R. K. Singh, R. Mishra, D. Demircol, T. Daim and Laxmi Pandit Vishwakarma a, "The Adoption of Social Robots in Service Operations: A Comprehensive Review," *Technology in Society*, 76, p. 102441, 2024.
111. Heather E. Gary, Solace Shen and Peter H. Kahn Jr, "Children's Social Relationships With Current and Near-Future Robots," *Child Development Perspective*, 7(1), pp. 32–37, 2012.
112. R. Bogue, "The Role of Robots in Entertainment," *Industrial Robot*, 49(4), pp. 667–671. 2022.
113. H. Yu, C. Shum, M. Alcorn, J. Sun and Z. He (2022). Robots can't Take My Job: Antecedents and Outcomes of Gen Z Employees' Service Robot Risk Awareness. *International Journal of Contemporary Hospitality Management*, *34*(8), 2971–2988.

114. Ryan C. Johnson, Kristin N. Saboe, Linda R. Elliott, Michael D. Coovert and Matthew S. Prewett, "Managing Workload in Human–Robot Interaction: A Review of Empirical Studies," *Computers in Human Behavior*, 26(5), pp. 840–856, 2010.
115. K. Dautenhahn, "Socially Intelligent Robots: Dimensions of Human–Robot Interaction," *Philosophical Transactions of the Royal Society B: Biological Sciences*, 362(1480), 2007.
116. E. Hargittai, "Second-Level Digital Divide: Mapping Differences in People's Online Skills," *Computers and Society, arXiv preprint cs/0109068,* 2001.
117. Z. Pei and Y. Nie, "Educational Robots: Classification, Characteristics, Application Areas and Problems." In *2018 Seventh International Conference of Educational Innovation through Technology (EITT)* (pp. 57–62). *IEEE*, 2018.
118. A. Kelemen, Y. Liang, C. Sik-Lanyi and Joan DiPietro, "Computer- and Robot-Assisted Therapies to Aid Social and Intellectual Functioning of Children With Autism Spectrum Disorder," *Medicina*, 55(8), p. 440, 2019.
119. M. N. Kamel Boulos and Torbjørn S. Dahl, "Robots in Health and Social Care: A Complementary Technology to Home Care and Telehealthcare?" *Robotics*, 3(1), pp. 1–21, 2014.
120. Mehdi Hellou, Ho Seok Ahn and Norina Gasteiger, "Factors for Personalization and Localization to Optimize Human–Robot Interaction: A Literature Review," *International Journal of Social Robotics*, 15(4), pp. 689–701, 2023.
121. J. Scholtz, "Theory and Evaluation of Human Robot Interactions," In 36th Annual Hawaii International Conference on System Sciences, 2003. Proceedings of the (pp. 1–20). IEEE. 2003.
122. Mohsen Soori, Behrooz Arezoo and Roza Dastres, "Artificial Intelligence, Machine Learning and Deep Learning in Advanced Robotics, a Review," *Cognitive Robotics*, pp. 54–70, 2023.
123. A. Cangelosi, A. Weinberger, E. Mazzoni, M. Benassi, M. Barbaresi, M. Orsoni and Martina Benvenuti a, "Artificial Intelligence and Human Behavioral Development: A Perspective on New Skills and Competences Acquisition for the Educational Context," *Computers in Human Behavior*, 148, p. 107903, 2023.
124. J.M. Box-Steffensmeier, J. Burgess and M. Corbetta et al. "The Future of Human Behaviour Research," *Nature Human Behaviour,* 6, pp. 15–24, 2022. https://doi.org/10.1038/s41562-021-01275-6
125. M. Javad Ranjbar Kalharoodi, M. Bagher Menhaj and H. Taheri, "Social robots: an open-source architecture for personal assistant robots," 2022 10th RSI International Conference on Robotics and Mechatronics (ICRoM), Tehran, Iran, Islamic Republic of, 2022, pp. 242–247, https://doi.org/10.1109/ICRoM57054.2022.10025253.
126. Editorial, "Top 5 robot personal assistants for your home in 2022," https://roboticsbiz.com/top-5-best-robot-personal-assistants-for-your-home-in-2019/, 2021
127. A. M. Panchea, F. Ferland and Pouyan Asgharian, "A Review on the Use of Mobile Service Robots in Elderly Care," *Robotics*, 11(6), p. 127. 2022.
128. N. Mavridis (2015). Ethical Ramifications of Social Robots. Science and Engineering Ethics, 21(3), pp. 553–570. https://doi.org/10.1007/s11948-014-9565-5
129. F. T. Tony Belpaeme, "Social Robots as Educators," OECD Digital Education Outlook 2021: Pushing the Frontiers with Artificial Intelligence, Blockchain and Robots, University of Plymouth, United Kingdom, 2021.
130. D. Jamshidnejad, "Autonomous robots protect the lives of emergency services in disasters," 2023. [Online]. Available: https://www.tudelft.nl/en/stories/articles/autonomous-robots-protect-the-lives-of-emergency-services-in-disasters.
131. Adam. S. Jacoff and Ann Virts, "Emergency Response Robots," National Institute of Standard Technology, US, 2020.

132. Carmen Lafuente-Ibañez, Natalia Fernandez Alvarez, Teresa Villace-Molinero and Laura Fuentes-Moraleda, "Willingness to Accept Social Robots in Museums: An Exploratory Factor Analysis According to Visitor Profile," *Laura Fuentes-Moraleda, Carmen Lafuente-Ibañez, Natalia Fernandez Alvarez, Teresa Villace-Molinero*, 40(4), pp. 894–913, 2021.
133. T. Report, "Social Robots: The Future of Personal Assistants and Companions," 23 JUNE 2023. [Online]. Available: https://techinsiderhq.com/article/social-robots-the-future-of-personal-assistants-and-companions/.
134. S. Saunderson, "https://theconversation.com/robots-can-be-companions-caregivers-collaborators-and-social-influencers-172215," 23 November 2021. [Online]. Available: https://theconversation.com/robots-can-be-companions-caregivers-collaborators-and-social-influencers-172215.
135. A. Corrales-Paredes, "User Experience Design for Social Robots: A Case Study in Integrating Embodiment," *Sensors*, 23(11), p. 5274, 2023.
136. P. Golata, "Social Impact of Robotics in the Workplace: The Arrival of the Perfect Coworker," 29 Sept 2020. [Online]. Available: https://www.wevolver.com/article/social-impact-of-robotics-in-the-workplace-the-arrival-of-the-perfect-coworker.
137. Daniel J. Rea, James E. Young, Ehud Sharlin and S. Banh, "Inspector Baxter: The Social Aspects of Integrating a Robot as a Quality Inspector in an Assembly Line," in HAI '15: Proceedings of the 3rd International Conference on Human-Agent Interaction, New YorkNYUnited States, 2015.
138. J. Chun, J. Lee, J. Kim and S. Lee, "An international systematic review of cyberbullying measurements," Comput. Human Behav, vol. 113, p. 106485, 2020.
139. Darragh, M., Ahn, H. S., MacDonald, B., Liang, A., Peri, K., Kerse, N., & Broadbent, E. (2017). Homecare robots to improve health and well-being in mild cognitive impairment and early stage dementia: results from a scoping study. Journal of the American Medical Directors Association, 18(12), 1099-e1.
140. Diakopoulos, N. (2017). Enabling accountability of algorithmic media: transparency as a constructive and critical lens. Transparent data mining for Big and Small Data, 25–43.

3 Strategies for Cyber-Resilience

Combatting Advanced Persistent Threats in the Digital Age

Mohd Haroon, Manish Madhava Tripathi, Afsaruddin Khan, Shish Ahmad, Muhammad Kalamuddin Ahmad, and Mohd Husain

3.1 INTRODUCTION

An advanced persistent threat (APT) is a sophisticated and stealthy cyberattack that is typically carried out by highly skilled and financially backed threat actors, such as nation-states, organized cybercrime groups, or state-sponsored hackers. Persistence, or the ability to enter a targeted network or system and remain hidden there for an extended period of time—often months or even years—is what sets APT attacks apart. Sophisticated, individualized hacking techniques are used by APT attackers to circumvent security measures and get to their targets. This could use social engineering strategies, zero-day vulnerabilities, or sophisticated malware. APT attackers typically invest a lot of time and energy into sustaining a prolonged presence within a compromised network because they are persistent and patient. They take care to avoid detection and keep access for as long as they can. APT attacks are designed to be undetectable; thus, it will be difficult for staff members and security systems to recognize them. This often requires the use of intricate evasive maneuvers. APT attacks are not random; instead, they target certain businesses, industries, or individuals who are of interest. These targets are specifically picked to align with the attackers' goals, which could involve espionage, data theft, disruption of critical infrastructure, or theft of intellectual property. APT attackers have an abundance of resources at their disposal, such as finance, skilled personnel, and state-of-the-art tools and technology [1].

APT attackers routinely keep a close eye on the networks of their targets in order to gather intelligence, retain access, and adjust their tactics as needed. One of the primary goals of APT attacks is to steal confidential information, such as bank account details, secret documents, or intellectual property. Attackers use a range of strategies to steal this data secretly. Enterprises must implement robust cybersecurity measures, such as network segmentation, recurring security audits, threat hunting, intrusion detection systems (IDSs), and employee education to spot and report suspicious

DOI: 10.1201/9781003409571-3

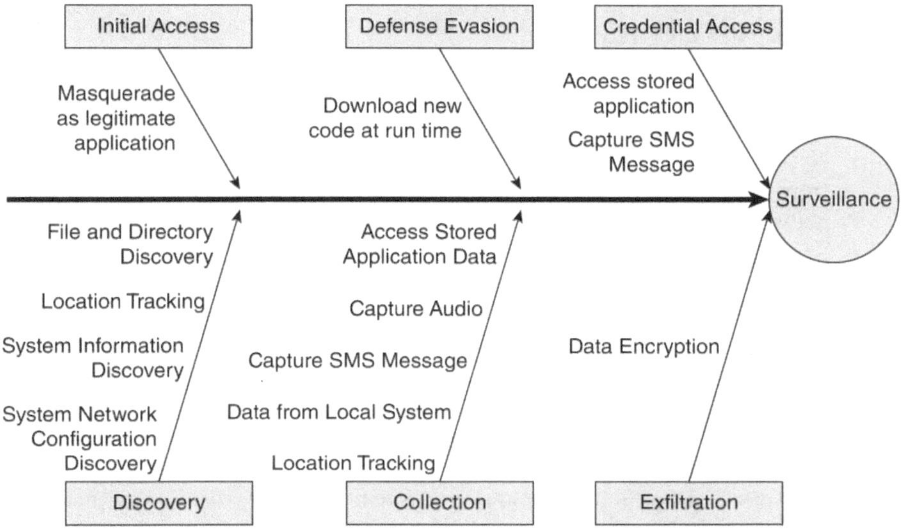

FIGURE 3.1 The MITRE framework-based life cycle of a frozen cell attack.

activities, in order to combat APTs. Cooperating with other organizations and governmental bodies, as well as exchanging threat intelligence, can help detect and neutralize APT assaults more successfully [2]. Figure 3.1 illustrates a vulnerability by organizing various attack patterns hierarchically based on the method used to exploit a vulnerability.

3.2 STRONG CYBERSECURITY PROTOCOLS

To protect an organization's data and digital assets from various online threats, including advanced persistent threats (APTs), robust cybersecurity defenses are essential. Some crucial components of effective cybersecurity measures are as follows:

- Firewalls and Intrusion Detection/Prevention Systems (IDS/IPS): Network traffic entering and leaving the system is inspected and filtered by firewalls and IDS/IPS systems, stopping known attack patterns and malicious activity.
- Access Control: Use strict access control policies and processes, such as role-based access control (RBAC), to ensure that only authorized people have access to vital systems and data.
- Regular Software Upgrades and Patch Management: Maintain all software, including operating systems, applications, and security tools, up to date with the latest upgrades and patches in order to address known vulnerabilities.
- Endpoint Security: Use endpoint security solutions (including antivirus, anti-malware, and host-based IDSs) to protect certain network devices.

- Encryption: When sensitive data is in transit or at rest, use encryption to prevent unauthorized access.
- Multi-Factor Authentication (MFA): Require multi-factor authentication (MFA) to access critical systems and apps, adding an extra layer of protection on top of passwords.
- Network Segmentation: Divide the network into segments to stop attackers from spreading laterally and to ensure that an intrusion in one part of the network does not impact the entire infrastructure.
- Security Awareness Training: Educate employees and users on cybersecurity best practices, including how to recognize and avoid social engineering and phishing frauds. To ensure a timely and well-coordinated response to security issues, including APTs, develop and periodically evaluate an incident response strategy.
- Threat Intelligence: You may stay informed about new dangers and APT activities by subscribing to threat intelligence services and sharing information with colleagues in the business and government organizations.
- Network and Endpoint Monitoring: To detect and handle any suspicious activity, continuously monitor network traffic and endpoint activity using security information and event management (SIEM) systems and other monitoring technologies.
- Regular Penetration Tests and Security Audits: Regular security audits, vulnerability assessments, and penetration tests are carried out to find weaknesses and vulnerabilities in the network and applications.
- Data Backup and Recovery: Conduct frequent data backups and evaluate the restoration process to ensure that crucial data can be restored in the case of a ransomware attack or data breach.
- Vendor Risk Management: Assess the security protocols adhered to by outside vendors and service providers with access to your company's systems or data.
- Screening of Employees and Contractors: Perform thorough security and background checks on employees and contractors who have access to confidential information.
- Security Rules and Processes: Establish comprehensive security rules and procedures that cover every aspect of internal cybersecurity for your business.
- Continuous Improvement: Cybersecurity maintenance is a continuous practice. Regularly review and update security measures to stay on top of evolving threats and vulnerabilities. By integrating technology, protocols, and human awareness, strong cybersecurity is a multi-layered approach that effectively protects an organization's assets from cyber threats, including APTs [3, 4].

3.3 THE GLOBAL MARKET FOR ADVANCED PERSISTENT THREATS

According to Advanced Persistent Threat Market Research, 2030, it was valued at \$5.9 billion in 2021 and is projected to grow at a compound annual growth rate (CAGR) of 20.5% to reach \$30.9 billion by 2030. Backdoors are used by malicious groups, also referred to as APTs, to obtain unauthorized access to servers or networks and remain undetected for an extended length of time. Businesses and governments are looking for advanced security to protect their databases because ransomware events and cybercrimes are growing quickly. APTs are challenging to detect and eliminate because they are purposefully built within a company's computer systems. In addition, the people who create these threats constantly monitor and control their actions, updating the code as necessary to evade detection and adopt new behaviors. It prevents the company from being able to keep an accurate eye on outward communication activity [5].

APT solutions are integrated for threat and malicious attack detection, defense, and potential repair. The solutions, services, and deployment mode sectors comprise the APT market. Since email and internet access are the most accessible platforms via which attacks might happen, providers need to offer solutions that logically gather and correlate data about a potential attack from a variety of sources and channels. These technologies have a wide range of applications, such as software, hybrid models, private or public clouds, and virtual or real appliances. To identify and neutralize APTs, network administrators, security companies, and individual users utilize a variety of security techniques, such as access restriction, bot-bot defense, and traffic monitoring. The primary factor driving the considerable rise of the APT security industry globally is the increasing frequency of APT-like intrusions. For instance, the Federal Bureau of Investigation (FBI) and the Cybersecurity and Infrastructure Security Agency (CISA) together alerted the public to the dangers of APT in October 2020. The Department of Justice (DOJ) revealed claims that six Russian nationals had worked with the Sandworm APT, a group well-known for utilizing TeleBots in very visible hacks. The majority of enterprises, including those in retail, healthcare, and information technology (IT), are using integrated security services to protect cloud environments from APT-style attacks. Over the course of the forecast period, this might potentially cause a spike in the need for APT protection solutions. Over the course of the forecast period, it is expected that the growing cost of APT protection platforms will hinder the growth of the global market. Additionally, the global market for APT defense may suffer from a lack of knowledge about APTs and their complexity [6].

Because of the increasing usage of cybersecurity platforms in the healthcare industry, biological research and medicine have witnessed tremendous increases in technological advancements in recent years. The platform is essential for locating and addressing any threats or vulnerabilities related to protected health information (PHI). Other benefits of this state-of-the-art platform include risk assessment, secure data transmission, authorization of security measures, and authentication of data storage. As a result, a lot of healthcare facilities have chosen to employ integrated security systems in recent years. For instance, the "HITRUST Cyber Threat XChange (CTX) Deceptive Program," which enables the deployment of honeypots throughout the healthcare ecosystem, was launched in October 2017 by the multinational American-Japanese cybersecurity software company Trend Micro Incorporated in partnership with the industry leader in IT and services, HITRUST Alliance. Such technological developments could eventually strengthen the APT protection market globally. Among the significant businesses included in this study are Cisco Systems, Inc., AO Kaspersky Lab., ESET spol. S r.o., Sophos Ltd., Forcepoint, VMware, Inc., Microsoft, Palo Alto Networks, McAfee, LLC, and F-Secure. The global market share of APTs is segmented according to deployment style, services, solutions, and geography. There are

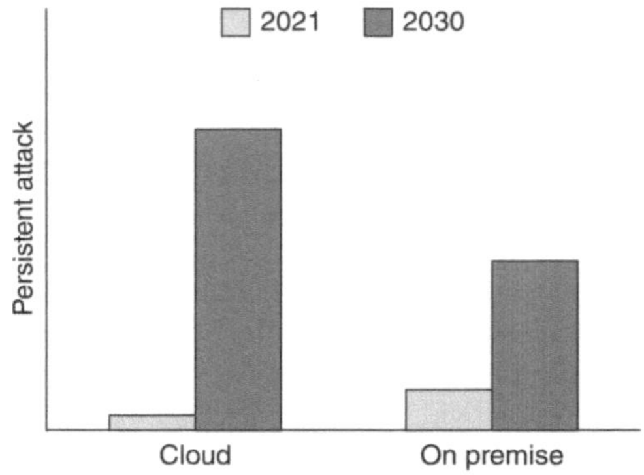

FIGURE 3.2 Advanced persistent threat market.

two types of deployment: on-premises and cloud. It is divided into several categories, including forensic analysis, endpoint security, next-generation firewall (NGFW), IDS/ intrusion prevention systems (IDSs/IPSs), sandboxing, and SIEM. Asia-Pacific, Latin America, Europe, and North America are the regions included in the market analysis [7]. Figure 3.2 presents a comparative study of advanced persistent attacks between cloud and on-premise environments from 2021 to 2030.

The on-premises subsegment led the market in 2019, but the cloud subsegment is expected to grow at the fastest rate throughout the projected period, depending on the deployment type. Because of its affordability, essential function in disaster recovery, expanding business concern about integrated safety standards, and accessibility of cloud security services for both small and large companies, cloud-based APT defense is gaining traction. Because of their greater dependability and ease of use, cloud services are becoming more and more popular among businesses, which is driving the segment's global revenue development [8]. The market for APT protection has changed over the last few years, moving from common, simple malware outbreaks to complex, zero-day threats. Regarding services, the managed services sub segment became the global leader in 2019; the professional services sub segment is expected to grow at the fastest rate during the projection period. Businesses now need to be aware of system vulnerabilities, which have led to a substantial rise in criminal activity across a number of industries, including IT and telecoms, retail, and healthcare. The significant increase in cyber hazards in both developed and emerging countries is one of the main factors contributing to the subsegment's growth throughout the projection period. Moreover, the widespread deployment of 5G infrastructure and a notable increase in technological support are some of the reasons that are expected to drive the demand for integrated security and consequently propel the subsegment market's growth. Highlights of the Advanced Persistent Threat Market Report: Aspects Details [9].

- *By Deployment Mode*: On-Premises Cloud
- Managed Services, Expert Services, Through Services
- Via Solutions Sandboxing, NGFW, Endpoint Security, SIEM, IDS/IPS, Forensic Analysis, and Other

- Europe (Germany, UK, France, Spain, Italy, and the Rest of Europe) as a Region
- Canada, Mexico, and the United States of America
- *Asia-Pacific*: comprising the remainder of the region as well as South Korea, Australia, China, Japan, and India
- LAMEA, which includes Saudi Arabia, Brazil, South Africa, and the United Arab Emirates.

Five Significant Players in the MarketMicrosoft, Raytheon Technologies Corporation, VMware, Inc., Cisco Systems, Inc., McAfee, LLC., FireEye, AO Kaspersky Lab., Sophos Ltd., Palo Alto Networks, and F-Secure.

3.3.1 COVID-19's Effect on the Global Advanced Persistent Threat Protection Market

Cybercriminals who are well-funded and well-organized launch highly competent cyberattacks known as advanced persistent threats, or APTs. During the COVID-19 pandemic, the market for APT defense has expanded dramatically. Globally, the pandemic has resulted in serious business challenges. Because of the COVID-19 disaster, a number of global market sectors, including retail, healthcare, and IT and communications, have become vulnerable to APT attacks. Consequently, most of the main companies are creating innovative products to help businesses throughout the epidemic. Palo Alto Networks, a global cybersecurity corporation with headquarters in the United States, for instance, debuted the first NGFW with machine learning capabilities in June 2020. The platform supports the proactive security of Internet of Things (IoT) devices, the prevention of cyberattacks, and the implementation of security policy recommendations. Due to their well-planned and creative product launches throughout the epidemic, these companies are having a positive effect on the worldwide APT protection market [10]. Figure 3.3 depicts the lifecycle of an attack, showing how an attacker gains unauthorized access to the network.

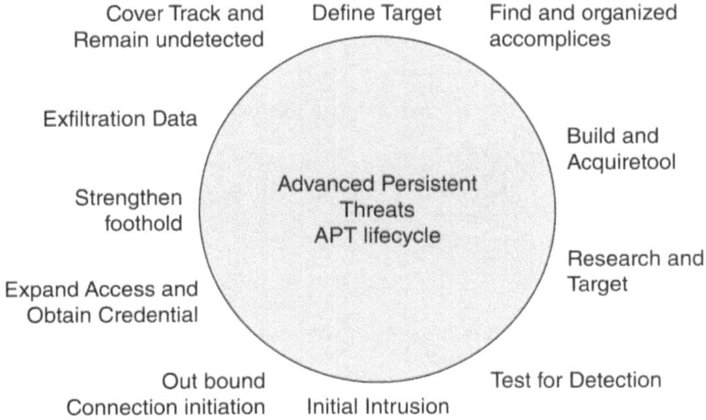

FIGURE 3.3 Advanced persistent threat's life cycle.

3.4 NETWORK SEGMENTATION FOR ADVANCED PERSISTENT THREATS

Network segmentation is a crucial security method for defending against advanced persistent attacks (APTs). By dividing your network into smaller, more isolated sections, you can limit an attacker's ability to move laterally within it. An attacker will find it more difficult to compromise crucial systems and steal confidential data as a result. For APT defense, the following network segmentation strategies and elements need to be considered: Network segmentation is the process of breaking your network up into many zones and giving each zone a particular set of security rules. The main purpose of network segmentation is to improve security and compliance management. Usually used to split traffic between network segments, virtual local area networks, or VLANs, offer an additional layer of security for data and application security. By dividing your bigger network into smaller ones, the devices, servers, and applications used by your business are kept apart from the rest of the network. A potential attacker is still restricted to the network segment they can access, even if they are able to get beyond your first barrier of defense [11]. Figure 3.4 describes a segmented network both with and without the presence of an advanced persistent threat. The segmented network helps prevent cyberattacks, while a non-segmented network allows the attacker to breach the security perimeter.

3.4.1 The Role of Network Segmentation in Protecting Sensitive Data and Systems

From the literature [12, 13], the following parameters have been identified for protecting sensitive data using network segmentation:

- *Identify Critical Assets*: Ascertain which assets, data, and systems are most crucial to your business. To protect these elements, the highest level of segmentation should be used.

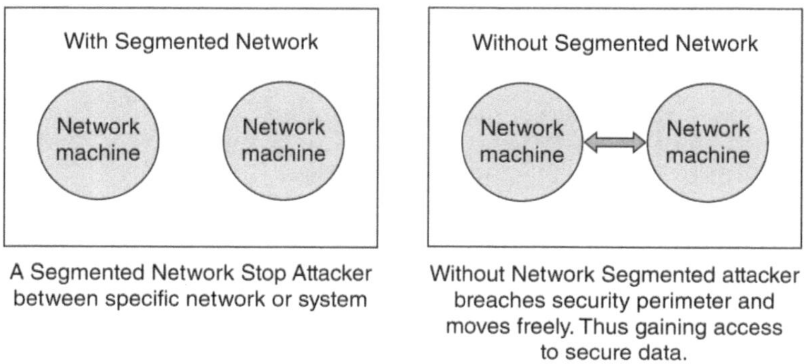

FIGURE 3.4 Network segmentation with and without advanced persistent threat.

- *Segmentation Zones*: Create segmentation zones based on the importance of specific systems and the delicate nature of data. Common zones include the demilitarized zone (DMZ), internal network, and high-security areas. Establish strict access controls and security system regulations among network segments. Only allow necessary traffic to pass across segments, and restrict access depending on user roles and permissions.
- *Sensitive Data Isolation*: Set apart sensitive data, such as databases holding client information or intellectual property, in a rigorous, highly separated area with restricted access.
- *Network Monitoring*: Ensure that continuous logging and monitoring are in place for each network segment. This makes it easier for you to spot unusual activities and potential APT intrusions early on.
- *Intrusion Detection and Prevention*: Position IDSs and IPSs (IDS/IPS) at key points across the network to recognize and thwart unwanted activity.
- *Locate Critical Assets*: Establish which information, assets, and systems are most crucial to your business. To protect these elements, the highest level of segmentation should be used.
- *Subnetting and Virtual LANs (VLANs)*: These strategies let various network segments be logically separated while yet making use of the same physical infrastructure.
- *Use a Security Model Based on Zero Trust*: No person or device should be trusted by default, regardless of where they are located within a network. Every network segment requires strict access controls and authentication protocols.
- *Internet of Things Device Segmentation*: Isolate IoT devices within a distinct network segment to prevent them from acting as a point of entry for hackers.
- *Remote Access Segmentation*: Set up separate segments for VPN connections and remote access to ensure that users from a distance have restricted access to internal resources.
- *Regular Testing and Audits*: Perform penetration tests and routine audits to verify the effectiveness of your network segmentation plan and spot any potential weaknesses. Make an incident response plan specifically for APTs that describes how to isolate compromised segments, contain the threat, and conduct a forensic investigation.
- *Least Privilege Concept*: Within each network segment, apply the least privilege concept to limit system and user access. Users and systems should only be given the minimal access necessary to do their jobs.
- *Encryption between Segments*: To protect sensitive data when it is being transferred among segmented networks, utilize encryption whenever it needs to go through several segments.
- *Consistent Updates*: Keep all network segmentation devices, like firewalls and routers, up to date with the latest security patches and configurations. Remember that network segmentation should only be utilized in conjunction with other security measures like endpoint security, user training, and threat intelligence as part of a comprehensive cybersecurity plan. You may strengthen your company's defenses against cyberattacks and APTs by integrating these elements.

3.5 INVESTIGATING THREATS AGAINST APTS AND ORGANIZATIONS

An initiative-driven cybersecurity strategy called "threat hunting" can help organizations find and stop sophisticated attacks, such as APTs, which might go undetected by traditional security measures [14]. It comprises aggressively searching for signs of hostile activity on a company's networks and systems. Businesses can identify threats from APTs in this way. Form a Team for Threat Hunting: Organize a specialized team or designate particular personnel from your security team to undertake threat-hunting duties. These individuals need to be very analytical and well-versed in the network and procedures of the company. Set Objectives: Give a clear explanation of your goals and the scope of your threat-hunting operations. Decide on your search criteria, such as specific APT indicators or peculiar behavior patterns. Gather and Centralize Information: Gather information from all relevant sources, including logs, threat intelligence feeds, network traffic statistics, and endpoint data. An organized data repository is necessary for effective analysis [15].

- *Create Hypotheses*: Make educated guesses or conjectures about potential APT activities. Threat intelligence, known attack routes, and historical data should all be used to bolster these beliefs. Formulating Hunting Questions and Strategies: Create original rules, search phrases, and detection techniques based on your hypotheses. These queries should be used to find any potential indications of APT activity, such as unusual network traffic patterns, anomalies in the system log, or file modifications. Constant Observation: To spot anomalies and potential threats rapidly, use automated and ongoing monitoring tools. This could shorten the time between detection and reaction. Analyze and Research Data: Review the details gathered from your searches and questions. Investigate any unexpected findings to see if there is a genuine risk involved. Lateral Movement Detection: Be vigilant for any signs of lateral movement within the network, as APTs often move surreptitiously among compromised systems. Watch out for unusual account access, privilege escalation, and data transfers.
- *Behavioral Analysis*: Take note of behavioral analysis rather than relying solely on signature-based detection. APTs frequently employ evasive techniques to prevent the setting of established signature-based alarms.
- Threat Intelligence Integration: When performing your hunt, utilize threat intelligence streams and external information sources. Stay up to date on the latest APT tactics, techniques, and indications of compromise (IoCs).
- *Reporting and Documentation*: Take meticulous notes on everything you find, including the affected systems, the order in which things happened, and the nature of the threat. Forward the findings to the incident response team so that they can take the necessary action.
- *Adapt and Develop*: Threat hunting is an iterative process. Make ongoing adjustments to your hunting techniques in light of new information, developing threats, and lessons learned from previous hunts. Encourage cooperation and communication among employees as well as with partners and

other professionals in the same field. Given that APTs usually target many firms, sharing threat intelligence can help detect bigger-scale attacks.

- *Training and Skill Development*: Invest in your threat-hunting team's training and skill development to ensure that they stay up to date on the latest techniques and technologies. Legal and Privacy Concerns: Recall legal and privacy concerns when conducting threat hunting, particularly when working with sensitive data. Respect for relevant regulations must be followed. The "threat hunting" method of proactive cybersecurity is intended to support more traditional security measures like firewalls, antivirus software, and IDSs. By continuously searching for hazards, organizations can reduce the amount of time that attackers remain on their networks and are better able to recognize and counter advanced threats.

3.6 ADVANCED PERSISTENT THREAT DETECTION SYSTEMS USING INTRUSION DETECTION

IDSs play a major role in the detection and mitigation of APTs. Identifying APTs with traditional security technologies may be challenging due to their hidden and persistent nature. Here's how to get IDS optimized for APT detection:

- *Detection Based on Signature*: Traditional IDSs use signature-based detection to find known attack patterns and signatures. For traditional attacks, this is important, but APTs often use customized attack techniques. Your IDS ought to contain current signatures for APT-related IoCs. It is advised to use anomaly-based detection in addition to signature-based detection. APTs often display peculiar network characteristics. Anomaly detection can help identify deviations from typical host and network activities. Employ behavioral analysis techniques to spot dubious patterns and anomalies in host and network activities. Monitoring unwanted user access, privilege expansion, and data transfers that can show APT agents relocating laterally are all part of this. Deep Packet Inspection: Examine the content of network communication using deep packet inspection. The detection of harmful payloads or communication patterns can be aided by deep packet inspection. APTs may use communication methods that are obscured or encrypted. Examine network data and system logs regularly to search for unusual or suspicious activity. This means monitoring for unusual port or protocol usage, behavioral changes in users, and attempts at unauthorized access [16].
- *Network Segmentation*: To stop APTs from migrating laterally, split up your network. Configure your IDS to monitor traffic among network segments and set up rules to recognize any unauthorized communication between them. Integration with SIEM: Integrate your IDS with an SIEM system to centralize and correlate security event data. This allows for more in-depth analysis and incident response. Integration of threat intelligence: Integrate threat intelligence feeds and sources that are unique to APTs into your IDS. This improves the system's ability to recognize known APT indicators,

like malicious file hashes, IP addresses, or domain names. Utilize machine learning and artificial intelligence (AI) approaches to conduct behavioral and statistical analysis to identify APTs. These instruments are able to spot subtle patterns that may point to APT activity. User and Entity Activity Analytics (UEBA): User and Entity Activity Analytics (UEBA) systems can help detect APTs by examining user and entity activity to spot deviations from normal trends. User and Entity Activity Analytics (UEBA): UEBA systems can help detect APTs by studying user and entity activity to find deviations from regular patterns [17].

- *Custom Rules and Alerts*: Set up custom rules and alerts in your IDS to specifically target APT-related actions and indicators. Customize notifications to help you prioritize and respond quickly to any APT incidents [18].
- *Integrating Incident Response*: Make sure that your IDS and incident response protocol are connected. Provide defined procedures for forensic analysis and the isolation of compromised systems in the event that APT activity is detected.
- *Regular Updates and Tuning*: Maintain your IDS tuned and current with the latest threat intelligence as threats evolve and to better match the specific needs of your organization. Turn on real-time monitoring and alerting so you can respond quickly to potential APT scenarios as soon as they appear.
- *Red Team Activities*: Red team activities or penetration testing can be used to assess how successfully your IDS recognizes attacks similar to APT.
- *Instruction and Record-Keeping*: Ensure that all IDS configurations, rules, and response methods are documented and that your security personnel has received the necessary training to operate the system.

3.6.1 Building a Resilient Defense against Advanced Persistent Threats (APTs)

By configuring your IDS to detect APTs using a combination of signature-based, anomaly-based, and behavioral analysis techniques, you may improve your organization's ability to recognize and neutralize these sophisticated and persistent threats. Frequent monitoring, upgrades, and interaction with other security solutions are also necessary for robust protection against APTs. Employees who receive training on spotting and reporting suspicious activity are better able to recognize and reduce APT threats. Employee training is a critical component of an organization's defense against APTs and other cybersecurity threats. Having employees recognize and report suspicious activities can significantly strengthen an organization's security posture since they are often the first to react. Staff members can be trained to recognize, report, and neutralize APT attacks in this way [19].

First and foremost, all employees should undergo basic cybersecurity awareness training. This should cover topics such as the importance of strong authentication, awareness of phishing, and password security. Give detailed instructions on APTs and their characteristics. Help employees understand the characteristics, tactics, and reasons why APTs are such a major risk to the organization. APTs usually begin with phishing attacks. Workers ought to be trained to recognize phishing emails, questionable links, and email attachments. Conduct simulated phishing activities to assess their ability to recognize phishing scams. APT actors may employ social engineering strategies to fool employees into divulging personal information.

3.7 A MULTI-LAYERED APPROACH TO CYBERSECURITY IN ORGANIZATIONS

Workers ought to receive training in declining requests for information, even if they appear to be valid, in order to prevent giving in to them. Teach employees the need to have unique, strong passwords and the risks associated with using the same one twice. Encourage the use of password managers to create and store complex passwords. Promote the use of MFA whenever it makes sense. Describe the ways in which MFA's demand for multiple factors of authentication can significantly enhance account security. Instruct employees on how to recognize unusual activity on their networks or PCs, including pop-up windows that don't seem to belong, peculiar system behavior, or peculiar system prompts. Employees should be trained on how to manage and classify sensitive data. Ensure that they understand how to follow security protocols and where to look for important information [20]. If your organization employs encryption to transport sensitive data, then employees should be trained on how to recognize encrypted emails and how to use them while sending sensitive data. Provide clear channels and procedures for reporting issues. Ensure that employees know where and how to report security breaches or questionable activities. Provide staff workers with ongoing training and information about the newest threats and developments in cybersecurity. APT tactics are dynamic; therefore, awareness needs to be too. Tailor training programs to the responsibilities and tasks of specific positions held by employees of the firm. Diverse organizational functions may necessitate distinct cybersecurity requirements. Red team drills or simulated APT scenarios can be used to gauge how well employees can respond to attacks similar to those seen in APT. This makes it easier to pinpoint areas that require development. Provide a reward and incentive program for employees who exhibit good cybersecurity practices, such as reporting suspicious activities.

Cybersecurity is a constantly evolving field. Encourage employees to attend relevant training sessions, engage in continuing education, and stay up to date on security innovations. Make sure top leadership shows that cybersecurity rules and procedures are important to them. If workers see leadership taking security seriously, they are more likely to do the same. Explain the organization's incident response plan and emphasize the need to promptly report incidents. Make certain that employees understand their roles in incident response procedures. Use tests, quizzes, or role-playing exercises to routinely assess staff members' cybersecurity knowledge and awareness to ensure that training is effective. Establish a feedback loop so that employees may submit ideas for security-related policies, procedures, and training materials. This makes training depending on needs easier. Remember that offering top-notch staff training requires ongoing work. It's imperative to continuously stress cybersecurity awareness and adapt training materials to cover new threats like APTs. By encouraging a culture of cybersecurity awareness and granting employees the latitude to remain vigilant, you can significantly reduce the organization's vulnerability to APT assaults. Create segmentation zones based on the importance of specific systems and the sensitive nature of the data. Common zones include the DMZ, internal network, and high-security areas. Establish strict access controls and security system regulations between network segments. Only allow necessary traffic to pass across segments, and restrict access depending on user roles and permissions. Sensitive Data Isolation: Store sensitive data, such as databases holding client information or intellectual property, in a strictly regulated, highly separated region with limited access [18, 19].

- *Network Monitoring*: Ensure that continuous logging and monitoring are in place for each network segment. This makes it easier for you to spot unusual activities and potential APT intrusions early on. Prevention and Detection of Intrusions: Install IDSs and IPSs (IDS/IPS) at key points across the network to detect and block malicious traffic [20].
- *Subnetting and Virtual LANs (VLANs)*: These methods enable the logical division of a network's sections while maintaining the same physical infrastructure.
- *Embrace a Security Model of Zero Trust*: No person or device should be trusted by default, regardless of where they are located within a network. Every network segment requires strict access controls and authentication protocols.
- *IoT Device Segmentation*: Isolate IoT devices within a distinct network segment to prevent them from acting as a point of entry for hackers.
- *Segmentation for Remote Access*: Use separate segments for VPN connections and remote access to ensure that users at a distance have restricted access to internal resources.
- *Regular Testing and Audits*: Perform penetration tests and routine audits to verify the effectiveness of your network segmentation strategy and find any potential weaknesses. Make an incident response plan specifically for APTs that describes how to isolate compromised segments, contain the threat, and conduct a forensic investigation.
- *Least Privilege Concept*: Within each network segment, apply the least privilege concept to limit system and user access. Users and systems should only be given the minimal access necessary to do their jobs. Encryption among Segments: Encrypt sensitive data as it travels among segmented networks to protect it during transmission across several segments.
- *Continual Updates*: Ensure that all network segmentation devices, including firewalls and routers, have the most recent security patches and configurations applied. Remember that network segmentation should only be utilized in conjunction with other security measures like endpoint security, user training, and threat intelligence as part of a comprehensive cybersecurity plan. You may strengthen your business's defenses against other cyberattacks and APTs by putting these elements together [20].

3.8 INVESTIGATING THREATS AGAINST APTS AND ORGANIZATIONS

An initiative-driven cybersecurity strategy called "threat hunting" can help organizations find and stop sophisticated attacks, such as APTs, which might go undetected by traditional security measures. It comprises aggressively searching for signs of hostile activity on a company's networks and systems. Businesses can identify threats from APTs in this way [21, 22].

- *Establish a Team for Threat Hunting*: Organize a specialized team or designate particular personnel from your security team to undertake threat-hunting duties. These individuals need to be very analytical and well-versed in the network and procedures of the company. Set Objectives: Give a clear explanation of your goals and the scope of your threat-hunting operations. Decide on your search criteria, such as specific APT indicators or peculiar behavior patterns.
- *Gather and Centralize Information*: Gather information from all relevant sources, including logs, threat intelligence feeds, network traffic statistics, and endpoint data. An organized data repository is necessary for effective analysis. Create Hypotheses: Make educated guesses or conjectures about potential APT activities. Threat intelligence, known attack routes, and historical data should all be used to bolster these beliefs.
- *Develop Hunting Methods and Questions*: Create original rules, search phrases, and detection techniques based on your hypotheses. These queries should be used to find any potential indications of APT activity, such as unusual network traffic patterns, anomalies in the system log, or file modifications.
- *Ongoing Surveillance*: To spot anomalies and potential threats rapidly, use automated and ongoing monitoring tools. This could shorten the time between detection and reaction.
- *Data Analysis and Research*: Review the details gathered from your searches and questions. Investigate any unexpected findings to see if there is a genuine risk involved. Lateral Movement Detection: Be vigilant for any signs of lateral movement within the network, as APTs often move surreptitiously among compromised systems. Watch out for unusual account access, privilege escalation, and data transfers. Behavioral Analysis: Take note of behavioral analysis rather than relying solely on signature-based detection. APTs frequently employ evasive techniques to prevent the setting of established signature-based alarms.
- *Using Threat Intelligence Integrated*: When performing your hunt, utilize threat intelligence streams and external information sources. Stay up to date on the latest APT tactics, techniques, and IoCs. Documentation and Reporting: Keep a thorough log of everything you find, including the affected systems, the order in which things happen, and the kind of danger. Forward the findings to the incident response team so that they can take the necessary action. Adapt and Develop: Threat hunting is an iterative process.

Make ongoing adjustments to your hunting techniques in light of new information, developing threats, and lessons learned from previous hunts. Working together: Promote cooperation and communication among staff members as well as with outside partners and coworkers in related fields. Given that APTs usually target many firms, sharing threat intelligence can help detect bigger-scale attacks. Training and Skill Development: Invest in your threat-hunting team's training and skill development to ensure that they stay up to date on the latest techniques and technologies.

- *Legal and Privacy Issues*: Be mindful of legal and privacy issues when doing threat hunting, particularly when working with sensitive data. Respect for relevant regulations must be followed. The "threat hunting" method of proactive cybersecurity is intended to support more traditional security measures like firewalls, antivirus software, and IDSs. By aggressively seeking out hazards, organizations can reduce the amount of time that attackers remain on their networks and are better able to recognize and counter advanced threats like APTs [22].

3.9 CATEGORIES OF APT OBJECTIVES

One type of cyber espionage is the theft of state secrets or intellectual property. Financially motivated cybercrimes Destroying Hacktivism To stop, recognize, and address an APT, you must be aware of its traits. Most APTs have a similar fundamental life cycle that includes infiltrating a network, gaining more access, and achieving the attack's goal, which is typically data extraction theft [23].

1. **Initial Phase: Stealth**
   Social engineering techniques are a common tool used by APTs to get access in the initial phase. APTs can be identified by phishing emails that frequently leverage material that has been stolen from other team members who have already been compromised and that specifically target high-level individuals, such as senior executives or technical leaders. The phrase "spear-phishing" refers to email attacks that target certain individuals. The email appears to be from a team member and may contain references to an ongoing project. If several executives report being the target of spear-phishing attacks, look for additional signs of an APT [24].

2. **Stage 2: Escalation and Lateral Movement**
   Once attackers have gained initial access to a company's network, they proceed to the second stage, expansion, by introducing malware into the network. They move laterally to map the network and gather account names and passwords for logging in to access critical company information. They may also set up a "backdoor," which is a scheme that allows them to surreptitiously join the network and conduct activities at a later time. Other access points are often set up in order to ensure that the attack can continue even in the event that a vulnerable point is identified and closed [6].

3. **Step 3: Persistency**
   Prepare for the third step: fraudsters typically store stolen data in a secured section of the network **until** sufficient information is obtained. Then they "exfiltrate," or extract, it without being noticed. They may use tactics like a denial-of-service (DoS) attack to distract the security team and occupy network workers while the data is being taken. The network may remain compromised and await the return of the crooks at any moment [9].

3.10 APT ATTACK CHARACTERISTICS

Because advanced persistent attacks use different techniques than ordinary hackers, they leave behind various indicators. Aside from spear-phishing attacks directed at company executives, unusual activity on user accounts, such as an increase in high-level logins late at night, are indicators of an APT attack. The widespread use of Trojan backdoors Unusual or unusual data bundles may indicate the collection of data in preparation for exfiltration. Examples of APTs [25].

The following are some notable APT examples that CrowdStrike found:

- CrowdStrike recognized GOBLIN PANDA (APT27), which was first discovered in September 2013, as an indication of an attack (IOA) on the network of a multisector technology company. Two Microsoft Word exploit papers with training-related topics prompted this Chinese opponent to dump malicious files when they were opened. Our whole APT profile is available on Goblin Panda [26].
- Based in Russia, the attacker FANCY BEAR (APT28) uses phishing messages and spoofs websites that closely resemble reliable ones to target desktop computers and mobile devices. For our comprehensive APT Group Profile, see Fancy Bear [27].
- Russian opponent Cozy Bear (APT29) is believed to be acting on behalf of the Foreign Intelligence Service of the Russian Federation. This adversary has been seen using enormous spear phishing tactics that disseminate a wide range of malware types to target political, scientific, and national security entities across numerous industries. See the whole APT Group Profile for Cozy Bear.
- With its base in Vietnam, targeted intrusion adversary Ocean Buffalo (APT32) is said to have been operational since at least 2012. The aforementioned adversary is known to employ a wide range of tactics, techniques, and procedures (TTPs), including the use of commercially available and proprietary tools in addition to spear-ploughing malware through strategic web compromise (SWC) operations.
- HELIX KITTEN (APT34) has been operational since late 2015 and is most likely based in Iran. The spear-phishing messages are organized, thoroughly researched, and very relevant to the individuals they are aimed at. It is directed toward businesses in the banking, energy, telecommunications, hotel, government, and aerospace industries. View HELIX KITTEN's whole APT profile by clicking this link.

- During the middle of the 2010s and the 2020s, APT41, also known as Wicked Panda, was one of the most active and effective opponents based in China. According to CrowdStrike Insight, Wicked Panda is a superset of companies comprising several contractors that function illegally and financially for the Chinese government, most likely with the tacit approval of CCP authorities. View the full APT profile of WICKED PANDA [28].

3.11 HOW CAN ONE DEFEND THEMSELVES FROM APT ATTACKS?

A multitude of cybersecurity and intelligence solutions are available to assist businesses in fortifying their defenses against attacks by APTs. The following are a few of the best tactics to employ:

- *Sensor Range*: Organizations must use technologies that provide defenders with 360-degree visibility of their environment if they want to stop blind spots from turning into safe havens for cybercriminals.
- *Technical Expertise*: By incorporating IOCs into an SIEM, you may use technical intelligence to enhance data. This makes it possible to apply additional intelligence to event correlation, highlighting network occurrences that could not have been seen in any other case.
- *Service Provider*: Working with an adaptable cybersecurity firm is essential. Organizations may need support in dealing with a highly sophisticated cyber threat in an inconceivable scenario. A web application firewall (WAF), a security solution that helps protect enterprises at the application level, filters, monitors, and analyzes hypertext transfer protocol (HTTP) and HTTP secure (HTTPS) traffic between a web application and the internet knowledge about potential dangers. Threat information makes it easier to track campaigns, malware families, and threat actor profiles. Threat intelligence is important because, in today's world, knowing an assault happened is more important than knowing its specifics.

3.12 CONCLUSION

Organizations around the world are facing a serious and constantly changing threat from advanced persistent threats, or APTs. These extremely persistent and sophisticated cyberattacks are frequently coordinated by highly competent and well-funded threat actors, such as state-sponsored hackers, nation-states, and organized crime groups. APTs are known for their patient and covert tactics. They seek to stay hidden within a company's network for extended periods of time as they carry out their goals, which can include data theft, espionage, or the sabotage of vital infrastructure. Organizations need to be prepared for the possibility of becoming targets of APTs in the current digital environment. Cyber-resilience encompasses not just the ability to fend against assaults but also the capacity to identify, address, and effectively recover from incidents. Organizations may increase cybersecurity and resilience by taking a proactive, comprehensive strategy.

REFERENCES

1. Zhang, Y., Geng, P., Sivaparthipan, C. B., & Muthu, B. A. (2021). Big data and artificial intelligence based early risk warning system of fire hazard for smart cities. Sustainable Energy Technologies and Assessments, 45, 100986.
2. Barja-Martinez, S., Aragüés-Peñalba, M., Munné-Collado, Í, Lloret-Gallego, P., Bullich-Massague, E., & Villafafila-Robles, R. (2021). Artificial intelligence techniques for enabling Big Data services in distribution networks: a review. Renewable and Sustainable Energy Reviews, 150, 111459.
3. Kibria, M. G., Nguyen, K., Villardi, G. P., Zhao, O., Ishizu, K., & Kojima, F. (2018). Big data analytics, machine learning, and artificial intelligence in next-generation wireless networks. IEEE Access, 6, 32328–32338.
4. Srivastava, S., Haroon, M., & Bajaj, A. (2013, September). Web document information extraction using class attribute approach. In 2013 4th International Conference on Computer and Communication Technology (ICCCT) (pp. 17–22). IEEE.
5. Minasny, B. The elements of statistical learning, Trevor Hastie, Robert Tishirani, Jerome Friedman, (2009).
6. Siddiqui, Z. A., & Haroon, M. (2023). Research on significant factors affecting adoption of blockchain technology for enterprise distributed applications based on integrated MCDM FCEM-MULTIMOORA-FG method. Engineering Applications of Artificial Intelligence, 118, 105699.
7. Najafabadi, M. M., Villanustre, F., Khoshgoftaar, T. M., Seliya, N., Wald, R., & Muharemagic, E. (2015). Deep learning applications and challenges in big data analytics. Journal of Big Data, 2(1), 1–21.
8. Khan, W., & Haroon, M. (2022). A Pilot Study and Survey on Methods for Anomaly Detection in Online Social Networks. In Human-Centric Smart Computing: Proceedings of ICHCSC 2022 (pp. 119–128). Singapore: Springer Nature Singapore.
9. Siddiqui, Z. A., & Haroon, M. (2022). Application of artificial intelligence and machine learning in blockchain technology. In Artificial Intelligence and Machine Learning for EDGE Computing (pp. 169–185). Academic Press.
10. Khan, R., Haroon, M., & Husain, M. S. (2015, April). Different techniques of load balancing in distributed system: A review paper. In 2015 Global Conference on Communication Technologies (GCCT), pp. 371–375. IEEE.
11. Xin, Y., Kong, L., Liu, Z., Chen, Y., Li, Y., Zhu, H., & Wang, C. (2018). Machine learning and deep learning methods for cyber-security. IEEE Access, 6, 35365–35381.
12. Khan, W., & Haroon, M. (2022). An unsupervised deep learning ensemble model for anomaly detection in static attributed social networks. International Journal of Cognitive Computing in Engineering, 3, 153–160.
13. Khan, W., & Haroon, M. (2022). An efficient framework for anomaly detection in attributed social networks. International Journal of Information Technology, 14(6), 3069–3076.
14. Ngiam, K. Y., & Khor, W. (2019). Big data and machine learning algorithms for healthcare delivery. The Lancet Oncology, 20(5), e262–e273.
15. Sarker, I. H., Kayes, A. S. M., Badsha, S., Alqahtani, H., Watters, P., & Ng, A. (2020). Cyber-security data science: an overview from a machine learning perspective. Journal of Big Data, 7, 1–29.
16. Haroon, M., Tripathi, M. M., Ahmad, T., & Afsaruddin. (2022). Improving the healthcare and public health critical infrastructure by soft computing: an overview. Pervasive Healthcare: A Compendium of Critical Factors for Success, Vol. 1, 59–71.
17. Husain, M. S., & Haroon, D. (2020). An enriched information security framework from various attacks in the IoT. International Journal of Innovative Research in Computer Science & Technology (IJIRCST) ISSN, Vol. 8, no. 3, 2347-5552.

18. Rawat, D. B., Doku, R., & Garuba, M. (2019). Cyber-security in the big data era: from securing big data to data-driven security. IEEE Transactions on Services Computing, 14(6), 2055–2072.
19. Khan, A. M., Ahmad, S., & Haroon, M. (2015, April). A comparative study of trends in security in cloud computing. In 2015 Fifth International Conference on Communication Systems and Network Technologies (pp. 586–590). IEEE.
20. Haroon, M., & Husain, M. (2013). Analysis of a dynamic load balancing in multiprocessor system. International Journal of Computer Science Engineering and Information Technology Research, 3(1).
21. Husain, M. S. (2020). A review of information security from a consumer'sperspective, especially in online transactions. International Journal of Engineering and Management Research, 10.
22. Sreedevi, A. G., Harshitha, T. N., Sugumaran, V., & Shankar, P. (2022). Application of cognitive computing in healthcare, cyber-security, big data and IoT: a literature review. Information Processing & Management, 59(2), 102888.
23. Khan, N., & Haroon, M. (2022). Comparative Study of Various Crowd Detection and Classification Methods for Safety Control System. Available at SSRN 4146666.
24. Petrenko, S. A., Petrenko, A. S., & Makoveichuk, K. A. (2017). Problem of developing an early-warning cyber-security system for critically important governmental information assets. Network, 4, 7–8.
25. Tripathi, M. M., Haroon, M., Khan, Z., & Husain, M. S. (2022). Security in digital healthcare system. Pervasive Healthcare: A Compendium of Critical Factors for Success, Vol. 1, 217–231.
26. Truong, T. C., Zelinka, I., Plucar, J., Čandík, M., & Šulc, V. (2020). Artificial intelligence and cyber-security: Past, present, and future. In Artificial intelligence and evolutionary computations in engineering systems (pp. 351–363). Springer Singapore.
27. Khan, W., Haroon, M., Khan, A. N., Hasan, M. K., Khan, A., Mokhtar, U. A., & Islam, S. (2022). DVAEGMM: dual variational autoencoder with gaussian mixture model for anomaly detection on attributed networks. IEEE Access, 10, 91160–91176.
28. Martínez Torres, J., Iglesias Comesaña, C., & García-Nieto, P. J. (2019). Machine learning techniques applied to cyber-security. International Journal of Machine Learning and Cybernetics, 10, 2823–2836.

4 Unveiling the Digital Trails

A Survey of Contemporary Digital Forensics Tools and Techniques

Wasim Khan, Mohammad Ishrat, and Monia Mohammed Al Farsi

4.1 INTRODUCTION

Criminals in today's increasingly digital world are increasingly turning to cutting-edge technologies to launch attacks on institutions of all kinds and individuals equally [1–3]. The most recent such incident occurred in Baltimore, Maryland, where hackers stole a National Security tool and paralyzed thousands of computers. Over the course of three weeks, the attack disabled many essential services, including email, real estate transactions, water billing, and health notifications. Although digital forensics techniques can be useful in civil and administrative matters, it is especially important in criminal investigations to make a strategic choice when deciding which tools to deploy. Human psychology is intrinsically linked to the field of digital forensics. Human behavior and psychological states can be revealed through forensic analysis. In the field of digital forensics, behavioral evidence analysis (BEA) is used to better comprehend the mental state of a suspect or criminal. The investigators utilize a wide variety of instruments to carry out forensic operations and collect the unavoidable evidence needed to bring the perpetrators of the crimes to justice [4]. When it comes to investigating and analyzing digital evidence to bolster judicial processes or handle cybersecurity problems, digital forensics (also called cyber forensics) is a specialized subfield of forensic science. It emerged as a response to the rapid technological advancements that led to the pervasive use of digital devices and networks in various aspects of human life [5]. The growing importance of digital evidence in both criminal and civil proceedings has prompted the creation of specialized techniques and approaches for its proper handling and interpretation [6]. When computers were first introduced, law enforcement agencies quickly realized the importance of digital data in solving crimes. This was the beginning of digital forensics [7, 8]. To investigate criminal activity, law enforcement and government agencies have used even the most fundamental digital techniques since the 1970s. Since this was a brand-new field at the time, efforts were initially concentrated on data retrieval and the identification of possible evidence.

DOI: 10.1201/9781003409571-4

To locate, preserve, extract, analyze, and interpret digital evidence is the domain of digital forensics, which is defined as the application of forensic principles and procedures. Over time, digital forensics has grown to include not just computers but also smartphones, tablets, servers, cloud storage, social media platforms, and even network traffic records [9, 10]. This widened focus was prompted by the growing quantity of digital data and the necessity of applying forensic methods to a wide range of digital resources. Digital forensics is a growing discipline due to the rapid development of computing technologies. New challenges have arisen in dealing with digital evidence as a result of the widespread use of computers, the internet, and mobile devices. New developments in areas, such as encryption, virtualization, cloud computing, the Internet of Things (IoT), and artificial intelligence (AI), have necessitated that investigators keep up. Effective data extraction and analysis from ever-evolving digital sources has been a driving force in the development of digital forensics over the course of its history [11, 12]. One critical milestone in the development of digital forensics was its recognition as a legitimate and admissible form of evidence in legal proceedings. Under specific guidelines, courts all around the world have begun to accept digital evidence in trials [6]. Guidelines for the admissibility of expert witness testimony and scientific evidence, including digital data, were created in the 1990s under the Daubert standard in the United States. Other countries have also implemented similar rules and procedures, further establishing digital forensics as an integral part of the court system. Computer science, criminal justice, information security, the law, and data analysis are just a few of the disciplines that feed into digital forensics [13]. It takes a mix of technical knowledge to handle and analyze digital evidence, investigative skills to reconstruct events and identify suspects, and a deep understanding of legal principles to ensure the admissibility and validity of evidence in court for a digital forensic investigation to be successful [9, 14, 15]. As a result of digital forensics' cross-disciplinary character, experts from other disciplines have begun working together to improve the subject.

Digital forensics' global influence goes beyond that of conventional law enforcement. Governments, businesses, and individuals are increasingly relying on digital forensic specialists to investigate and respond to cybersecurity issues, preserve digital evidence, and identify criminals in light of the rise of cybercrime, data breaches, and digital fraud [16]. The importance of digital forensics in today's society has been highlighted by its use in fields as varied as national security, business investigations, intellectual property protection, and civil litigation [10, 17]. The ever-evolving nature of technology presents various difficulties for digital forensics. Data capture and analysis are made more difficult by the prevalence of new devices, operating systems, and software. Digital evidence is difficult to access and analyze because of encryption and other data security techniques. Investigative needs must be balanced with the rights and interests of individuals and organizations in digital forensic investigations, making privacy and data protection crucial ethical considerations [6, 18, 19]. Our motivation for the study is the rapidly evolving landscape of digital forensics, shaped by the continuous advancement in digital technologies and the increasing sophistication of cybercrimes. Our aim is to provide a comprehensive and up-to-date survey of the field, highlighting not only the tools and techniques but also the critical role digital forensics plays in legal and cybersecurity contexts. Our contribution through this chapter is threefold. First, we offer an extensive analysis of contemporary digital forensic tools, addressing both their capabilities and limitations. Second, we delve into the field's obstacles, covering anything from technological issues to ethical and legal concerns, and offer insights into the ways these challenges are being addressed. Finally, we take a comprehensive look at digital forensics by highlighting its multidisciplinary character, which incorporates technological, legal, and psychological viewpoints. The purpose of this review is to offer a thorough analysis of digital forensics research, covering all pertinent works, methodologies, tools, and advancements. By conducting this review, we intend to achieve many objectives. Providing a cohesive overview of digital forensics' evolution and history is the major objective of this study. If we look at where digital forensics came from and how it developed, we can better grasp its relevance in our technological age. Digital forensics will be defined, and its rapid ascent to prominence as an effective investigation tool will be detailed for readers. Researchers can better prioritize their efforts by weighing the pros and cons of existing approaches. The technical and ethical challenges faced by digital forensics professionals will also be discussed. Encrypted data, virtualized systems, and cloud storage are just a few of the new challenges that have emerged as a result of rapid technological advancements. Solving these problems is crucial to developing the field's potential and keeping it current in the face of technological advancements.

This chapter was written with the hope that it will provide helpful information to anyone working in the field of digital forensics. It will help subsequent studies and developments in the subject by compiling existing knowledge and highlighting research needs. Researchers and professionals alike can use this book to learn more about digital forensics and find new directions for their work. Finally, we believe that this study will help spread the word about the value of digital forensics in contemporary investigation and cybersecurity among policymakers, legal professionals, and the general public. Just and fair legal proceedings in the digital age depend on a thorough understanding of the value of digital evidence and the role of digital forensics in maintaining its integrity. In conclusion, this review paper on digital forensics aims to provide a thorough and informative analysis of this emerging topic. By delving into its history, practical applications, toolkit, difficulties, and potential, we hope to progress the field and help more people recognize its relevance in the modern world. As a whole, the digital forensics community and the general public stand to profit from the efforts that this review article makes to promote collaboration, knowledge-sharing, and progress in the field.

4.2 DIGITAL FORENSICS: DEFINITION AND EVOLUTION

The scientific and methodical examination of digital data and electronic devices for the goals of discovering, storing, retrieving, evaluating, and presenting digital evidence in the context of legal or investigative processes is the essence of computer forensics, sometimes called cyber forensics [20, 21]. Reconstructing what happened, finding evidence of possible digital crimes, and providing proof that can stand up in court are the main goals of digital forensics. Electronic data, including files, documents, emails, databases, social media activity, internet history, and other digital artifacts, are the focus of this subfield of forensics, which expands on the classic definition of the term [17, 22, 23]. Experts in digital forensics, often called digital forensic analysts or digital forensic examiners, follow a rigorous scientific process to ensure the integrity of the digital data they review [24]. Forensic investigations include the collection and storage of digital evidence according to strict protocols, the identification of important data through careful analysis, and the presentation of findings in an unambiguous report. Law enforcement, cybersecurity, corporate investigations, and intelligence collection are just some of the many fields that might benefit from digital forensics [25–27]. Along with the development and sophistication of digital devices and networks, the field of digital forensics has expanded and matured. Mobile devices, tablets, IoT devices, cloud storage, and even network logs are now included in the scope of digital forensics, which was formerly limited to computers and other fixed media [12, 28]. As the area of digital forensics expands, forensic investigators will undoubtedly face novel challenges and opportunities. The term "digital forensics" is defined as "the scientific and ethical examination of digital evidence to support legal proceedings and uphold justice," which highlights the importance of the field in the present day [29].

Law enforcement agencies have been using digital evidence since the dawn of computing. The field of digital forensics had its start at this point. When computers were accessible to the masses in the 1970s, a new field of study called "digital forensics" emerged [30–32]. During this period, investigators focus mostly on restoring deleted files and performing rudimentary computer probing for evidence of criminal behavior. Because of the explosion of personal computers in the 1980s and 1990s, innovative techniques in digital forensics had to be created. Specialized procedures to extract and analyze digital evidence from various storage media, such as hard drives and floppy discs, were developed through collaboration between law enforcement agencies and computer professionals. At this time, the first digital forensic tools and techniques emerged, albeit on a much more primitive level than those used today [33, 34]. A major transition occurred in digital forensics in the late 1990s and early 2000s. Cybercrime increased as the internet became more widely used for both communication and commerce. More specialized skills and techniques for investigating digital evidence relating to internet use, email correspondence, and online transactions have been developed in response to the rise in cybercrime cases. The early 2000s saw more changes to the digital landscape as a result of the proliferation of mobile devices and the exponential growth of technology. Experts in digital forensics had to adapt to new difficulties in processing data from mobile devices and online storage services. New forms of digital evidence have emerged with the rise of social media and other online communication tools [35–37].

Thanks to developments in computing hardware, software, and networks, digital forensics have grown tremendously since the year 2000. The development of specialized digital forensics tools and software has greatly improved the efficiency with which investigators can obtain, analyze, and interpret digital evidence. Encryption, data hiding methods, and the ever-increasing complexity of digital storage systems were all addressed by these instruments. Both technological and legal/regulatory changes have contributed to digital forensics' ongoing progress [38, 39]. The significance of digital evidence and the necessity of standardized forensic practices to protect the credibility of evidence presented in court have been acknowledged by courts all over the world. Thus, groups and governments have set up certificates and regulations to guarantee the skills and integrity of digital forensics experts. In recent years, digital forensics has grown to include applications outside of law enforcement. It has become increasingly important in fields, such as cybercrime investigations, corporate audits, civil lawsuits, and intelligence gathering. Challenges for digital forensics experts arise with the introduction of new technologies like the IoT, AI, and blockchain [35, 40]. Another factor that makes digital forensic investigations more difficult is the widespread adoption of encryption and other privacy safeguards [41, 42]. In conclusion, the growth and change of digital forensics over time are indicative of the interplay between technological progress and the evolution of policing techniques. From its humble beginnings in data recovery to its present status as a highly developed and indispensable field, digital forensics has been crucial in helping authorities solve crimes, protect digital evidence, and ensure justice is served in the digital age. To keep up with the demands of an ever-evolving digital ecosystem, digital forensics professionals must keep current with technology changes and embrace novel approaches.

4.3 LITERATURE REVIEW

The field of digital forensics includes cybercrime, which refers to any crime that involves the internet or other forms of virtual communication. The variety of cybercrimes has grown over the years, especially in this age of smart devices and technology [43, 44]. Data preservation, extraction, identification, analysis, and report generation are the standard procedures in digital forensics. There are several digital investigative technologies that might make a forensic investigator's job easier these days [45]. To aid investigators or users in applying composite digital forensics (DF) tackles for their investigation, researchers in [46] analyzed various DF tools and outlined their attributes, limitations, and uses. Finding a way to restore the destroyed data is the only focus of the team. Using programs like Autopsy, OS Forensics, WIRESHARK, TRUECRYPT, FTK Imager, SANS SIFT, X-WAYS, and Hacker Combat, we can get back the lost data. Forensic evidence in several situations includes a weapon found at the scene of the crime, as well as bloodstains on the weapon and the perpetrator's clothing, which are further investigated using chemical testing [47]. The perpetrator mostly deletes, overwrites, or otherwise damages files via damaging hard drives, etc. Finding a way to restore the destroyed data is the sole focus of the team. Hacker Combat, Autopsy, Operating System Forensics, WIRESHARK, TrueCrypt, FTK Imager, SANS SIFT, and X-WAYS are among the technologies that can be used to retrieve lost data.

An alternative approach [48] was suggested to aid investigators in their analysis by coordinating evidence uncovered by several forensic instruments. The method was implemented using a system that could add semantic assertion to the data generated by forensics tools during extraction. It is conceivable, according to these assertions, to have better memory and reasoning abilities and to have easier access to relevant knowledge. Through the incorporation of AF (Anti-forensic) approaches that impact different stages of a digital forensic process [49], validate Digital forensic models (DFMs) and present an abstract framework for digital forensics. Applying this validation concept to improve current DFMs allows for principled detection and countering of AF techniques, which may then be employed in a real-time investigation. In order to be prepared for digital forensics, to monitor, plan, and formulate ahead of time before any possible security incidents occur [50], suggested a new technique that offers simultaneous support for essential functions like public auditing, data integrity, confidentiality, an unbounded number of audits, blockless verification, stateless verification, and digital forensic readiness. In order to keep track of the current status of all virtual machines (VMs) in the cloud, [51] installed a reliable cloud forensic system that periodically transmits a snapshot to TCS. By locating the forensic server in the cloud and making use of the computing power of the cloud, they developed a model that could solve the problems associated with cloud-based forensic investigations and provide forensics as a service.

4.4 CUTTING-EDGE TECHNIQUES IN DIGITAL FORENSICS

Understanding the methodology and approaches used by investigators to collect and analyze digital evidence requires a thorough review of the many techniques utilized in digital forensics. When it comes to digital media and data, digital forensics involves a wide range of methods. These methods are shown in Figure 4.1. This analysis clarifies the most important methods used in digital forensics:

Keyword Search and Data Filtering: The use of keyword searches and data filtering is a cornerstone of digital forensics. Investigators use keyword searches to locate records, correspondence, and other data that may be useful to their inquiries. Using this method, they can rapidly whittle down the mountain of data and zero in on the most promising pieces of evidence [52–54].

- *File Signature Analysis*: In order to recognize specific file types, even if file extensions have been changed, file signature analysis includes locating distinct patterns, or signatures, within files. Data carving, in which investigators try to retrieve lost or hidden files by tracing their headers and footers, makes excellent use of this method [55, 56].

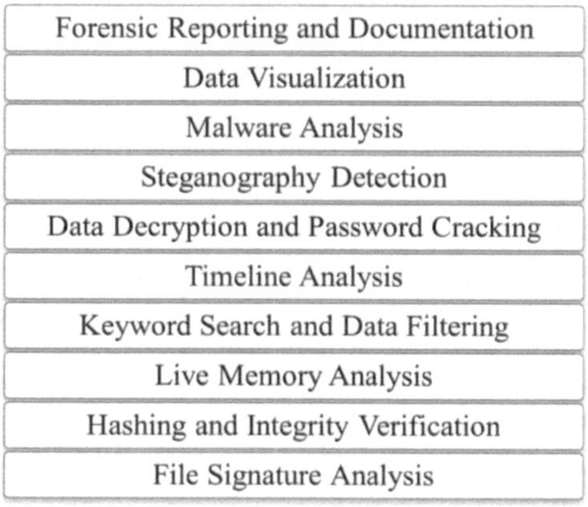

FIGURE 4.1 List of digital forensics techniques.

- *Hashing and Integrity Verification*: In the field of digital forensics, hashing is employed to ensure the reliability of digital evidence. To confirm that no changes were made to the data throughout the investigation, investigators calculate hash values and compare them with known hash values [57].
- *Timeline Analysis*: The purpose of timeline analysis is to rebuild and display a timeline of events using digital evidence such as timestamps and metadata. Timelines help investigators piece together what happened by showing the sequence of events that a user or system took [58–60].
- *Live Memory Analysis*: Performing a live memory analysis entails inspecting the RAM of a functioning system to detect processes, network connections, and malicious actions. This method aids in the detection of artifacts stored in memory that could otherwise be missed by conventional disk-based analysis [61].
- *Data Decryption and Password Cracking*: If digital forensics specialists come across encrypted data, they may use data decryption methods to gain access to the hidden files and messages. Recovery of passwords used to unlock encrypted data or password-protected files is another common use for password-cracking tools [62].
- *Steganography Detection*: Steganography refers to the practice of secretly encoding information within otherwise unobtrusive media. Experts in digital forensics use steganography detection tools and techniques to unearth information that has been cloaked within another file [63, 64].
- *Malware Analysis*: Malware analysis methods are used to investigate and assess potential malware infections or cyber-attacks. Analysis of malware lets investigators learn more about the threat's actions, capabilities, and effects on compromised machines [65].
- *Data Visualization*: In order to help investigators and stakeholders better comprehend the digital evidence at hand, data visualization techniques are used to present the data in a graphical format [66].
- *Forensic Reporting and Documentation*: To convey the results of a digital investigation in a clear and simple manner, it is crucial to employ effective forensic reporting and documentation methodologies. Throughout an investigation, investigators should keep meticulous records of all of their methods, techniques, and gathered evidence [67, 68].

In conclusion, the wide variety of resources and approaches utilized in digital forensics to unearth and analyze digital evidence has been revealed through this investigation. Digital forensics professionals have a wide variety of tools at their disposal, including but not limited to: keyword searches, file signature analysis, memory analysis, decryption, and data visualization. Investigators can help bring digital investigations to a close by properly employing these tools to reconstruct events, identify criminals, and provide persuasive evidence in judicial processes. To stay up with the ever-changing demands of the digital world, experts in the field of digital forensics must devote themselves to constant study and experimentation.

4.5 ANALYZING DIGITAL FORENSICS TOOLS

In order to properly investigate and analyze digital evidence, digital forensics tools are required. These specialized software and hardware solutions are designed to assist digital forensics practitioners in acquiring, preserving, and examining digital data from various sources. These tools are displayed in Figure 4.2. An overview of commonly used digital forensics tools provides insights into the capabilities and functionalities that aid investigators in efficiently and effectively handling digital evidence:

- *Forensic Imaging Tools*: Forensic imaging tools play a central role in creating bit-by-bit copies (forensic images) of digital storage media, ensuring the preservation of data integrity [51, 69]. Widely used imaging tools, such as "dd" in Linux environments or commercial solutions like EnCase and FTK Imager, enable investigators to acquire exact replicas of hard drives, solid-state drives, USB drives, and other media while maintaining the chain of custody.
- *Data Carving Tools*: These tools are vital for recovering deleted or fragmented files from digital storage media. These tools employ advanced algorithms to scan unallocated space on storage devices and reassemble fragmented files or retrieve deleted data. PhotoRec, Scalpel, and foremost are three of the most well-known data carving tools, and they can restore a broad variety of file formats [56, 70].
- *File Analysis Tools*: The content of files can be analyzed with the help of file analysis tools, which can then be used to extract important information

FIGURE 4.2 Categories of digital forensic tools in cyber investigations.

for investigators. Documents, pictures, movies, and emails are just some of the forms that may be read and analyzed by these programs. ExifTool and Metadata Extractor are two examples of programs that may be used to remove metadata from photos, and OLETools can be used to examine object linking and embedding (OLE) compound files [46, 71].

- *Registry Analysis Tools*: The Windows registry is a central storage location for configuration settings and system data, and it is the primary focus of registry analysis tools. Experts in digital forensics can make better use of user activity, system settings, and program installations with the aid of these technologies. Common tools for reading and understanding registry hives are RegRipper and Registry Explorer.
- *Internet History Analysis Tools*: Researching online behavior and uncovering evidence related to it can be done with the help of internet history analysis tools. Information such as search terms, downloaded files, and websites visited can all be gleaned through these instruments. ChromeCacheView and Mozilla Firefox's SQLite Manager are two examples of such tools.
- *Mobile Forensics (MF) Tools*: With the increasing prevalence of smartphones and mobile devices, MF tools are indispensable in acquiring and analyzing data from these platforms. Tools like Cellebrite UFED, Oxygen Forensic Detective, and Magnet AXIOM are capable of extracting data from mobile devices, recovering deleted information, and providing valuable insights into user activities and communication.
- *Memory Analysis Tools*: Memory analysis tools are used to examine the volatile memory (RAM) of a live system or a memory dump file. These tools are essential in identifying running processes, open network connections, and potential malware presence in memory [72, 73]. Volatility, Redline, Encase, BEC, and Rekall are prominent memory analysis tools commonly used by digital forensics experts.
- *Network Forensics Tools*: Network forensics involves proactive monitoring and analysis of extremely dynamic and volatile network traffic in order to trace attacks back to their initial point of entry and collect evidence of such attacks [74, 75]. The switch, router, firewall, intrusion detection system, and IoT devices are all integral parts of any network. LogRhythm, Wireshark, Zeek (formerly known as Bro), PLIXER, XPLiCO, and NetworkMiner are popular network forensics tools that facilitate the investigation of network-based incidents [76].

The overview of commonly used digital forensics tools highlights the diverse capabilities that aid digital forensics investigators in their work. From forensic imaging and data carving to file analysis, registry examination, internet history analysis, and mobile and memory forensics, these tools empower investigators to extract and analyze digital evidence efficiently, contributing to effective and reliable forensic examinations. As the field of digital forensics continues to evolve, new tools and advancements will further enhance the ability to uncover crucial evidence from the ever-expanding digital landscape.

4.6 DIGITAL FORENSIC CHALLENGES

Many technical, legal, and ethical obstacles must be surmounted before digital crimes may be effectively investigated and prosecuted. Investigators in the field of digital forensics must be prepared to deal with a complicated and ever-changing environment, which calls for cutting-edge skills and knowledge. Different challenges are shown in Figure 4.3. Digital forensic investigations and the outcomes of legal procedures that rely on digital evidence depend on identifying and overcoming these difficulties. Here we'll take a look at some of the biggest difficulties encountered by forensic digital investigators. Relevant research focusing on the technical difficulties has discovered significant hurdles [77, 78].

4.6.1 Technical Challenges

Unique challenges arise from the technological aspects of digital forensics investigations. It is difficult for digital forensic investigators to keep up with the newest technological trends due to the constant production of new devices, operating systems, and apps [12, 79]. Encryption and other data protection methods can also make it difficult to obtain and examine digital evidence. Another difficulty is the time and effort required to analyze the massive amounts of data. Last, it can be challenging to trace the origin of security events or data breaches, as well as the

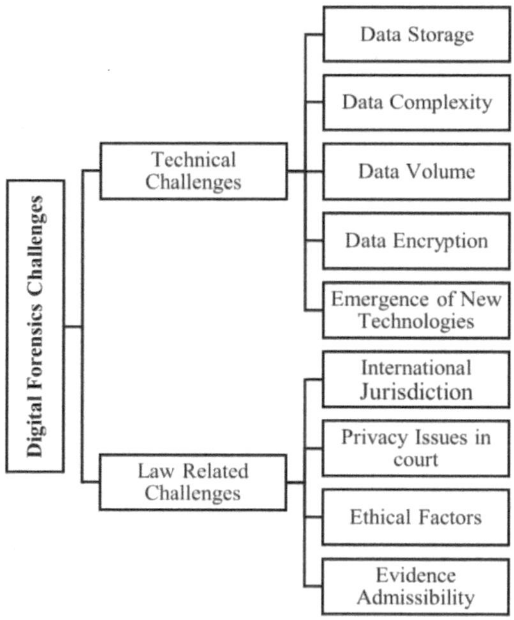

FIGURE 4.3 Taxonomy of challenges in digital forensics.

degree of the harm due to the complexity of digital systems and networks. As a whole, these technical obstacles show how important it is to keep improving digital forensics tools and methods.

- *Data Storage*: Local and distant servers, the cloud, and physical devices are all viable options for storing digital data. Finding and compiling all pertinent data from these various sources presents a significant difficulty for investigators. Local and distant servers, the cloud, and physical devices are all viable options for storing digital data. Identifying and collecting all relevant evidence from these different sources can be a substantial difficulty for digital forensic investigators. The storage media itself may also pose problems, as certain devices may be broken or inaccessible, while others may need particular expertise to access them. To properly gather and analyze evidence from any storage media, digital forensic investigators need to be familiar with a wide range of technologies and storage systems and flexible in their approach [12, 80].
- *Data Complexity*: Digital data analysis is sometimes complicated and time-consuming due to the fact that it often contains metadata, system logs, hidden or deleted files, and a variety of data kinds (structured and unstructured). Digital data may include a wide variety of information types, including structured and unstructured data, metadata, system logs, and even hidden or deleted files. Because of the need to review each piece of data to establish its relevance to the study, this can make analysis complicated and time-consuming. Further complicating efforts to glean useful information from digital devices is the growing prevalence of encryption and other data protection mechanisms. Expertise in data analysis, cryptography, and information security are just a few of the technical areas that digital forensic investigators need to successfully analyze and understand complicated digital evidence [62, 81].
- *Data Volume*: When many devices are involved, the sheer volume of digital data that must be analyzed during a digital forensic investigation can be daunting. Investigators may need more time and money due to the complexity and number of data they must process. In addition, the greater the volume of data, the less likely it is that investigators will be able to uncover the evidence they need because they will have to sift through more of it to get to the important bits. The timely and efficient organization and analysis of massive volumes of data is thus a significant problem for digital forensics. In order to overcome this obstacle, researchers will need to employ data analysis and processing techniques developed by experts and develop ways for extracting insights from big data sets [82, 83].
- *Data Encryption*: Despite its effectiveness in preventing unauthorized access, investigators may find it more challenging to obtain and assess digital data due to encryption. Encryption is a major obstacle for digital forensics when dealing with sensitive data that must remain hidden. It can be very challenging, if not impossible, to decode encrypted data so that it can

be read and analyzed. Therefore, investigators need to use a wide range of methods and instruments in an effort to crack encryption and gain access to encrypted data. Methods that aim to circumvent encryption and obtain unencrypted data are included in this category [81, 84, 85].

- *Emergence of New Technologies*: Digital forensic investigators must continually adapt and evolve to keep up with the high rate of technical innovation in terms of new devices, operating systems, and applications. As new devices, operating systems, and applications become available at an ever-increasing rate, digital forensic investigators must constantly adapt and change to keep up. This calls for an in-depth familiarity with cutting-edge tech and the agility to pick up and use new methods and tools. As a corollary, investigators need to be able to spot new weaknesses and dangers as well as create proactive plans to counter them [84, 86].

4.6.2 Law-Related Challenges

Issues occur when trying to collect, analyze, and present digital evidence in court within the confines of the legal and regulatory frameworks in place. These problems can have a major bearing on whether or not digital evidence can be used in court and on the integrity of that evidence. Key law-related challenges in digital forensics include the following:

- *International Jurisdiction*: Obtaining evidence and sharing information during a digital forensic investigation that spans many jurisdictions can be difficult due to the differences in the applicable laws and regulations. It can be challenging to collect and analyze digital evidence in investigations that span numerous jurisdictions with different legal systems [87, 88]. Information and data interchange can be made more difficult by the absence of a standardized legal framework among nations. As a result, when investigating across international borders, sleuths need to be cognizant of and careful to adhere to any applicable legal limits.
- *Privacy Issues in Court*: Digital forensic investigations may involve the collection and analysis of personally identifiable information, which raises concerns with data protection and privacy laws. The EU General Data Protection Regulation (GDPR) and other privacy and data protection laws have made it harder for investigators to obtain personal data. Data processing is severely limited by these regulations, which can hamper digital forensic investigations [89, 90].
- *Evidence Admissibility*: Digital evidence's admissibility in court might be challenged for reasons, including validity, reliability, and chain of custody. There are a number of challenges that digital evidence faces when trying to be admitted into court, including questions of validity, accuracy, and reliability. Prior to its presentation in court, the evidence must have been collected, kept, and examined in accordance with predetermined processes. In general, courts demand that evidence be relevant, credible, and collected legally.

- *Ethical Factors*: Among the many ethical considerations that digital forensic investigators must keep in mind are the individuals whose data is being examined and their right to privacy. When conducting their work, investigators must maintain the utmost professionalism, act ethically, and comply with all local, state, and federal regulations.

4.7 DISCUSSION AND ROAD AHEAD IN DIGITAL FORENSICS

Rapid technological breakthroughs and growing problems characterize the environment of digital forensics tools and procedures, as surveyed in this chapter. We have recognized significant advancements in the discipline, such as the growth of cloud and MF, which mirror the evolving character of digital evidence and crime. As a whole, our results support the idea that forensic methodologies need updating to stay up with technology developments. Additionally, our research highlights the intricate relationship between technological capacity and legal-ethical factors. There have to be frameworks that can accommodate both security and individual rights to effectively balance privacy issues with digital investigations. Discussions surrounding data privacy and encryption, topics that are becoming more important in the modern digital era, highlight this contradiction. Additionally, our research sheds light on how multidisciplinary methods can improve digital forensic procedures. An exciting new direction for digital forensics is the merging of technological investigations with psychological and behavioral assessments. This viewpoint not only adds to the resources available to forensic experts but also creates opportunities for further study and teamwork. The outcomes of our survey also reveal gaps in the current digital forensics literature, particularly in the integration of diverse forensic domains. Our systematic research approach has cataloged a wide range of methods and tools, yet it also highlights the need for more comprehensive, cross-disciplinary studies that can provide a more cohesive understanding of the digital forensics ecosystem. In reflecting on these findings, it becomes evident that the field of digital forensics is at a critical juncture. As technology continues to advance at a rapid pace, the methods and tools of digital forensics must evolve accordingly. This evolution is not just a matter of technical advancement but also involves a deeper understanding of the legal, ethical, and psychological dimensions of digital investigations.

Forensic analysis of network data is an established field with many resources at its disposal. Since most traffic is encrypted, the growth in the number of protocols utilized and the sheer volume of data being transmitted are the primary areas where improvement is needed. The amount of network traffic that must be studied to complete a forensics investigation grows proportionally with the prevalence of computer systems and the internet. There is a significant need for techniques that can quickly and accurately examine massive network traffic traces. It's also more labor to gather proof from everywhere because network traffic protocols are diverse.

Researchers have taken an interest in file system, memory, and data storage forensics because these areas provide a wealth of digital evidence. The sheer quantity of files and data within these domains is a significant challenge, as we saw in Section III-E. To get useful inquiry data out of the sea of seemingly unrelated or redundant digital items, the focus should be on big data analysis and data mining approaches.

The question of whether or not forensic analysis should be performed on the cloud becomes more complicated in situations involving distributed filesystems, databases, or data repositories. The latter situation is more difficult because it necessitates not just specific tools and processes but also close interaction with cloud service providers. MF is a branch of digital forensics concerned with the recovery of digital evidence from handheld and mobile devices. There are essentially three stages to MF processes: gathering, storing, and analyzing. Sound forensics methods have significant challenges in the MF field due to the large number of embedded operating systems, the short device life cycles, and the worldwide proliferation of smartphone manufacturers. The ability to perform live forensics requires fully functional mobile devices. The last step is for an investigator to get into the device, which, because of the security features included in most current mobile devices, will almost certainly change the data.

Data accessibility and operational excellence have both benefited from the broad deployment of IoT devices and IoT-related apps, but this has given rise to new security and forensic concerns. Another challenge in developing effective IoT-related forensics techniques is the time and effort required to compile comprehensive reports outlining all findings and their appropriate representation. IoT forensic investigators have additional challenges due to the increasing use of cryptographically secured storage systems, which already provide a substantial obstacle to effective digital forensic analysis. There are a lot of security storage restrictions related to IoT forensics evidence, and there are a lot of interoperability and availability issues because of all the linked IoT devices. Big Data is also an issue with IoT forensics data. Finally, there are several regulatory problems in the IoT forensics space, especially concerning cloud data ownership as defined by local regulations.

Since blockchain technology has only been around for a decade, blockchain forensics is a young field. It is important to realize that the demand for blockchain forensics techniques will likely increase over the next few years. Developing effective methods to carry out this sort of study is a significant obstacle. Public ledgers have ever-increasing databases, but their storage formats vary widely. Methods and tools for efficient data analysis on the most popular blockchain systems need to be developed.

4.8 CONCLUSION

Despite the numerous benefits, new worries regarding privacy and safety have emerged as a result of our increasingly digital existence. The motivation behind this manuscript is to fill a need in the literature by suggesting a review of digital forensics. The absence of a comprehensive overview of the ecosystem is mainly because different topics are examined separately, making the need for this review all the more pressing. Using a methodical research technique, we produced our list of the main digital forensics subjects. We compiled a list of all the digital forensics methodologies and tools that are now available to make a taxonomy. In conclusion, we discussed the article's main ideas and pointed out some of the future difficulties that digital forensics may face. That being said, we did offer some suggestions on how to avoid or circumvent such problems but we must not lose sight of the need to develop effective, cross-domain solutions; doing so would guarantee faster and more robust results, which should mitigate the effects of illegal acts. Forensics teams might use our study as a benchmark against which to measure the many toolkits available to them. Our literature review also covers future trends and open questions in digital forensics, as well as the fact that many of the problems that were first raised a long time ago are still unanswered, so it should be a good starting point for future studies in this area.

REFERENCES

1. C. Iwendi, Z. Jalil, A.R. Javed, T. Reddy, R. Kaluri, G. Srivastava, et al., Keysplitwatermark: Zero watermarking algorithm for software protection against cyber-attacks, IEEE Access 8 (2020), pp. 72650–72660.
2. Rehman Javed, Abdul, Zunera Jalil, Syed Atif Moqurrab, Sidra Abbas, and Xuan Liu. "Ensemble adaboost classifier for accurate and fast detection of botnet attacks in connected vehicles." Transactions on Emerging Telecommunications Technologies 33, no. 10 (2022): e4088.
3. N. Khadam, N. Anjum, A. Alam, Q.A. Mirza, M. Assam, E.A.A. Ismail, et al., How to punish cyber criminals: a study to investigate the target and consequence based punishments for malware attacks in UK, USA, China, Ethiopia & Pakistan, Heliyon (2023).
4. B.E. Turvey, Forensic Victimology: Examining Violent Crime Victims in Investigative and Legal Contexts, Academic Press, 2023.
5. A. MacDermott, T. Baker and Q. Shi, IoT forensics: Challenges for the IoA era, in 2018 9th IFIP International Conference on New Technologies, Mobility and Security (NTMS), 2018, pp. 1–5.
6. F. Casino, T.K. Dasaklis, G.P. Spathoulas, M. Anagnostopoulos, A. Ghosal, I. Borocz, et al., Research trends, challenges, and emerging topics in digital forensics: A review of reviews, IEEE Access 10 (2022), pp. 25464–25493.
7. A. Alazab, A. Khraisat and S. Singh A Review on the Internet of Things (IoT) Forensics: Challenges, Techniques, and Evaluation of Digital Forensic Tools [Internet]. Digital Forensics - Challenges and New Frontiers [Working Title]. IntechOpen; 2023.
8. K. Choi, S. Back and M.M. Toro-Alvarez, Digital Forensics & Cyber Investigation, Cognella Academic, 2022.
9. M.I. Alghamdi, Digital forensics in cyber security—recent trends, threats, and opportunities, Cybersecurity Threats with New Perspectives (2021).
10. T.J. Holt, A.M. Bossler and K.C. Seigfried-Spellar, Cybercrime and Digital Forensics: An Introduction, Routledge, 2022.
11. K. Barmpatsalou, T. Cruz, E. Monteiro and P. Simoes, Current and future trends in mobile device forensics: A survey, ACM Computing Surveys (CSUR) 51 (2018), pp. 1–31.
12. M. Stoyanova, Y. Nikoloudakis, S. Panagiotakis, E. Pallis and E.K. Markakis, A survey on the Internet of Things (IoT) forensics: Challenges, approaches, and open issues, IEEE Communications Surveys & Tutorials 22 (2020), pp. 1191–1221.
13. A. Harisha, A. Mishra and C. Singh, Advancements in Cybercrime Investigation and Digital Forensics, CRC Press, 2023.
14. S. Costantini, G. De Gasperis and R. Olivieri, Digital forensics and investigations meet artificial intelligence, Annals of Mathematics and Artificial Intelligence 86 (2019), pp. 193–229.
15. J.-P.A. Yaacoub, H.N. Noura, O. Salman and A. Chehab, Advanced digital forensics and anti-digital forensics for IoT systems: Techniques, limitations and recommendations, Internet of Things 19 (2022), p. 100544.
16. D. Beardall, Unveiling the Digital Shadows: Cybersecurity and the Art of Digital Forensics, (2023).

17. G. Gogolin, Digital Forensics Explained, CRC Press, 2021.
18. S. Krishnan, B. Zhou and M.K. An, Smartphone forensic challenges, (2019).
19. Etow, Tambue Ramine. "Impact of anti-forensics techniques on digital forensics investigation." (2020). https://www.diva-portal.org/smash/get/diva2:1453485/FULLTEXT02.pdf
20. D.K.V.K. Santhy and A.S. Padmanabhan, A review on the changing dimensions of digital forensics in criminal investigations, SVP National Police Academy Journal, Vol. LXXI, No. 1-2, 160-183 (2022).
21. W. Khan, M. Ishrat, M. Haleem, A.N. Khan, M.K. Hasan and N.A. Farooqui, An Extensive Study and Review on Dark Web Threats and Detection Techniques, in Advances in Cyberology and the Advent of the Next-Gen Information Revolution, IGI Global, 2023, pp. 202–219.
22. V. Roussev, Digital Forensic Science: Issues, Methods, and Challenges, Morgan & Claypool Publishers, 2016.
23. M. Rasool and W. Khan, Big data: Study in structured and unstructured data, 2015.
24. J. Brunty, Validation of forensic tools and methods: A primer for the digital forensics examiner, Wiley Interdisciplinary Reviews: Forensic Science 5 (2023), p. e1474.
25. D. Quick and K.-K.R. Choo, Impacts of increasing volume of digital forensic data: A survey and future research challenges, Digital Investigation 11 (2014), pp. 273–294.
26. W. Khan, H. Ansari and A.A. Shaikh, Log files utility for software maintenance, International Journal of Advanced Research in Computer Engineering & Technology (IJARCET) 4 (2015).
27. A. Årnes, Digital Forensics, John Wiley & Sons, 2017.
28. R. Ayers, S. Brothers and W. Jansen, Guidelines on mobile device forensics (draft), NIST Special Publication 800 (2013), pp. 101.
29. G. Horsman, Defining principles for preserving privacy in digital forensic examinations, Forensic Science International: Digital Investigation 40 (2022), p. 301350.
30. J. Kävrestad, Fundamentals of Digital Forensics, Springer, 2020.
31. P.S. Shaji, Understanding the digital forensics framework of cloud computing-cloud forensics, An Open Access Journal from The Law Brigade (Publishing) Group 6 (2020), pp. 10–37.
32. W. Khan and L. BITM and M.S. Awasthi, Matlab Based Implementation and Comparative Analysis of De-Blocking Algorithms, 2017.
33. W. Khan and M. Haroon, A Pilot Study and Survey on Methods for Anomaly Detection in Online Social Networks, in Human-Centric Smart Computing: Proceedings of ICHCSC 2022, Springer, 2022, pp. 119–128.
34. B. Nelson, A. Phillips and C. Steuart, Guide to Computer Forensics and Investigations, Cengage Learning, 2014.
35. I. Yaqoob, I.A.T. Hashem, A. Ahmed, S.M.A. Kazmi and C.S. Hong, Internet of things forensics: Recent advances, taxonomy, requirements, and open challenges, Future Generation Computer Systems 92 (2019), pp. 265–275.
36. R.K. Gibson, Party change, social media and the rise of 'citizen-initiated' campaigning, Party Politics 21 (2015), pp. 183–197.
37. W. Khan and M. Haroon, An efficient framework for anomaly detection in attributed social networks, International Journal of Information Technology 14 (2022), pp. 3069–3076.
38. L.L. Dhirani, N. Mukhtiar, B.S. Chowdhry and T. Newe, Ethical dilemmas and privacy issues in emerging technologies: A review, Sensors 23 (2023), p. 1151.
39. V. Galaz, M.A. Centeno, P.W. Callahan, A. Causevic, T. Patterson, I. Brass, et al., Artificial intelligence, systemic risks, and sustainability, Technology in Society 67 (2021), pp. 101741.
40. S. Li, T. Qin and G. Min, Blockchain-based digital forensics investigation framework in the Internet of Things and social systems, IEEE Transactions on Computational Social Systems 6 (2019), pp. 1433–1441.

41. A.M. Balogun and S.Y. Zhu, Privacy impacts of data encryption on the efficiency of digital forensics technology, arXiv preprint arXiv:1312.3183 (2013).
42. W. Khan and M. Haroon, An unsupervised deep learning ensemble model for anomaly detection in static attributed social networks, International Journal of Cognitive Computing in Engineering 3 (2022), pp. 153–160.
43. D. Rawtani and C.M. Hussain, Modern Forensic Tools and Devices: Trends in Criminal Investigation, John Wiley & Sons, 2023.
44. A.M. Alenezi, Digital Forensics in the Age of Smart Environments: A Survey of Recent Advancements and Challenges, arXiv preprint arXiv:2305.09682 (2023).
45. K.M. Mahdi Salih and N.B. Ibrahim, Digital forensic tools: A literature review, Journal of Education & Science 32 (2023).
46. F. Amato, G. Cozzolino, V. Moscato and F. Moscato, Analyse digital forensic evidences through a semantic-based methodology and NLP techniques, Future Generation Computer Systems 98 (2019), pp. 297–307.
47. S. Sachdeva, B.L. Raina and A. Sharma, Analysis of digital forensic tools, Journal of Computational and Theoretical Nanoscience 17 (2020), pp. 2459–2467.
48. F. Amato, A. Castiglione, G. Cozzolino and F. Narducci, A semantic-based methodology for digital forensics analysis, Journal of Parallel and Distributed Computing 138 (2020), pp. 172–177.
49. D. Mothi, H. Janicke and I. Wagner, A novel principle to validate digital forensic models, Forensic Science International: Digital Investigation 33 (2020), p. 200904.
50. I. Jayaraman and A. Stanislaus Panneerselvam, A novel privacy preserving digital forensic readiness provable data possession technique for health care data in cloud, Journal of Ambient Intelligence and Humanized Computing 12 (2021), pp. 4911–4924.
51. E.E.-D. Hemdan and D.H. Manjaiah, An efficient digital forensic model for cybercrimes investigation in cloud computing, Multimedia Tools and Applications 80 (2021), pp. 14255–14282.
52. G. Pozsgai, G.L. Lövei, L. Vasseur, G. Gurr, P. Batáry and J. Korponai, et al., Irreproducibility in searches of scientific literature: A comparative analysis, Ecology and Evolution 11 (2021), pp. 14658–14668.
53. N.A. Adderley, Graph-based temporal analysis in digital forensics, (2019).
54. W. Khan, An exhaustive review on state-of-the-art techniques for anomaly detection on attributed networks, Turkish Journal of Computer and Mathematics Education (TURCOMAT) 12 (2021), pp. 6707–6722.
55. E.A. Azeem, The data carving-the art of retrieving deleted data as evidence, International Journal for Electronic Crime Investigation 6 (2022), p. 8.
56. R.R. Ali, K.M. Mohamad, S. Jamel and S.K.A. Khalid, A review of digital forensics methods for JPEG file carving, Journal of Theoretical and Applied Information Technology 96 (2018), pp. 5841–5856.
57. H.-J. Ke, J. Liu, S.-J. Wang and D. Goyal, Hash-algorithms output for digital evidence in computer forensics, in 2011 International Conference on Broadband and Wireless Computing, Communication and Applications, 2011, pp. 399–404.
58. S. Bhandari and V. Jusas, An abstraction based approach for reconstruction of timeline in digital forensics, Symmetry (Basel) 12 (2020), p. 104.
59. M. Huber, M. Mulazzani, M. Leithner, S. Schrittwieser, G. Wondracek and E. Weippl, Social snapshots: Digital forensics for online social networks, in Proceedings of the 27th annual computer security applications conference, 2011, pp. 113–122.
60. W. Khan, M. Haroon, A.N. Khan, M.K. Hasan, A. Khan, U.A. Mokhtar, et al., DVAEGMM: Dual variational autoencoder with Gaussian mixture model for anomaly detection on attributed networks, IEEE Access 10 (2022), pp. 91160–91176.

61. N.R. Mistry and M.S. Dahiya, Signature based volatile memory forensics: A detection based approach for analyzing sophisticated cyber attacks, International Journal of Information Technology 11 (2019), pp. 583–589.
62. E.A. Vincze, Challenges in digital forensics, Police Practice and Research 17 (2016), pp. 183–194.
63. N. Alsaidi, M. Alshareef, A. Alsulami, M. Alsafri and A. Aljahdali, Digital steganography in computer forensics, International Journal of Computer Science and Information Security (IJCSIS) 18 (2020), pp. 54–61.
64. H. Kundra, W. Khan, M. Malik, K.P. Rane, R. Neware and V. Jain, Quantum-inspired firefly algorithm integrated with cuckoo search for optimal path planning, International Journal of Modern Physics C 33 (2022), p. 2250018.
65. F. Aldauiji, O. Batarfi and M. Bayousef, Utilizing cyber threat hunting techniques to find ransomware attacks: A survey of the state of the art, IEEE Access 10 (2022), pp. 61695–61706.
66. K. Börner, A. Bueckle and M. Ginda, Data visualization literacy: Definitions, conceptual frameworks, exercises, and assessments, Proceedings of the National Academy of Sciences 116 (2019), pp. 1857–1864.
67. G. Johansen, Digital Forensics and Incident Response: Incident Response Techniques and Procedures to Respond to Modern Cyber Threats, Packt Publishing Ltd, 2020.
68. A.A. Mughal, A comprehensive study of practical techniques and methodologies in incident-based approaches for cyber forensics, Tensorgate Journal of Sustainable Technology and Infrastructure for Developing Countries 2 (2019), pp. 1–18.
69. P. Sanda, D. Pawar and V. Radha, An insight into cloud forensic readiness by leading cloud service providers: A survey, Computing 104 (2022), pp. 2005–2030.
70. N.I.S. Ramli, S.I. Hisham and G. Badshah, Analysis of file carving approaches: A literature review, in Advances in Cyber Security: Third International Conference, ACeS 2021, Penang, Malaysia, August 24–25, 2021, Revised Selected Papers 3, 2021, pp. 277–287.
71. M. Zakariah, M.K. Khan and H. Malik, Digital multimedia audio forensics: Past, present and future, Multimedia Tools and Applications 77 (2018), pp. 1009–1040.
72. H. Nyholm, K. Monteith, S. Lyles, M. Gallegos, M. DeSantis, J. Donaldson, et al., The evolution of volatile memory forensics, Journal of Cybersecurity and Privacy 2 (2022), pp. 556–572.
73. M. Parekh and S. Jani, Memory forensic: Acquisition and analysis of memory and its tools comparison, Communication, Integrated Networks & Signal Processing 5 (2018), pp. 90–95.
74. A.R. Jayakrishnan and V. Vasanthi, Empirical survey on advances of network forensics in the emerging networks, International Journal of Cyber-Security and Digital Forensics 7 (2018), pp. 38–46.
75. S. Rizvi, M. Scanlon, J. Mcgibney and J. Sheppard, Application of artificial intelligence to network forensics: Survey, challenges and future directions, IEEE Access 10 (2022), pp. 110362–110384.
76. A.R. Javed, W. Ahmed, M. Alazab, Z. Jalil, K. Kifayat and T.R. Gadekallu, A comprehensive survey on computer forensics: State-of-the-art, tools, techniques, challenges, and future directions, IEEE Access 10 (2022), pp. 11065–11089.
77. J. Wiles, Technosecurity's Guide to e-Discovery and Digital Forensics: A Comprehensive Handbook, Elsevier, 2011.
78. X. Lin, X. Lin and Lagerstrom-Fife, Introductory Computer Forensics, Springer, 2018.
79. Z.A. Baig, P. Szewczyk, C. Valli, P. Rabadia, P. Hannay, M. Chernyshev, et al., Future challenges for smart cities: Cyber-security and digital forensics, Digital Investigation 22 (2017), pp. 3–13.

80. T.R. Sree and S.M.S. Bhanu, Data collection techniques for forensic investigation in cloud, Digital Forensic Science, Vol. 1, pp. 101–157 (2020).
81. A.M. Alenezi, Digital and Cloud Forensic Challenges, arXiv preprint arXiv:2305.03059 (2023).
82. A. Jarrett and K.R. Choo, The impact of automation and artificial intelligence on digital forensics, Wiley Interdisciplinary Reviews: Forensic Science 3 (2021), p. e1418.
83. V.S. Harichandran, F. Breitinger, I. Baggili and A. Marrington, A cyber forensics needs analysis survey: Revisiting the domain's needs a decade later, Computers & Security, 57 (2016), pp. 1–13.
84. T. Janarthanan, M. Bagheri and S. Zargari, IoT forensics: An overview of the current issues and challenges, in Digital Forensic Investigation of Internet of Things (IoT) Devices, Advanced Sciences and Technologies for Security Applications. Springer, 2021, pp. 223–254.
85. H. Arshad, A. Bin Jantan and O.I. Abiodun, Digital forensics: Review of issues in scientific validation of digital evidence, Journal of Information Processing Systems 14, vol. 14, no. 2 pp. 346–376. (2018).
86. D. Hayes and M. Kyobe, The adoption of automation in cyber forensics, in 2020 Conference on Information Communications Technology and Society (ICTAS), 2020, pp. 1–6.
87. A. Singh, R.A. Ikuesan and H. Venter, Secure storage model for digital forensic readiness, IEEE Access 10 (2022), pp. 19469–19480.
88. R. Montasari and R. Hill, Next-generation digital forensics: Challenges and future paradigms, in 2019 IEEE 12th International Conference on global security, safety and sustainability (ICGS3), 2019, pp. 205–212.
89. A. Nieto, R. Rios, and J. Lopez, Privacy-Aware Digital Forensics , Security and Privacy for Big Data, Cloud Computing and Applications, 2019. NICS Lab. Publications: https://www.nics.uma.es/publications
90. R.I. Ferguson, K. Renaud, S. Wilford and A. Irons, PRECEPT: A framework for ethical digital forensics investigations, Journal of Intellectual Capital 21 (2020), pp. 257–290.

5 Issues and Challenges in Emerging Cyber Forensics and Traditional Cyber Forensics Methods

Harold Andrew Patrick, Praveen Gujjar J, Guru Prasad M S, Rishi Kumar, and Naveen Kumar H N

5.1 INTRODUCTION

In digital cyberspace, the digital hobby is each a boon and a bane. It has spread out plenty of possibilities for communication, commerce, and innovation. Cyberspace has given an upward push to complex and complex cybercrime [1]. The digital era continues to grow, and so too does the need for effective and advanced strategies for investigating cybercrimes. This is where the sector of cyber forensics plays a crucial function. Cyber forensics also referred to as virtual forensics, is the science and art of collecting, reading, and retaining electronic evidence in a way that is admissible in a courtroom of regulation. It has emerged as an essential subject in our more and more interconnected world, wherein facts breaches, hacking incidents, identification robbery, and different cybercrimes have grown to be distressingly commonplace [2]. Traditional cyber forensics strategies were instrumental in figuring out and prosecuting cybercriminals; however, they're now confronted with unparalleled demanding situations [3–5].

This chapter serves as an exploration into the multifaceted global of cyber forensics, delving into both the installed methodologies and the rising issues that cast a shadow over the subject [6]. We will embark on a journey that takes us from the rules of traditional cyber forensics to the frontiers of cutting-edge techniques, all while addressing the complexities and dilemmas that investigators come upon along the way [7]. The rapid tempo of technological innovation poses one of the most pressing challenges for cyber forensics experts. The gadgets and structures we use to speak and store records are constantly evolving, and keeping pace with those changes isn't any small feat. The diversification of garage mediums, from traditional tough drives to cloud storage and Internet of Things (IoT) gadgets, has exponentially increased the scope of cyber forensics investigations. The globalization of cybercrime has blurred the bounds of jurisdiction, making it increasingly hard to prosecute offenders who perform throughout international borders. The criminal and ethical issues surrounding cyber forensics have ended up extra intricate, elevating questions about privacy, civil liberties, and the restrictions of investigative powers. Striking the proper stability in safety and personal rights is an impressive undertaking in this virtual age. This

DOI: 10.1201/9781003409571-5

dive into cyber forensics is driven through the digital panorama's double individual as an advantage supplying possibilities and a deadly disease containing complicated cybercrimes. Considering our growing reliance on digital devices, the want for efficient ways to identify and deal with cyber risks has by no means been more. Cyber forensics, as the dominant actor in this zone, is critical in fixing digital crimes and keeping the integrity of our related global. The bankruptcy covers the fast technical progress, emphasizing the model vital to combat cybercrime. This chapter highlighted the globalized man or woman of cybercrime, in addition to criminal and ethical troubles, to create an extra comprehensive view. We hope to provide readers with insights into the complexity of cyber forensics, developing a broader expertise of the subject's critical role in ensuring a safe digital environment.

5.2 CHALLENGES IN TRADITIONAL CYBER FORENSICS

The challenges in traditional cyber forensics are broadly classified into five different categories are as follows:

- Proliferation of Data
- Encryption and Privacy
- Jurisdictional Complexities
- Ephemeral Data
- Complex Attack Vectors

The large volume of information stored on various devices and platforms makes it challenging to identify and collect relevant evidence. This is categorized as the proliferation of data. Encrypted communication channels can hinder the ability to access and analyze data. Hence encryption and privacy are considered as one of the challenges in traditional cyber forensics. Traditional cyber forensics is bounded by jurisdictional limits, making it difficult to identify cybercrimes with an international scope. If the data resides in the main memory then the understanding of cyberattacks is relatively easy by cyber forensics but capturing ephemeral data and analyzing is one of the challenges in traditional cyber forensics. Cybercrimes usually occur with sophisticated attacks; traditional forensics may not be well-equipped to identify such attacks in real time.

5.3 CHALLENGES IN EMERGING CYBER FORENSICS

The challenges in emerging cyber forensics are classified into six main categories which are as follows:

- Technological Advancement.
- Big data Analysis
- Artificial intelligence (AI) Bias
- Ethical and Legal Dilemmas
- Lack of Standardization
- Resource Intensiveness

Technological advancements are dynamic in nature; cyber forensics must continuously adapt to the technological landscape such as new hardware, new software, and other protocols. A large volume of data has to be processed and analyzed for this specialized gear, and techniques are required to handle massive statistics efficiently. Machine learning and AI bias may hinder the efficiency of cyber forensics. AI must offer superior analytical capabilities, which do not persist any bias. Ethical and legal dilemmas are also one of the challenges in emerging cyber forensics. The framework has to be prepared for identification of the activity which outlines the boundaries for public and private data. The lack of standardization can lead to inconsistencies in methods and effects. Emerging cyber forensics should have professional experts with great sources of information and knowledge and any of these gaps may lead to a cybersecurity skill gap.

5.4 FUNDAMENTAL PRINCIPLES OF TRADITIONAL CYBER FORENSICS

Traditional cyber forensics, also known as digital forensics, is a discipline rooted in well-established principles and practices for investigating and analyzing digital evidence in a legally sound manner. The fundamental principles of traditional cyber forensics have been shown in Figure 5.1. Fundamental principles of traditional cyber forensics include the following points.

- *Preservation of Evidence*: The objective of cyber forensics is to keep digital evidence in a manner that maintains its integrity. This entails growing a chain of custody to record who had been admitted to the proof always to save you from tampering or contamination [8].
- *Legal Admissibility*: The cyber forensics processes ought to adhere to the laws and guidelines of the jurisdiction wherein the investigation is carried out.
- *Evidence Identification*: Cyber forensics investigators must as it should be identify and record potential sources of digital evidence [9].
- *Collection and Acquisition*: Once the proof is diagnosed, it has to be collected and purchased in a forensically sound manner to save your records from alteration or contamination.
- *Documentation and Analysis*: Thorough and well-maintained documentation is essential. Investigators must record the details of their actions, including the identification of evidence, collection procedures, and any changes made during the investigation.
- *Recovery of Deleted Data*: Cyber forensics experts often need to recover deleted or hidden data, as criminals often attempt to cover their tracks. Specialized tools and techniques are used to identify and retrieve such data while maintaining its integrity.
- *Timeline Reconstruction*: Creating a timeline of events is essential in lots of cyber forensics investigations. It helps set up a chain of movements and presents the context for the evidence. Timeline reconstruction may be essential in information on the who, what, while, and how of a cybercrime.

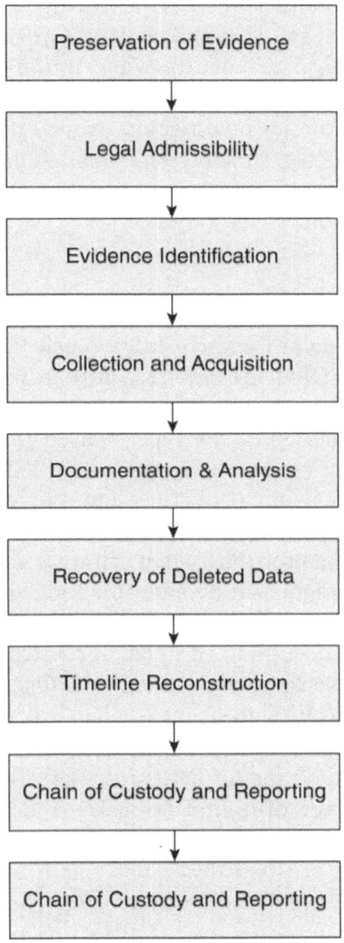

FIGURE 5.1 Fundamental principles of traditional cyber forensics.

- *Chain of Custody and Reporting:* Maintaining a secure chain of custody is paramount. This record tracks the possession and handling of evidence from the moment it is identified until it is presented in court. A well-documented chain of custody ensures the integrity and authenticity of the evidence. Investigators are expected to produce comprehensive reports that detail their findings, methodologies, and any relevant conclusions. These reports are essential for legal proceedings and may need to be presented to attorneys, law enforcement, or other relevant parties. Cyber forensics professionals can be known to testify in the courtroom as professional witnesses. They should be organized to give an explanation for their findings, methodologies, and the significance of the evidence to judges and juries.

The field of cyber forensics is constantly evolving due to advances in the era and modifications in cyber threats [10]. Cyber forensics professionals should interact in ongoing education and training to live contemporary with new gear, techniques, and exceptional practices. These fundamental ideas guide cyber forensics investigators as they navigate the complexities of digital proof and cybercrime investigations. Adhering to those standards guarantees that the proof gathered is reliable, admissible in court docket, and able to withstand legal scrutiny.

5.5 LITERATURE REVIEW

There are numerous technical issues in IoT forensic science. Technical difficulties such as establishing what IoT devices, how to professionally acquire data and secure your chain of control among other unexplored territories, including supplies needed for instruction or the type of used forensics tools [11]. A comprehensive comprehension of the challenges identified in the literature will aid academics in identifying new areas of study and provide a few recommendations to aid forensics investigators. The digital imprint of our daily lives is now enormous and consequently, the probability that criminal acts will leave digital traces is very high. Due to the increased demand for forensic investigators, numerous academic education and certification programs related to digital forensics have emerged [12, 13]. In the discipline of cyber forensics, "Data," also interpreted as "Information," has come to play a crucial role. One of the most vital circumstances in which data must be significant is when it is used as evidence in cybercrimes. In numerous instances involving crime scenes, these offenses can occur in the domains of electronic media and networks. Crime and forensic investigators require the assistance of digital forensics to determine whether a victim has committed a crime or not. Therefore, an investigator must utilize a suitable, accurate, affordable, and trustworthy cyber forensic instrument for forensics investigations conducted in relation to crimes [14]. The repeated increase in new types of cybercrime demands adaptive investigation process models, innovative technology, and cutting-edge techniques to combat these incidents. In terms of portability and efficiency, modern technology is more advanced. With billions of devices connecting to the internet, a vast quantity of details is generated that must be stored and accessed, posing significant challenges in sustaining the authenticity and reliability of digital proof for its acceptability in court. Due to the fact that digital evidence is latent, transient, fragile, can cross jurisdictional borders rapidly and easily, and in many cases can be time/machine dependent, their management presents unique challenges. In a digital society, ensuring the legality and authenticity of the methods and protocols used to collect and transmit evidence is a significant challenge [15–18]. As AI methods advance, digital image modification and the creation of false images become simpler and cheaper. Digital image manipulation can be harmless (e.g., to improve the quality or appearance of an image) or malicious (e.g., to alter the semantic meaning of the image to spread erroneous information about a specific individual, group, or event, or to establish an alibi in an illegal activity investigation). There are implications for judicial systems, the global economy, financial stability, and even territory and national security resulting from the spread of false images [19]. In recent years, the digital forensics and image forensics communities have shown interest in detecting intentional manipulation of digital pictures. According to recent research, the commercial sector has also shown interest [20]. The increasing number of mobile devices has given birth to complex cybercriminal activities, demanding the development of sophisticated mobile forensic techniques. This discussion discusses contemporary techniques for conducting mobile forensics and the unique limitations faced by forensic investigators. Additionally, it highlights the importance of identifying and exploring potential research that can improve the efficiency and effectiveness of mobile forensic technologies [21–23].

5.6 COMPARISON BETWEEN TRADITIONAL CYBER FORENSICS AND EMERGING CYBER FORENSICS

Emerging cyber forensics is characterized by its broader scope, dealing with larger volumes of data from diverse sources, reliance on advanced technologies, a proactive approach, and the need for interdisciplinary collaboration. It addresses the evolving landscape of cyber threats and digital evidence. The comparison between emerging and traditional cyber forensics is shown in Table 5.1.

5.7 EMERGING CYBER FORENSICS

According to Guru Prasad et al. [24], emerging cyber forensics, sometimes referred to as advanced or next-generation cyber forensics, builds on the core ideas of traditional cyber forensics while tackling the particular difficulties brought on by developing technology and cyber threats. Emerging cyber forensics are shown in Figure 5.2. The below paragraph outlines the emerging cyber forensics. The emerging cyber forensics are broadly classified into different categories which are as follows:

- Real-Time and Near-Real-Time Analysis
- Big Data and Cloud Forensics
- IoT Device Forensics
- Machine Learning and AI
- Blockchain Forensics
- Network Forensics
- Endpoint Detection and Response (EDR)
- Threat Intelligence Integration

Emerging cyber forensics frequently addresses cyber occurrences and threats that are changing quickly. In order to detect and address ongoing cyberattacks or data breaches, investigators need to be prepared to perform analysis in real time or very close to it.

Because of the widespread use of cloud computing and the exponential rise in data, a new field called cyber forensics is focused on developing methods and instruments

TABLE 5.1
Comparison between Traditional Cyber Forensics and Emerging Cyber Forensics

Sl. No.	Parameter	Traditional Cyber Forensics	Emerging Cyber Forensics
1	Focus and scope	Primarily focused on collecting and analyzing digital evidence related to cybercrimes such as hacking, data breaches, and malware attacks.	Cyber forensics includes emerging technologies like IoT, cloud computing, blockchain, and AI.
2	Data volume	Dealing with relatively smaller volumes of data, which were manageable with conventional tools and techniques.	Facing massive data volumes due to the proliferation of connected devices and cloud-based services, necessitating advanced analytics and machine learning for efficient analysis.
3	Data sources	Focused on data from traditional computing devices like computers and servers.	Handling data from a diverse range of sources, including IoT devices, wearables, smart appliances, and cloud-based platforms.
4	Complexity of evidence	Dealing with well-known digital evidence types, such as logs, emails, and files.	Addressing complex and non-traditional evidence types like blockchain transactions, telemetry data from IoT devices, and AI-generated content.
5	Timeliness	Focused on post-incident analysis and evidence collection after a cyberattack has occurred.	Emphasizing real-time or near-real-time monitoring and response to detect and mitigate threats as they happen, shifting from a reactive to a proactive approach.
6	Tools and techniques	Relied on established tools and methodologies for data collection and analysis.	Utilizing cutting-edge technologies like machine learning, artificial intelligence, automation, and advanced analytics for faster and more accurate analysis.
7	Challenges	The main challenges included dealing with encrypted data, volatile memory analysis, and anti-forensic techniques.	Faces additional challenges like securing IoT devices, tracing cryptocurrency transactions, and handling AI-generated deepfake content.
8	Legal and privacy concerns	Operated within established legal frameworks and privacy regulations.	Facing evolving legal and ethical dilemmas related to the use of AI, privacy in IoT, and blockchain's anonymity features.
9	Interdisciplinary approach	Primarily within the domain of cybersecurity and digital forensics.	Requires collaboration across diverse fields, including cybersecurity, data science, AI, law enforcement, and legal experts.
10	Globalization	Typically conducted within national or regional boundaries	Requires international cooperation and collaboration due to the global nature of cyber threats and digital evidence.

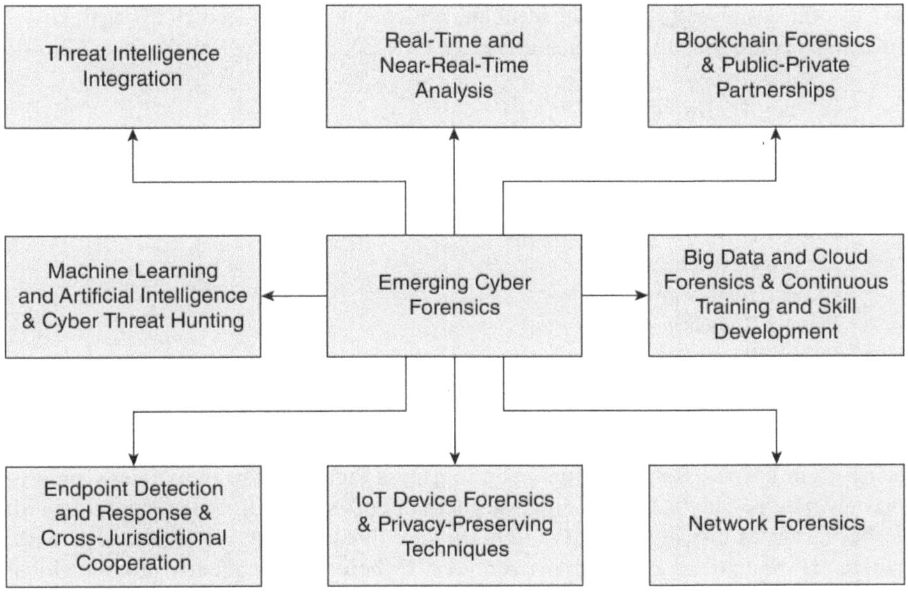

FIGURE 5.2 Emerging cyber forensics.

for quickly analyzing massive datasets and looking into data stored in the cloud [25]. A network of connected devices known as the IoT may contain important forensic evidence. For the purpose of extracting and analyzing data from the IoT devices, which could have limited processing and storage capacity—cyber forensics requires specific knowledge and tools. Emerging cyber forensics leverages machine learning and AI algorithms to automate the analysis of large datasets, detect anomalies, and identify patterns indicative of cyberattacks [24]. Blockchain forensics techniques can trace cryptocurrency transactions and identify illicit activities, such as money laundering or ransomware payments. Investigating network traffic and communication patterns is a fundamental aspect of emerging cyber forensics. EDR solutions are used to monitor and collect data from endpoints (e.g., computers and smartphones) in real time [26]. Emerging cyber forensics incorporates threat intelligence feeds and databases to enrich investigations. These feeds provide context about known threats, attacker tactics, techniques, and procedures (TTPs), which can aid in attributing cyberattacks.

5.8 LEGAL AND ETHICAL DILEMMAS FACED BY CYBER FORENSICS INVESTIGATORS

Cyber forensics investigators face numerous criminal and ethical dilemmas in their work, frequently stemming from the complex nature of virtual investigations and they want to stabilize regulation enforcement targets with individual rights and societal expectations. The below paragraphs outline the legal and ethical dilemmas faced by cyber forensics investigators. The legal and ethical dilemmas faced by cyber forensics investigators are broadly classified into eight different categories which are as follows:

- Privacy vs. Investigation
- Chain of Custody
- Jurisdiction
- Encryption
- Handling Sensitive Information
- Zero-Day Exploits
- Expert Testimony
- Privacy Invasion
- Public Perception
- Retention and Disposal

Cyber forensics investigators must appreciate the privacy of people while gathering digital proof. Balancing the need to access facts with an individual's right to privacy may be tough. Maintaining the chain of custody for digital evidence is vital. Cyber forensics investigators have to make sure that evidence is not tampered with, which can be hard in a virtual environment. Cybercrime frequently crosses global borders. Determining jurisdiction and cooperating with overseas law enforcement can be legally and ethically complicated. Decrypting encrypted records for investigative functions increases questions on individuals' right to privacy and the potential for authorities' overreach. Investigators may additionally encounter touchy or confidential facts unrelated to the case. Ethical dilemmas arise in determining the way to cope with this data and whether to document it. Cyber forensics investigators discover 0-day vulnerabilities at some point in research, they need to decide whether to file them to the software seller or take advantage of them for investigative purposes. Providing professional testimony requires making sure accuracy and objectivity. The use of surveillance strategies like adware or hacking gear increases moral questions on privacy invasion, even within the pursuit of crook suspects. Investigating insider threats within groups can contain tracking personnel's virtual sports, which may also infringe upon their privacy rights. Maintaining public acceptance as true is crucial. Public perception of cyber forensics investigations may be inspired by the aid of media insurance, which could have moral implications. The use of forensics equipment for each valid purpose and potential misuse can create ethical dilemmas. This equipment may be used by regulation enforcement but additionally via hackers or authoritarian regimes. Determining how lengthy to hold virtual evidence and how to safely eliminate it is important to each felony compliance and ethical considerations.

5.9 ADHERING TO LEGAL FRAMEWORKS AND ETHICAL GUIDELINES

Complying with legal frameworks and ethical recommendations is of utmost importance in investigations, especially in the context of digital forensics and ever-evolving cybercrime. This commitment now emphasizes not only standards of fairness and accountability but also the need to protect character rights and preserve public input in research procedures. Legal frameworks and ethical guidelines are designed to protect fundamental individual rights and freedoms. Respect for those considerations assures that the investigation does not violate privacy, due process, and different rights. Fairness and justice are enforced through adherence to established norms of law and ethics. Legal frameworks that help ensure that investigations are conducted without prejudice or discrimination, and that evidence is collected and processed in a manner that supports the rule of law, provide clear guidelines for accountability. Investigators working within the boundaries of the law often gather admissible evidence that can withstand forensic scrutiny. Law enforcement agencies and investigative agencies need to consider the public. Compliance with ethical requirements and legal recommendations can be taken into account by demonstrating transparency and responsible practices. The discipline of investigation, particularly in cybercrime and digital forensics, is dynamic. Rules and guidelines are being developed to address emerging threats and technologies. Recognizing these changes ensures that researchers are prepared to face new and challenging situations while supporting ethical requirements. International cooperation in research is forever important. Knowledge and compliance with global criminal treaties and protocols enable effective cross-border cooperation. Failure to adhere to prison and ethical standards can have criminal implications for investigators and exclude evidence from court, possibly jeopardizing prosecutions. In an era of growing concerns, it is often necessary to adhere to written confidentiality, ethical guidelines, and prison policies to ensure that investigators recognize the right of individuals to privacy, especially in digital analysis. Strict adherence to criminal and ethical requirements helps you protect against abuse of power and assures that investigators act within the limits of their authority. Investigators who adhere to high ethical standards and adhere to the law are viewed as credible and honest professionals, which is important in any law enforcement and private investigation. Maintaining awareness of and compliance with changing legal guidelines and policies is essential in a quickly changing virtual environment where the boundaries of generation and character rights may be hazy. It demands cooperation between technical and criminal experts, continual education and training for investigators, and a commitment to preserving moral principles as generation after generation comes of age. Ultimately, when investigations are conducted with the highest integrity and respect for the law, the pursuit of justice and the safety of a person's rights are greatly served.

5.10 COMPLEXITIES SURROUNDING EMERGING AREAS OF CYBER FORENSICS

Emerging areas of cyber forensics convey particular complexities and challenges due to the constantly evolving nature of the era and cyber threats. Cyber forensics investigators should adapt to these emerging complexities by continuously updating their abilities, participating with experts in associated fields, and staying abreast of technological improvements and criminal trends. Additionally, interdisciplinary techniques and the mixing of superior tools and methodologies are important to efficiently cope with those demanding situations. IoT devices range significantly in terms of architecture, operating structures, and communication protocols, making it hard to analyze and secure them. Cloud environments distribute information across a couple of servers and places, making it tough to collect and examine virtual evidence. Determining the physical area of cloud records and which jurisdiction applies to them could be complex in pass-border instances. Investigating cryptocurrency transactions may be tough because of the pseudonymous nature of blockchain generation, which may prevent the identity of people concerned with cybercrimes. Smart contracts on blockchain systems introduce complexities in tracing and studying agreement-primarily based cybercrimes. The use of AI via cybercriminals to automate assaults and evasion approaches complicates forensic evaluation, as investigators have to apprehend and counter AI-driven threats. AI may be used for forensic analysis, however demanding situations like bias in algorithms and the need for interpretability should be addressed. Detecting deepfakes and manipulated media calls for specialized forensic techniques, as content can be rather convincing and hard to distinguish from true materials. Investigating assaults on important infrastructure and cyber-physical structures includes knowledge of the interaction among digital and physical components, making it greater complicated than conventional virtual forensics. The creation of quantum computing threatens modern-day encryption techniques, which could affect the confidentiality of virtual proof and require new forensic techniques. The proliferation of statistics generated by using 5G and part computing technology will increase the quantity of capability digital evidence, posing challenges in records series, garages, and evaluation. Attackers may additionally use adverse system learning techniques to evade detection, requiring forensic specialists to understand and counter those threats. Investigating cybercrimes often requires cooperation between law enforcement organizations throughout borders, necessitating harmonization of prison frameworks and sharing of facts. Compliance with statistics privacy rules like General Data Protection Regulation (GDPR) can complicate digital evidence series and processing at the same time as safeguarding people's rights. As technology evolves, the ethical implications of using advanced tools and strategies in cyber forensics emerge as greater, traumatic, responsible, and obvious practices.

5.11 CONCLUSION

The fields of rising cyber forensics and traditional cyber forensics techniques are critical in our increasingly virtual world, but they each encompass their very own precise sets of issues and demanding situations. Traditional cyber forensics strategies have been essential in investigating virtual crimes for many years, but they face ongoing troubles in preserving pace with swiftly evolving generations and complicated cyber threats. The challenges surrounding information privacy, jurisdictional problems, and the want for proper evidence management hold to call for cautious attention and version inner this established concern. On the other hand, rising areas of cyber forensics, which incorporates IoT forensics, cloud forensics, blockchain analysis, and AI-driven investigations, provide thrilling opportunities to cope with novel cyber threats and technology. They introduce complexities that require specialized expertise, equipment, and methodologies. As we navigate the uncharted territories of these growing fields, troubles related to statistics fragmentation, the pseudonymous nature of blockchain, the speedy growth of interconnected devices, and the ethical implications of advanced technology become more and more giant. In this dynamic landscape, cyber forensics experts ought to encompass a multidisciplinary technique, collaborate throughout sectors, and continually update their talents to successfully address the issues and demanding situations. They need to moreover adhere to ethical principles, felony frameworks, and fine practices to ensure the integrity of virtual evidence at the same time as respecting people's rights to privacy. In the face of ever-evolving cyber threats and the fast-paced advancement of the era, cyber forensics remains a crucial pillar in keeping virtual safety and consideration.

REFERENCES

1. Lone, Auqib Hamid, and RoohieNaaz Mir. March 2019. "Forensic-Chain: Blockchain Based Digital Forensics Chain of Custody with PoC in Hyperledger Composer." Digital Investigation 28. Elsevier BV: 44–55. doi:10.1016/j.diin.2019.01.002.
2. Janarthanan, Tharmini, Maryam Bagheri, and Shahrzad Zargari 2020. "IoT Forensics: An Overview of the Current Issues and Challenges." In Advanced Sciences and Technologies for Security Applications, 223–254. doi:10.1007/978-3-030-60425-7_10.
3. Guru Prasad, M.S., H.N. Naveen Kumar, and K. Raju et al. 2023. "Glaucoma Detection Using Clustering and Segmentation of the Optic Disc Region from Retinal Fundus Images." SN Computer Science 4: 192. https://doi.org/10.1007/s42979-022-01592-1.
4. Prasad, Guru, Praveen Gujjar, HN Naveen Kumar, M. Anand Kumar, and S. Chandrappa. "Advances of Cyber Security in the Healthcare Domain for Analyzing Data." In Cyber Trafficking, Threat Behavior, and Malicious Activity Monitoring for Healthcare Organizations, pp. 1–14. IGI Global, 2023.
5. Kumar, M. A., A. H. Pai, J. Agarwal, S. Christa, G. M. S. Prasad, and S. Saifi, "Deep Learning Model to Defend against Covert Channel Attacks in the SDN Networks," 2023 Advanced Computing and Communication Technologies for High Performance Applications (ACCTHPA), Ernakulam, India, 2023, pp. 1–5, doi: 10.1109/ACCTHPA57160.2023.10083336.
6. J, P. G., M. H. Sharief, U. Mukherjee, M. Anand Kumar, and G. P. M S, "Malicious Host Detection in Software Defined Networks Using Machine Learning Algorithms," 2023 Third International Conference on Advances in Electrical, Computing, Communication and Sustainable Technologies (ICAECT), Bhilai, India, 2023, pp. 1–6, doi: 10.1109/ICAECT57570.2023.10118020.
7. Guru, P. M. S., G. J. Praveen, R. Dodmane, T. H. Sardar, A. Ashwitha, and A. N. Yeole, "Brain Tumor Identification and Classification using a Novel Extraction Method based on Adapted Alexnet Architecture," 2023 6th International Conference on Information Systems and Computer Networks (ISCON), Mathura, India, 2023, pp. 1–5. doi: 10.1109/ISCON57294.2023.10112075.
8. Chandrappa, S., M.S. Guruprasad, and H.N.N. Kumar et al. 2023. "An IOT-Based Automotive and Intelligent Toll Gate Using RFID." SN Computer Science 4: 154. https://doi.org/10.1007/s42979-022-01569-0.
9. Anand Kumar, M., N. Abirami, M. S. Guru Prasad, and M. Mohankumar, "Stroke Disease Prediction based on ECG Signals using Deep Learning Techniques," 2022 International Conference on Computational Intelligence and Sustainable Engineering Solutions (CISES), Greater Noida, India, 2022, pp. 453–458, doi: 10.1109/CISES54857.2022.9844403.
10. Avinash, S., H.N. Naveen Kumar, and M.S. Guru Prasad et al. 2023. "Early Detection of Malignant Tumor in Lungs Using Feed-Forward Neural Network and K-Nearest Neighbor Classifier." SN Computer Science 4: 195. https://doi.org/10.1007/s42979-022-01606-y.
11. Kirubasri, G., S. Sankar, and M.S. Guru Prasad et al. 2023. "LQETA-RP: Link Quality Based Energy and Trust Aware Routing Protocol for Wireless Multimedia Sensor Networks." International Journal of System Assurance Engineering and Management. https://doi.org/10.1007/s13198-023-01873-9.

12. Casino, Fran et al. January 2022. "Research Trends, Challenges, and Emerging Topics in Digital Forensics: A Review of Reviews." IEEE Access 10: 25464–25493. doi:10.1109/access.2022.3154059.
13. Overill, Richard E., and Jan Collie. 2021. "Quantitative Evaluation of the Results of Digital Forensic Investigations: A Review of Progress." Forensic Sciences Research 6, no. 1. Taylor & Francis: 13–18. doi:10.1080/20961790.2020.1837429.
14. Fernando, Vihara, "Cyber Forensics Tools: A Review on Mechanism and Emerging Challenges," 2021 11th IFIP International Conference on New Technologies, Mobility and Security (NTMS), April 2021. doi:10.1109/ntms49979.2021.9432641.
15. Ariffin, Khairul Akram Zainol, and FarisHanif Ahmad. June 2021. "Indicators for Maturity and Readiness for Digital Forensic Investigation in Era of Industrial Revolution 4.0." Computers & Security 105. Elsevier BV: 102237. doi:10.1016/j.cose.2021.102237.
16. Purnaye, Prasad, and Vrushali Kulkarni. 2021. "A Comprehensive Study of Cloud Forensics." Archives of Computational Methods in Engineering 29, no. 1. Springer Science+Business Media: 33–46. doi:10.1007/s11831-021-09575-w.
17. Waseem, Quadri, Sultan S. Alshamrani, Kashif Nisar, Wan IsniSofiah Wan Din, and Ahmed Saeed Al Ghamdi. 2021. "Future Technology: Software-Defined Network (SDN) Forensic." Symmetry 13, no. 5. Multidisciplinary Digital Publishing Institute: 767. doi:10.3390/sym13050767.
18. Bhatia, Manjot Kaur, Palak Gambhir, Shefali Sinha, and Saksham Kumar Singh. 2022. "A Comparative Analysis of OS Forensics Tools." International Journal for Research in Applied Science and Engineering Technology 10, no. 11): 494–502. doi:10.22214/ijraset.2022.47346.
19. Nowroozi, Ehsan, Ali Dehghantanha, Reza M. Parizi, and Kim-Kwang Raymond Choo. January 2021. "A Survey of Machine Learning Techniques in Adversarial Image Forensics." Computers & Security 100. Elsevier BV: 102092. doi:10.1016/j.cose.2020.102092.
20. Information Security, Audit and Control Association (ISACA). July 2000. COBIT 3rd Edition Control Objectives.
21. Chernyshev, Maxim, Sherali Zeadally, Zubair A. Baig, and Andrew Woodward. 2017. "Mobile Forensics: Advances, Challenges, and Research Opportunities." IEEE Security & Privacy 15, no. 6. Institute of Electrical and Electronics Engineers: 42–51. doi:10.1109/msp.2017.4251107.
22. Urias, Vincent, Curtis Hash, and Lorie M. Liebrock. 2009. "Consideration of Issues for Parallel Digital Forensics of RAID Systems." Journal of Digital Forensic Practice 2, no. 4. Taylor & Francis: 196–208. doi:10.1080/15567280903140953.
23. Köhn, Michael, Martin S. Olivier, and Jan H. P. Eloff. January 2006. "Framework for a Digital Forensic Investigation." Information Security for South Africa, 1–7. http://martinolivier.com/open/dfframe.pdf.
24. Guru Prasad, M. S., J. Agarwal, S. Christa, H. Aditya Pai, M. A. Kumar, and A. Kukreti, "An Improved Water Body Segmentation from Satellite Images using MSAA-Net," 2023 International Conference on Machine Intelligence for GeoAnalytics and Remote Sensing (MIGARS), Hyderabad, India, 2023, pp. 1–4, doi: 10.1109/MIGARS57353.2023.10064508.
25. Prasad, Guru, Amith K. Jain, Prithviraj Jain, and H. R. Nagesh. 2019. "A Novel Approach to Optimize the Performance of Hadoop Frameworks for Sentiment Analysis." International Journal of Open Source Software and Processes (IJOSSP) 10, no. 4: 44–59.
26. N, H. R, G. P. M. S, S. B. G, D. Jain, P. B. R, and M. Anadkumar, "E-Voting System Using Blockchain Technology," 2022 4th International Conference on Advances in Computing, Communication Control and Networking (ICAC3N), Greater Noida, India, 2022, pp. 2106–2111, doi: 10.1109/ICAC3N56670.2022.10074164.

6 Optimized Approaches for Cyber Forensics Tools Applications Case Studies

Shubham Kumar and Sarfaraz Alam

6.1 INTRODUCTION

In the contemporary landscape, digital crime and computer-supported criminal activities have become lucrative enterprises, with criminals, fraudsters, and terrorists exploiting every opportunity [1]. This necessitates the role of effective digital forensic investigators to identify criminals and crime scenes. The investigator's primary responsibilities include data recovery [2], system stability restoration, process rollback, vulnerability tracing, and the identification of cyber-attack paths, leading to legal proceedings against the perpetrators [3]. The investigator also implements intrusion detection and prevention mechanisms for effective computing. To facilitate meticulous investigation, digital forensic investigators rely on advanced tools, encompassing both hardware and software, specifically tailored for digital crime scenes [3]. An essential tool in this domain is Sleuth Kit Autopsy, which offers a user-friendly environment and incorporates critical strategies such as data runtime process backup, data cloning, extraction, analysis, and visualization. This comprehensive toolset ensures not only reduced investigation time but also correctness in case proceedings and data segregation. A complementary focus in the literature centers on the increasing importance of digital forensics in law enforcement, given the extensive use of computers and mobile devices. Forensic data analysis, conducted through digital techniques, is critical in this context. The chapter emphasizes the role of pattern recognition systems, positioning them within the Analysis Phase of Digital Forensics. This phase involves two integral processes: analysis and recognition. Attributes derived from pattern analysis, such as faces and fingerprints, are utilized to develop forensic tools for collecting evidence. These tools prove instrumental in solving specific types of crimes, providing crucial assistance to examiners during the analysis phase by identifying relevant data.

Furthermore, the surge in digitalization across organizations and individuals has led to an increased reliance on digital systems for various tasks. However, this digital transformation has also given rise to a higher rate of cybercrime. To address this, the chapter advocates for the installation of cyber security software to prevent cyber-attacks. It underscores the importance of awareness regarding available methods to handle attacks. Importantly, the chapter directs attention to the role of computers and

DOI: 10.1201/9781003409571-6

digital forensics in responding to and mitigating cyber threats, positioning them as essential components when tackling potential attacks. Collectively, these perspectives underscore the critical role of digital forensics in addressing the challenges posed by digital crimes, advocating for advanced tools, pattern recognition systems, and proactive cyber security measures. The integration of these approaches enhances the overall efficacy of digital forensic investigations in our increasingly digitized and interconnected world.

6.2 DIGITAL FORENSICS

In an era defined by digital transformation, where every facet of our lives is influenced by technology, the role of digital forensics has emerged as a linchpin in unraveling complex investigations and securing justice in the digital age [1]. The proliferation of smartphones, computers, cloud storage, and the internet has left a trail of digital footprints that, when analyzed with precision, can unlock invaluable insights and answers [4]. As cyber threats evolve, the need for optimized and advanced approaches to digital forensics tools becomes imperative. Motivated by the growing challenges in cybercrime investigations, this study delves into the development and application of optimized approaches for cyber forensics tools, with a particular emphasis on the widely used Autopsy platform. Our motivation stems from the recognition that current digital forensics methodologies face limitations in terms of processing time and accuracy, especially in the context of complex and rapidly evolving cyber threats.

The authors' contribution lies in proposing innovative methodologies that significantly enhance the efficiency and effectiveness of digital forensics investigations, as demonstrated through comprehensive case studies [5]. By addressing the shortcomings of existing tools and leveraging advanced techniques, our research aims to provide a substantial contribution to the ongoing evolution of cyber forensics, empowering investigators with tools that not only keep pace with technological advancements but also surpass traditional methodologies. This study seeks to bridge the gap between theoretical advancements and practical application, offering tangible solutions to the ever-growing challenges faced in the field of cyber forensics.

6.2.1 Importance of Digital Forensics in Today's World

In today's world, digital forensics holds a paramount position in investigations, be they criminal, civil, corporate, or related to cyber security incidents [5]. The ubiquity of digital devices and the extent to which we rely on them for communication, data storage, and transactional activities underscore the significance of digital evidence. Every text message sent, every email received, every financial transaction made, and every file saved leaves behind a digital trail—an electronic testament to our actions, intentions, and interactions. As society evolves, so too do the methods of committing crimes and concealing illicit activities. Criminals have become increasingly adept at exploiting digital channels to perpetrate offenses [2], making it imperative for law enforcement, legal professionals, and cyber security experts to harness the power of digital forensics. It is within the digital realm that crucial evidence is hidden, waiting to be uncovered by skilled investigators.

6.2.2 Significance of Digital Evidence in Investigations

Digital evidence, often referred to as electronic evidence or e-evidence, encompasses a vast spectrum of information stored in digital formats. This can range from text messages, emails, and social media posts to images, videos, and log files. The integrity and accuracy of digital evidence are paramount in legal proceedings, as it serves as a factual record of events. In courtrooms around the world, digital evidence has played a pivotal role in establishing guilt or innocence, shedding light on cybercrimes, and providing the means to trace illicit activities back to their source. Digital evidence is not confined to criminal cases alone. In the corporate world, it is instrumental in investigations related to intellectual property theft, fraud, and employee misconduct. Additionally, in the realm of cyber security, digital evidence helps organizations understand the nature and scope of data breaches, enabling them to bolster their defenses and mitigate future risks.

6.2.3 Autopsy: A Powerful Digital Forensics Tool

At the heart of effective digital forensics lies the need for robust, user-friendly tools capable of sifting through the mountains of digital data. One such tool, which stands as a beacon of excellence in the field, is Autopsy [6]. Developed as an open-source, cross-platform digital forensics platform, Autopsy offers an array of features and capabilities that are indispensable to investigators. It provides the means to analyze file systems, perform keyword searches, create timelines of activities, examine registries, scrutinize web artifacts, and recover data hidden within digital landscapes. Autopsy's power lies not only in its feature set but also in its accessibility. Its open-source nature ensures that it is available to a global community of forensic practitioners, free of cost. Moreover, its user-friendly interface and extensible architecture make it an ideal choice for both novice and seasoned investigators.

In an age where technology permeates every aspect of our lives, digital forensics has emerged as a pivotal discipline in the realm of investigations [7]. It is the science and art of collecting, analyzing, and preserving digital evidence to uncover the truth and establish the facts in legal, criminal, or corporate matters. This section delves into the fundamental aspects of digital forensics, from its definition and role in investigations to the types of cases where it is indispensable and the methodology guiding the process.

6.2.4 Define Digital Forensics and Its Role in Investigations

Digital forensics, often referred to as computer forensics, is the application of forensic science techniques to examine and recover information stored electronically [1]. It encompasses a wide range of digital devices, from computers and mobile phones to servers and cloud-based platforms. The primary objective of digital forensics is to retrieve, preserve, and analyze digital evidence in a manner that maintains its integrity and admissibility in a court of law.

The role of digital forensics in investigations is multifaceted. It serves as the means to:

- *Uncover Digital Evidence*: In a world where much of our interactions and transactions occur digitally, crucial evidence is often hidden within the electronic footprint. Digital forensics enables investigators to locate and extract this evidence.
- *Establish Facts*: Whether in criminal cases, civil litigation, or corporate investigations, digital evidence can help establish the facts surrounding an incident or dispute. It provides a factual record of actions, communications, and transactions.
- *Attribute Actions*: Digital forensics can attribute actions and activities to specific individuals or entities. This is particularly vital in cases of cyber-crimes, data breaches, and unauthorized access.
- *Support Legal Proceedings*: Digital evidence is admissible in court when collected and handled following established forensic procedures. It plays a critical role in legal proceedings, helping to prove or disprove allegations.

6.2.5 Types of Cases where Digital Forensics Is Crucial

Digital forensics is indispensable in a wide range of cases, including:

- *Criminal Investigations*: In criminal cases, digital forensics can uncover evidence of computer-related crimes such as hacking, identity theft, child exploitation, and cyberbullying [7]. It can also retrieve data from suspects' devices, shedding light on their activities.
- *Civil Litigation*: In civil cases, digital forensics is used to gather evidence related to intellectual property disputes, fraud, employment issues, and more. Email communications, financial records, and digital documents are often key sources of evidence.
- *Corporate Investigations*: Organizations turn to digital forensics when dealing with insider threats, data breaches, and employee misconduct. It helps them understand how a breach occurred, who was responsible, and what data was compromised.
- *Cyber Security Incidents*: When a data breach or cyber security incident occurs, digital forensics is instrumental in identifying the attack vector, the extent of the breach, and the compromised data. It assists in incident response and recovery efforts.

6.2.6 Digital Forensic Process and Methodology

The digital forensic process is a systematic and meticulous approach to handling digital evidence [5]. It involves several key steps:

- *Identification*: The process begins with the identification of potential sources of digital evidence. This includes identifying devices, storage media, or online platforms relevant to the investigation.

- *Preservation*: Once identified, digital evidence must be preserved to maintain its integrity and admissibility in court. This involves creating a forensic copy, often referred to as a "bit-by-bit" or "forensic image," of the original data. This copy is used for analysis while the original remains unchanged.
- *Collection*: With preserved copies in hand, investigators collect digital evidence. This may include data from hard drives, mobile devices, cloud services, network logs, and more. Chain of custody protocols are critical to maintaining the integrity of the evidence.
- *Examination and Analysis*: In this phase, forensic experts use specialized tools and techniques to examine and analyze the digital evidence. They may recover deleted files, examine file metadata, conduct keyword searches, and reconstruct timelines of activities.
- *Documentation*: All actions taken during the examination and analysis phase are meticulously documented. This documentation is crucial for creating a clear and defensible trail of the investigation process.
- *Reporting*: Investigators generate comprehensive reports that detail their findings, the methodology used, and the relevance of the evidence to the case. These reports are often presented in court.
- *Presentation*: In legal proceedings, investigators may be called upon to present their findings and provide expert testimony regarding the digital evidence they have uncovered.

Digital forensics is a dynamic and evolving field that demands not only technical expertise but also a deep understanding of legal and ethical considerations [2]. The methodologies employed in digital forensics continue to advance to keep pace with the rapidly changing landscape of technology. As we delve further into this chapter, we will explore one of the most potent tools in the digital forensics toolkit: Autopsy. This open-source platform empowers investigators with a wide range of features and functionalities designed to unlock the secrets hidden within digital evidence.

6.3 INTRODUCTION TO AUTOPSY AND FEATURES

In the realm of digital forensics, the quest for tools that empower investigators and analysts to navigate the labyrinth of digital evidence has led to the emergence of powerful solutions. Among these, Autopsy stands as a beacon of excellence—an open-source digital forensics platform that has revolutionized the way experts examine and dissect digital artifacts. In this section, we embark on a journey into the world of Autopsy, tracing its history, elucidating its open-source nature, and uncovering its primary purpose and versatile use cases [6].

- History and Background of Autopsy Digital Forensics Tool
  The story of Autopsy as a digital forensics tool began with its development by Brian Carrier in the early 2000s. Carrier, a renowned expert in digital forensics, recognized the need for a comprehensive, extensible, and user-friendly platform to assist investigators in their quest to extract insights

from digital evidence. Drawing inspiration from his work on The Sleuth Kit (TSK), a widely respected open-source toolkit for digital forensics, Carrier embarked on the journey to create Autopsy [6]. Autopsy was designed to serve as a graphical interface for TSK, thus making it more accessible and user-friendly for a broader audience. Carrier's vision was clear—to provide a robust tool that would streamline the digital forensic process, making it more efficient and effective for both experienced practitioners and newcomers to the field.

- Open-Source Nature and Availability
  One of the defining features of Autopsy is its open-source nature. This characteristic embodies the ethos of collaboration and knowledge sharing that underpins the field of digital forensics. By being open-source, Autopsy is freely available to digital forensics professionals, law enforcement agencies, cyber security experts, and anyone with a vested interest in uncovering the truth hidden within digital data. The open-source model ensures that Autopsy remains accessible and adaptable. Users can scrutinize the source code, contribute to its development, and customize it to meet their specific needs. This level of transparency and community involvement has not only fostered the growth of Autopsy but also contributed to its reputation as a trustworthy and dependable tool. The availability of Autopsy across multiple platforms, including Windows, macOS, and Linux, further enhances its accessibility. Investigators can deploy Autopsy on their preferred operating systems, ensuring that it seamlessly integrates into their workflow.
- Primary Purpose and Use Cases
  At its core, Autopsy is a digital forensics platform designed to streamline the examination of digital evidence [3]. Its primary purpose is to assist investigators in uncovering, analyzing, and documenting digital artifacts efficiently and comprehensively. Autopsy achieves this through an array of features and functionalities that cater to a broad spectrum of use cases.
- Interface of Autopsy
  In an autopsy features application, the interface for managing new cases and open cases plays a vital role in organizing and tracking the progress of forensic examinations [6]. Here's how these sections might be structured:
  - New Cases Management
    - *Create New Case*: This feature allows users to create a new case entry by providing essential information such as case ID, date of examination, case description, and relevant details about the deceased individual.
    - *Import Data*: Option to import data from external sources such as medical imaging files (CT scans, MRI scans), forensic databases, or digital evidence.
    - *Initial Examination Tools*: Tools for conducting the initial examination of the body or body part, including 3D visualization, measurements, annotations, and documentation of preliminary findings.

- Open Cases Management
  - *Case Dashboard*: A dashboard displaying all open cases with essential details such as case ID, status, date of examination, and assigned forensic experts.
  - *Case Details View*: Access to detailed information about each open case, including examination notes, findings, progress updates, and any pending tasks or follow-ups.
  - *Collaboration Tools*: Features for team collaboration, such as discussion threads, shared annotations, and file attachments within each open case.
  - *Task Management*: Ability to assign tasks, set deadlines, and track the progress of forensic examinations, ensuring efficient workflow management.
  - *Evidence Tracking*: Tools for managing and tracking digital evidence, lab results, medical records, and other relevant documentation associated with each open case.
- File System Analysis
  At the core of digital forensics lies the examination of file systems—as shown in Figure 6.1, the organized structures that store digital data on storage devices. Autopsy excels in file system analysis, offering a range of features that enable investigators to delve deep into digital landscapes [6].
  - *File Listing and Metadata*: Autopsy provides a detailed listing of files shown in Figure 6.2, complete with metadata such as file size, timestamps, and permissions. This information is crucial for establishing the timeline of events and understanding user activities.

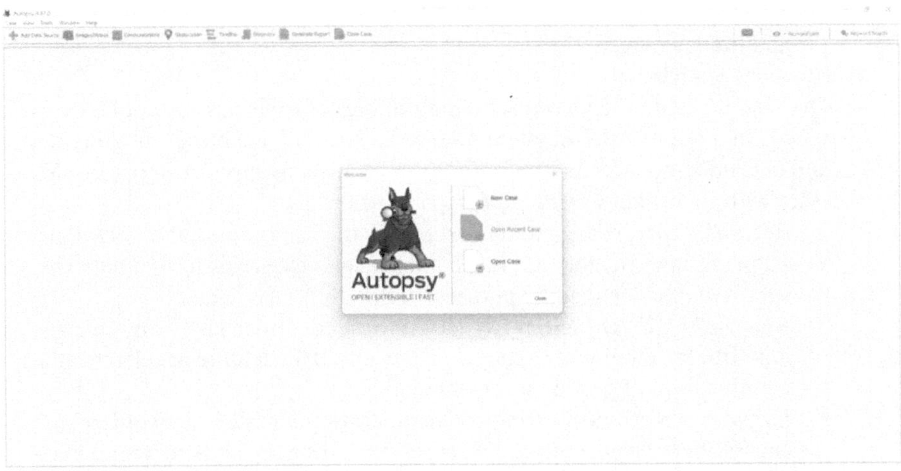

FIGURE 6.1 Interface of autopsy.

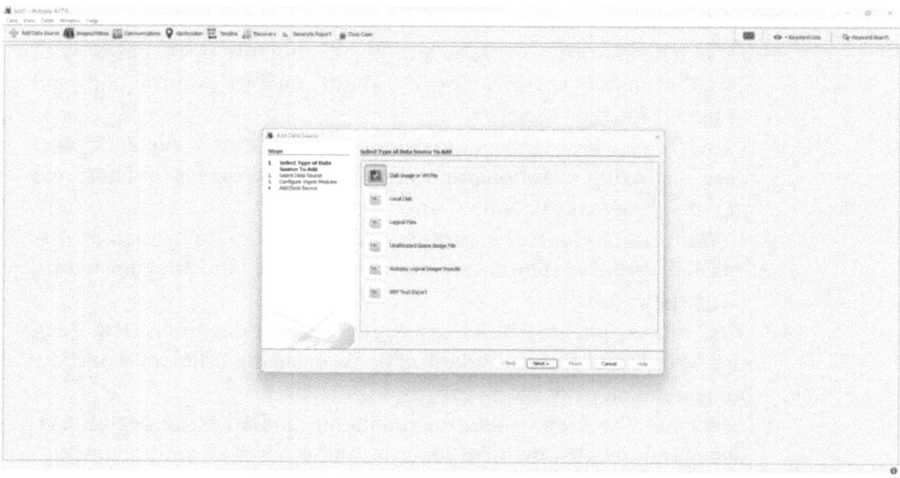

FIGURE 6.2 File system analysis.

    - *File Viewing*: Investigators can view the content of various file types directly within Autopsy, eliminating the need for external applications. This feature facilitates the quick review of documents, images, and more.
    - *File Timeline*: Autopsy constructs a chronological timeline of file activity, allowing investigators to track the creation, modification, and deletion of files. This timeline can provide critical insights into user actions.
    - *File Analysis*: Through the use of file analysis modules, Autopsy can identify potentially suspicious files based on signatures, hashes, and other attributes. This aids in the detection of malware, illegal content, and hidden data.
- Keyword Search
  The significance of keyword searching in digital forensics cannot be overstated. In a sea of data, keywords serve as anchors, enabling investigators to locate specific information relevant to an investigation. Autopsy equips users with robust keyword search capabilities.
    - *Regular Expressions*: Autopsy supports the use of regular expressions, allowing investigators to craft complex search queries to find patterns, sensitive data, or specific phrases within digital artifacts.
    - *Keyword Hit Highlighting*: When Autopsy identifies keyword matches, it highlights them within the search results. This feature accelerates the identification of pertinent information.
    - *Integrated Searching*: Autopsy seamlessly integrates keyword search capabilities across various data types, including file content, email messages, chat logs, and more. This comprehensive approach streamlines the search process.

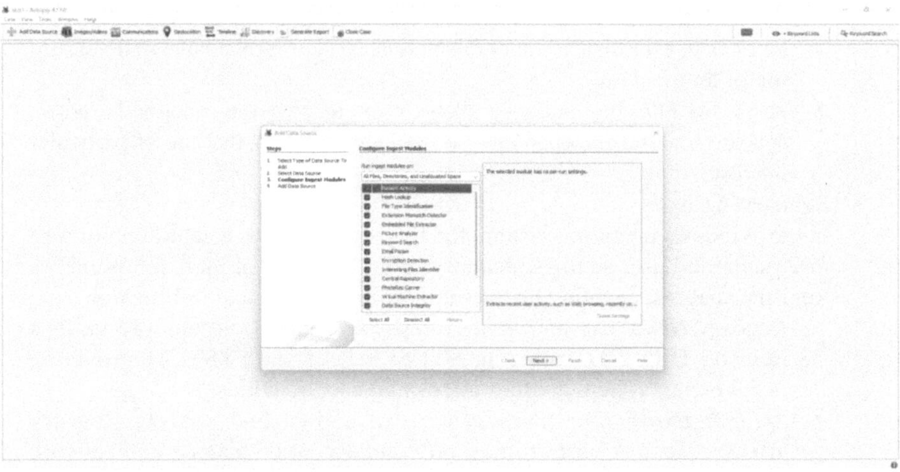

FIGURE 6.3 Configure ingest modules.

- Timeline Analysis
  In many investigations, understanding the chronological order of events is pivotal. Autopsy excels in timeline analysis, constructing a visual representation of activities and interactions based on digital artifacts shown in Figure 6.3.
  - *Artifact Timeline*: Autopsy compiles a timeline that includes a wide range of digital artifacts shown in Figure 6.4, such as file creations, document access, email communications, and web browsing history. This holistic view offers insights into user actions.

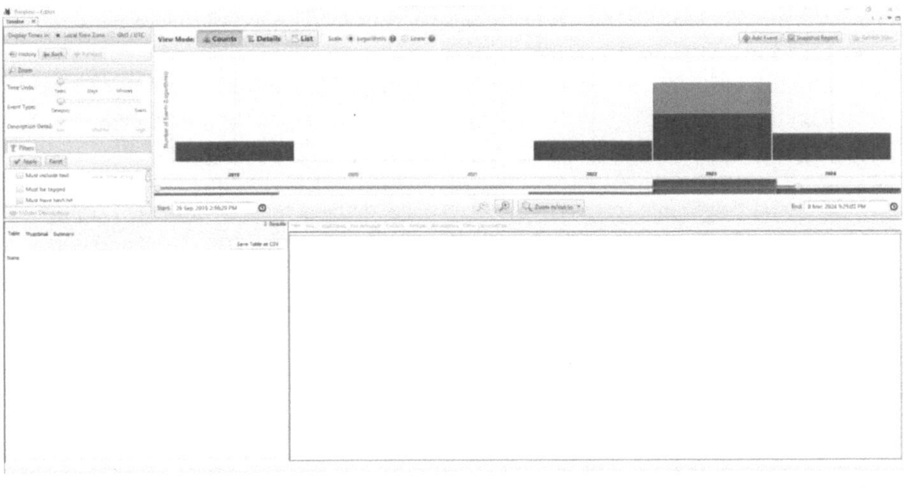

FIGURE 6.4 Timeline analysis.

- *Event Correlation*: Investigators can correlate events from different sources to establish connections and gain a comprehensive understanding of an incident.
- *Temporal Filtering*: Autopsy allows users to filter the timeline based on specific timeframes, enabling the isolation of events that are of particular interest.

- Registry Analysis
  In the Windows operating system, the Registry serves as a central repository for system and user settings. Autopsy offers specialized tools for Windows Registry analysis, enabling investigators to uncover crucial information.
  - *Registry Hives Examination*: Autopsy can parse and analyze various Registry Hive files, including NTUSER.DAT and SYSTEM, providing insights into system configurations and user activities.
  - *Artifact Extraction*: Investigators can extract and analyze Registry artifacts, such as user profiles, program execution history, and network configurations, to piece together the puzzle of a digital investigation.
- Web Artifact Analysis
  The internet is a treasure trove of digital evidence, from web history and cookies to downloads and online communications. Autopsy equips investigators to scrutinize this digital footprint effectively.
  - *Web History Examination*: Autopsy parses web browser history data to reveal a user's online activities, including visited websites, search queries, and timestamps.
  - *Cookie Analysis*: By analyzing cookies, Autopsy unveils information about a user's online interactions, preferences, and potential access to websites.
  - *Download Tracking*: Autopsy tracks downloaded files and associated metadata, aiding in the identification of downloaded content and its source.
  - *Email Analysis*: Email artifacts are a critical part of web artifact analysis. Autopsy allows investigators to examine email communications, attachments, and message details within the context of a case.
- Carving and Data Recovery
  Digital artifacts can be fragmented, damaged, or hidden. Autopsy's data carving and recovery features come to the rescue, enabling investigators to reassemble the puzzle of digital data.
  - *Data Carving*: Autopsy employs data carving techniques to extract files and fragments from unallocated space, deleted files, or damaged storage media. This capability is instrumental in recovering lost or deleted data.
  - *File Carving*: Investigators can use Autopsy to carve files based on file signatures or known header/footer patterns, even when file system metadata is missing.
- File and Image Carving:
  File and image carving in Autopsy are specialized techniques used to reconstruct files and images when their original structures are damaged or inaccessible.
  - *File Carving*: Autopsy's file carving capabilities are designed to recover files, even when the file system is severely damaged or absent. It identifies file signatures and reconstructs files based on patterns.

    - *Image Carving*: In cases where digital images are fragmented or incomplete, Autopsy can extract and reconstruct images, providing a comprehensive view of the visual evidence.
- Illustrating Autopsy's Features: Real-World Examples
  To illustrate Autopsy's capabilities in action, let's consider a few real-world examples:
    - *Criminal Investigation*: Autopsy's file system analysis assists investigators in reconstructing a suspect's file activity timeline, revealing when specific files were created, accessed, or deleted. Keyword searches help identify relevant communications or documents, and timeline analysis correlates these activities to establish a timeline of events.
    - *Corporate Data Breach*: In a corporate investigation, Autopsy can be used to analyze email communications and web artifacts to uncover the source and extent of a data breach. Carving and data recovery features help in piecing together fragmented evidence, such as deleted documents.
    - *Cyber Security Incident Response*: Autopsy's web artifact analysis can aid in understanding the tactics and techniques used by threat actors. Timeline analysis can help incident response teams identify the initial entry point and lateral movement within a compromised network.

In each of these scenarios, Autopsy's diverse feature set proves invaluable in the systematic examination of digital evidence, contributing to successful investigations and the pursuit of justice. In summary, Autopsy's comprehensive set of features, ranging from file system analysis to keyword search, timeline analysis, Registry examination, web artifact analysis, and data recovery, empowers digital forensics professionals to dissect and interpret digital artifacts with precision. Its role in investigations spans criminal, civil, corporate, and cyber security domains, making it an indispensable tool in the pursuit of truth within the digital age.

6.4 AUTOPSY PLUGINS

Autopsy, renowned for its robust and versatile digital forensics capabilities, becomes even more potent with the addition of plugins. These extensions, often created by the digital forensics community, provide specialized functionalities that cater to the diverse needs of investigators. In this section, we delve into the world of Autopsy plugins, exploring their concept, popular examples, and how they enhance Autopsy's functionality, making it an even more indispensable tool in the field of digital forensics.

Autopsy's plugin system follows a simple yet powerful concept—extendability. It recognizes that no single tool can encompass the vast spectrum of scenarios encountered in digital forensics. Plugins act as modular extensions that augment Autopsy's core features, allowing users to tailor the tool to their specific investigative requirements. In essence, Autopsy plugins are additional sets of tools, scripts, or modules that seamlessly integrate with the Autopsy framework. They can be developed by the Autopsy community, forensic experts, or organizations to enhance Autopsy's capabilities for a range of forensic scenarios. This extensible architecture ensures that Autopsy remains adaptable to emerging challenges and diverse investigative needs.

6.4.1 Popular Autopsy Plugins and Their Functionality

- *Artifact Exchange*: This plugin streamlines the sharing of forensic artifacts among investigators and organizations. It allows for the import and export of artifacts, making it easier to collaborate and share findings in multi-team investigations.
- *Hash Lookup*: The Hash Lookup plugin facilitates the identification of known files by comparing their hashes against a database of known file hashes. This aids in the rapid identification of common files, reducing the time needed for analysis.
- *GeoIP Lookup*: GeoIP Lookup adds geolocation information to IP addresses found in digital artifacts. This can be invaluable for tracing the physical location of a device or user, shedding light on the geographical context of an investigation.
- *Email Viewer*: In email-centric investigations, the Email Viewer plugin offers enhanced capabilities for parsing and viewing email artifacts. It supports a variety of email formats and provides a more detailed view of email metadata and content.
- *File Type Identification*: This plugin specializes in identifying the file types of unknown or ambiguous files. It uses file signatures and magic numbers to determine file types, helping investigators make sense of unclassified data.
- *Volatility Integration*: The Volatility plugin integrates the Volatility Framework—a renowned memory forensics tool—with Autopsy. This enables the analysis of volatile memory dumps, providing insights into running processes, open network connections, and more.
- *Registry Viewer*: For cases involving Windows systems, the Registry Viewer plugin offers advanced capabilities for parsing and examining Windows Registry hives. It provides a comprehensive view of registry keys and values, aiding in the analysis of system configurations and user activities.
- *Yara Rule Manager*: Yara is a powerful tool for creating and applying custom rules to identify patterns or specific content within files. The Yara Rule Manager plugin integrates Yara rules into Autopsy, enabling users to create and apply custom rulesets for their investigations.

6.4.2 Enhancing Autopsy's Functionality Through Plugins

The integration of plugins significantly enhances Autopsy's functionality in several ways:

- *Specialized Capabilities*: Plugins extend Autopsy to handle specialized tasks or data types that may not be covered by the core functionality. This means investigators can access the right tools for the specific challenges they face.
- *Time Efficiency*: Plugins often automate and streamline tasks that would otherwise be time-consuming. For example, hash lookup plugins can quickly identify known files, saving investigators valuable time.

- *Customization*: Autopsy's extensible nature allows investigators to build a customized workflow for their investigations. They can choose and activate plugins that align with the case requirements, creating a tailored approach to analysis.
- *Community Collaboration*: The Autopsy plugin ecosystem thrives on community collaboration. Digital forensics professionals and organizations contribute plugins to address emerging needs and share them with the community. This collaborative spirit ensures that Autopsy remains adaptable and up-to-date.
- *Cross-Platform Compatibility*: Many Autopsy plugins are cross-platform, meaning they can be used on different operating systems, enhancing Autopsy's versatility for investigators who work in diverse environments.

6.4.3 Examples Of Enhanced Functionality Through Plugins

Consider a scenario where an investigator is dealing with a complex case involving a compromised server. In this situation, Autopsy's core functionality may not provide all the necessary tools to extract and analyze specific server logs. However, by integrating a custom server log parsing plugin, the investigator can:

- Efficiently parse server logs, identifying potential indicators of compromise (IOCs).
- Correlate log entries with file system data and network traffic.
- Generate detailed reports specific to server activity, aiding in incident response and attribution.
- In another example, when dealing with a case involving a suspected data exfiltration attempt via email, a custom email analysis plugin can:
- Extract email headers, attachments, and embedded objects.
- Analyze email metadata to track communication patterns and potential collaborators.
- Utilize custom Yara rules to identify malicious email content.
- Produce comprehensive reports summarizing email communication patterns and any identified threats.

In conclusion, Autopsy's plugin system is a testament to the flexibility and extensibility of this renowned digital forensics tool. Plugin enriches Autopsy's functionality, offering investigators a diverse toolkit to tackle complex digital investigations. Through community collaboration and customization, Autopsy remains at the forefront of digital forensics, providing professionals with the means to uncover digital evidence effectively and comprehensively.

6.5 CASE STUDIES, CHALLENGES, AND FUTURE TRENDS

6.5.1 Case Study 1: Uncovering a Cyber Extortion Scheme

In a high-stakes case of cyber extortion, Autopsy played a pivotal role in unraveling a complex digital crime that targeted a multinational corporation [7]. The threat actors had infiltrated the organization's network and encrypted critical data, demanding a substantial ransom for its release. The corporation turned to digital forensics experts to investigate and resolve the breach.

- Autopsy Features and Tools in Action
    - *File System Analysis*: Autopsy's file system analysis capabilities allowed investigators to examine the extent of the data encryption and identify the affected files and directories. This provided a clear understanding of the scope of the breach.
    - *Timeline Analysis*: Autopsy's timeline analysis feature was instrumental in reconstructing the sequence of events leading up to the breach. It revealed the initial point of entry, lateral movement within the network, and the execution of ransomware.
    - *Data Recovery*: Using Autopsy's data recovery tools, investigators were able to salvage encrypted files and partially recover some of the critical data. This helped in mitigating the impact of the breach and reducing the ransom demand.
    - *Keyword Search*: Autopsy facilitated a targeted keyword search for communication logs related to the extortion attempt. This led to the discovery of email exchanges between the threat actors and the victim, providing valuable leads for further investigation.

Through the combined use of Autopsy's features, investigators successfully identified the threat actors, their tactics, and the means used to infiltrate the network [3]. This information was invaluable in assisting law enforcement agencies in pursuing legal action against cybercriminals.

6.5.2 Case Study 2: Unmasking an Insider Threat

In a corporate environment, Autopsy played a critical role in unmasking an insider threat that was compromising sensitive company data [2]. An employee, with legitimate access to the organization's systems, had been exfiltrating confidential documents and intellectual property over an extended period.

- Autopsy Features and Tools in Action
- *File System Analysis*: Autopsy's file system analysis helped investigators track the unauthorized access and copying of sensitive files. It revealed patterns of file movement and modifications.
- *Keyword Search*: Investigators used Autopsy's keyword search capabilities to identify documents related to the organization's proprietary technology and confidential projects. This led to the discovery of files that had been surreptitiously copied.
- *Timeline Analysis*: Autopsy's timeline analysis allowed investigators to create a comprehensive timeline of the employee's activities. This timeline revealed a consistent pattern of unauthorized data access coinciding with the exfiltration of sensitive documents.
- *Data Carving*: Autopsy's data carving feature helped recover deleted files and remnants of documents that the insider threat had attempted to conceal.

By leveraging Autopsy's capabilities, investigators were able to build a compelling case against the insider threat. The evidence gathered through Autopsy's tools and analysis was presented to law enforcement and legal authorities, leading to the arrest and prosecution of the employee responsible for the data breaches.

These case studies exemplify Autopsy's crucial role in digital investigations. Its diverse features and tools empower investigators to dissect digital evidence, reconstruct timelines, recover deleted or hidden data, and ultimately contribute to the successful resolution of digital crimes, from cyber extortion schemes to insider threats within organizations. Autopsy continues to be an indispensable asset in the pursuit of truth and justice in the digital age.

6.5.3 Challenges in Digital Forensics

Digital forensics, while a powerful tool for uncovering digital evidence, faces several challenges [2]:

- *Encryption*: The widespread use of strong encryption techniques by individuals and organizations can make it challenging to access and decrypt data. Investigators must stay updated on encryption methods and employ advanced techniques to overcome this hurdle.
- *Evolving Technology*: Rapid technological advancements result in an ever-expanding array of digital devices and platforms. Keeping pace with these changes and adapting forensic tools and methods accordingly is a constant challenge.
- *Data Volume*: The exponential growth of digital data poses a significant challenge in terms of data storage, processing, and analysis. Investigators must efficiently sift through vast amounts of data to identify relevant evidence.
- *Cloud Computing*: The migration of data and applications to cloud platforms complicates digital investigations. Investigators need to navigate legal and technical challenges associated with accessing cloud-stored data.
- *Data Privacy and Legal Issues*: Striking a balance between privacy rights and the needs of investigations is a complex challenge. Laws and regulations governing data privacy vary across jurisdictions, adding further complexity to cross-border investigations.

6.5.4 Emerging Trends and Technologies

The field of digital forensics is continuously evolving to address these challenges and harness emerging trends [1]:

- *Machine Learning and artificial intelligence (AI)*: Machine learning and AI are increasingly employed to automate and enhance aspects of digital forensics, from data analysis to threat detection and pattern recognition.
- *Blockchain Analysis*: As cryptocurrencies gain prominence, blockchain analysis becomes critical in tracking illicit transactions and activities. Forensic tools and techniques for blockchain analysis are on the rise.

- *Internet of Things (IoT) Forensics*: The proliferation of IoT devices presents new challenges and opportunities. IoT forensics is emerging as a specialized field to investigate connected devices and their data.
- *Cloud Forensics*: With the growing use of cloud services, cloud forensics is becoming essential. It involves techniques for preserving and analyzing data stored in remote servers.
- *Quantum Computing Challenges*: While quantum computing promises remarkable advancements, it also poses challenges to encryption methods. Digital forensics must adapt to this changing landscape.
- *Privacy-Preserving Forensics*: As privacy concerns rise, techniques for conducting investigations while preserving individuals' privacy rights are being developed, such as homomorphic encryption and secure multi-party computation.
- *Cyber Threat Intelligence*: The proactive gathering and analysis of threat intelligence data help organizations prepare for cyber threats. Digital forensics is increasingly integrated with threat intelligence to identify and mitigate risks.

In summary, digital forensics faces challenges related to encryption, evolving technology, data volume, cloud computing, and legal issues. To overcome these challenges, the field is embracing emerging trends and technologies such as machine learning, blockchain analysis, IoT forensics, and quantum computing adaptation. The future of digital forensics is marked by innovation and adaptation to an ever-changing digital landscape.

6.6 CONCLUSION

In this comprehensive chapter, we embarked on a journey through the dynamic world of digital forensics, exploring the challenges, trends, and the invaluable role of Autopsy—a digital forensics powerhouse. Let's recap the key points discussed and reiterate the significance of digital forensics and Autopsy as indispensable tools in the modern age. We began by emphasizing the paramount importance of digital forensics in today's world, where virtually every aspect of our lives is intertwined with digital technology. Digital evidence has become the cornerstone of investigations, ranging from cybercrimes and corporate breaches to criminal cases and incident responses. Understanding the role of digital forensics in investigations, we highlighted its critical importance in unraveling complex cases and maintaining the integrity of evidence. The systematic digital forensic process, including evidence acquisition, analysis, and reporting, was outlined, emphasizing its pivotal role in the pursuit of justice. Introducing Autopsy, we unveiled a tool that stands as a beacon of excellence in the field of digital forensics. Developed by Brian Carrier and nurtured by a thriving open-source community, Autopsy has emerged as a versatile, user-friendly, and extensible platform that empowers investigators to dissect digital artifacts with precision and insight. As we delved into Autopsy's features, we witnessed how it excels in file system analysis, keyword search, timeline analysis, registry examination, web artifact analysis, data recovery, and file and image carving.

Through real-world case studies, we saw Autopsy in action, playing a pivotal role in solving digital crimes, from cyber extortion schemes to insider threats within organizations.

Moreover, we explored the realm of Autopsy plugins, which extend its capabilities to address specialized investigative needs. These plugins, developed by a collaborative digital forensics community, enhance Autopsy's functionality, from hash lookup and GeoIP analysis to email parsing and blockchain investigation. In contemplating the future of digital forensics, we acknowledged the challenges posed by encryption, evolving technology, data volume, and cloud computing. Yet, we also embraced emerging trends, from AI and blockchain forensics to IoT investigation and quantum computing preparedness. In conclusion, the importance of digital forensics cannot be overstated, and Autopsy shines as an indispensable tool in this field. Its capabilities, coupled with the collaborative spirit of the digital forensics community, empower investigators to uncover the truth concealed within the digital landscape. We encourage further exploration and learning in this ever-evolving field of digital forensics. As technology continues to advance, the need for skilled digital investigators and innovative tools like Autopsy will only grow. Together, we navigate the digital frontier, ensuring that justice prevails in our increasingly digitalized world.

REFERENCES

1. Shraddha Vedre, Waman Parulekar, "Digital Forensic and Role of Computers in Digital Forensic", IJCRT, 10 (2022): a772–a778.
2. Ghonge M M, "Cyber Security and Digital Forensics Challenges and Future Trends", 1st Edition, John Wiley, 2022
3. Shaweta Sachdeva, Bansi Lal Raina, Avinash Sharma, "Analysis of Digital Forensic Tools", Journal of Computational and Theoretical Nanoscience, 17.6 (2020): 2459–2467.
4. Gehlot A, "Digital Forensics and Internet of Things Impact and Challenges", 1st Edition, John Wiley, 2022
5. Greg Gogolin, "Digital Forensics Explained", 2nd New edition, Taylor & Francis Ltd, 2021
6. Ashok Kumar Mohan, Paul Selwin, "Digital forensic investigation using sleuth kit autopsy", National Conference on Information, Communication and Cyber Security, 2016
7. Gerard Johansen, "Digital Forensics and Incident Response", 2nd edition, Packt Publishing Limited, 2020

7 A Study of Cyber Threats and Security Frameworks in Cloud Computing

Kashif Asad and Mohd. Faizan Farooqui

7.1 INTRODUCTION

Cloud computing has significantly transformed the manner in which enterprises and individuals manage, store, and utilize data and applications. Its significance in modern computing cannot be overstated, as it offers unparalleled accessibility, scalability, flexibility, and cost-efficiency in comparison to traditional on-premises systems [1]. Cloud computing technology implies new security concerns, as well as new difficulties [1]. The importance of security in these cloud systems is growing as cloud computing continues to influence the digital world. The exponential growth in the use of cloud services in recent years has made data and apps housed in the cloud attractive targets for cybercriminals. Cloud computing security requires more than just protecting against standard cybersecurity risks. It also entails managing complicated shared responsibility structures, difficult data isolation situations, and ecosystem-specific compliance issues. It is impossible to exaggerate the significance of security in cloud systems. Cloud infrastructure vulnerabilities have given rise to new cyber threats that may cause data loss, financial consequences, and reputational harm to a business. Ensuring strong security measures is a joint obligation between cloud users and providers since organizations and people entrust their sensitive information to these entities. Consequently, for successful risk mitigation, it is essential to comprehend the varied nature of cloud security. Solving security issues is crucial to getting businesses and organizations to use cloud computing [2, 3]. The use of cloud services is rapidly growing in popularity. Cyber threats are expanding at the same rapid rate as the number of users of these sites [4]. The confidentiality, availability, and integrity of sensitive information are all jeopardized as a result. There is an immediate need to learn about and effectively counter the many cyber risks prevalent in cloud computing environments, which is why this research is being conducted [5]. Businesses must rapidly identify, investigate, and eliminate these risks as they increasingly migrate critical operations to the cloud. This research is the first step toward a comprehensive analysis of the many cyber risks associated with cloud computing. When it comes to safeguarding digital assets, it's critical to be strategic and knowledgeable. Using industry best practices and real-world case studies, the article provides a comprehensive picture of the evolving cloud computing safety landscape [4]. A better knowledge of the difficulties we face and better approaches to mitigate cyber dangers may be achieved by integrating scientific discoveries with helpful ideas [5]. The authors believe that this will help practitioners, decision-makers, and

DOI: 10.1201/9781003409571-7

the larger research community [3]. Cloud computing and cybersecurity researchers will find this study useful in their pursuit of understanding the interplay between the two. In order to keep data secure in modern IT systems, it provides crucial, up-to-date information [1].

The goal of this study is to examine the complex world of cloud computing security frameworks and cyber threats. Through the analysis of diverse cyber threats aimed at cloud settings and the examination of security frameworks and countermeasures used to mitigate these risks, this research aims to provide significant perspectives for scholars and practitioners alike. The following are the main goals of this study:

- To recognize and classify online threats that might endanger cloud computing settings.
- To examine and assess the well-known security standards and frameworks that are applied to cloud security.
- To investigate the variety of security methods and technologies used to safeguard cloud apps and data.
- To handle the particular security issues that come with cloud computing, such as compliance and shared accountability.
- To provide advice and top techniques for improving cloud security.
- To foresee new developments in the ever-evolving world of cloud security, including trends and threats.

7.2 CLOUD COMPUTING CYBER THREATS

Although cloud computing has many benefits, it is not impervious to a variety of cyber threats that might jeopardize the security and integrity of data and applications housed in cloud settings. This section examines the many kinds of cyber threats that are directed at cloud computing, including information on the dangers and weaknesses of cloud-based systems. Furthermore, case studies and real-world examples highlight how serious these concerns are.

7.2.1 Classification of Cyber Threat

Many kinds of cyber threats can seriously damage data integrity, privacy, and system uptime in cloud computing. Some of these threats are data leaks, which happen when someone gets unauthorized access to private data kept on cloud computers and steals or shares that data. Malware, such as viruses and ransomware, can get into cloud settings damage data integrity, and stop operations. Distributed denial of service (DDoS) attacks flood cloud services with too much data, which shuts them down and stops their services from working. Insider threats come from inside a company when workers or other trusted people abuse their access to cloud resources or change them in a bad way. Phishing and social engineering attacks try to trick users into giving up their log in information or other private data, allowing unauthorized users to access their cloud accounts. Insecure application programming interfaces (APIs) and incorrectly set up cloud infrastructure make these risks even worse. This shows how important strong security measures and effective risk management strategies are

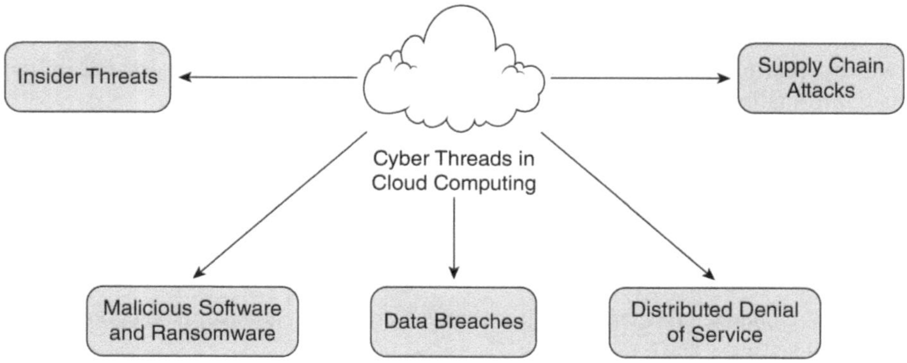

FIGURE 7.1 Cloud computing cyber thread types.

in cloud computing settings. Figure 7.1 illustrates the various cyber threats associated with cloud computing. It categorizes specific types of threats, including insider threats, supply chain attacks, malicious software and ransomware, data breaches, and DDoS.

- *Malicious Software and Ransomware*: Via compromised files or links, malicious software, such as viruses, worms, and ransomware, may enter cloud systems. Once inside, they might demand ransoms to be paid before they can be allowed to leave, disrupt operations, or steal data [4].
- *Data Breaches*: Gaining unauthorized access to confidential information is a common risk. Misconfigurations, lax authentication, or insider threats may all result in data breaches that allow for theft or leakage of data [3].
- *DDoS*: It aims to interrupt cloud services by flooding them with excessive traffic. Attackers are able to execute huge DDoS assaults that impact many clients by making use of cloud resources [5].
- *Insider Threats*: Cloud security may be jeopardized by staff members or authorized users with bad intentions. They could willfully release information or misuse authority to get access to private data [6].
- *Supply Chain Attacks*: Cloud service companies often use software and parts from outside sources. By breaching these supply chains, attackers may introduce malware or get illegal access to cloud computing infrastructure [7].

7.2.2 Literature Review

Millions of people were affected by the Capital One Data Breach in 2019, in which a former employee of a cloud service provider (CSP) used a badly set up web application firewall to access and steal private data [6]. It has been reported that businesses have accidentally shared data because they set up their Amazon S3 files wrong [8]. It was these hacks that showed how weak Internet of Things (IoT) devices in the cloud were and caused service interruptions. Jensen et al. [1] found that enhancing cloud computing security necessitates enhancing the security capabilities of both web browsers and web service frameworks. The most effective way to do this is by integrating the latter into the former. In their study, Jensen et al. [4] focus on a particular kind of denial of service assault that targets network-based services. This attack involves flooding victims with a large number of fake requests in order to overload them. The text provides comprehensive information on both well-recognized and more recent attacks, along with discussing parallels and approaches to security measures. According to Fox et al. [7], it is recommended to virtualize resources to hide the way they are combined and distributed. Cloud has had outages that have shown an effect on organizations that significantly depend on cloud services [9]. These actual cases highlight how serious cyber threats are to cloud computing and how important it is to have strong security protocols and watchful procedures in place to safeguard cloud-based assets [3]. This research chapter's subsequent parts will examine the security frameworks, technologies, and best practices required to successfully counter these attacks. In order to fully meet these goals, this research chapter will consult a broad variety of academic sources, case studies, and professional opinions [10]. This will eventually lead to a deeper knowledge of the always-changing environment of cyber risks and security in cloud computing. According to the outcomes of a report by the CIO Research Center in Figure 7.2, security remains the primary worry of CIOs with regard to the use of cloud computing. As a result, 66 percent of CIOs have concerns about the level of security offered by cloud computing [11].

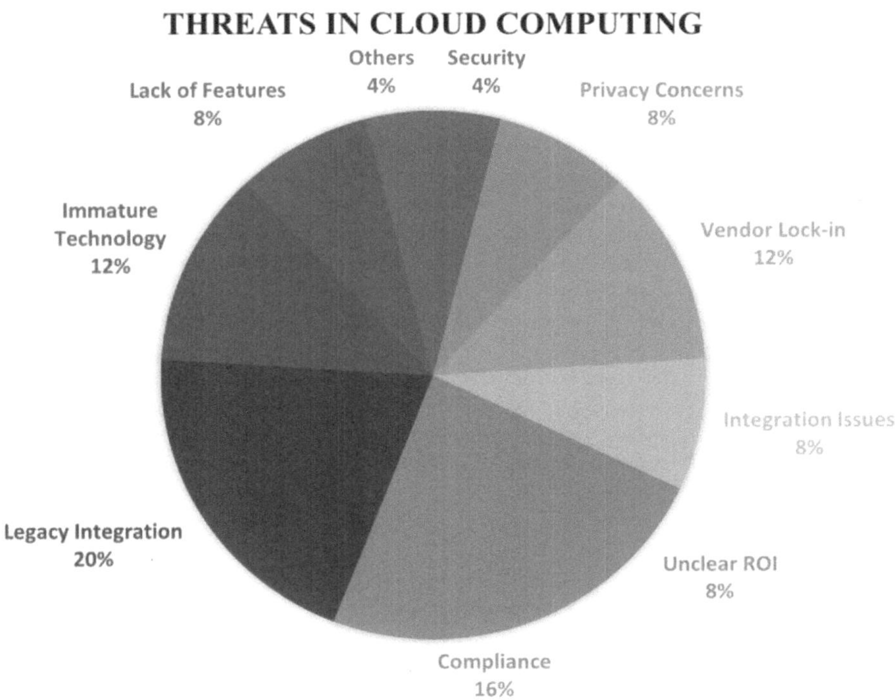

FIGURE 7.2 CIO Research Center report security threats [11].

7.3 SECURITY FRAMEWORKS IN CLOUD COMPUTING

Cloud computing security frameworks are collections of standards, best practices, and controls that businesses may adhere to improve the safety of the cloud-based infrastructure, apps, and data they manage. These frameworks provide companies with assistance in addressing the one-of-a-kind security difficulties and dangers posed by cloud computing.

7.3.1 In-Depth Examination of Major Security Frameworks

Table 7.1 provides an overview and analysis of three key security frameworks: ISO 27001, NIST Cybersecurity Framework, and CSA Security Guidance, with a focus on their relevance to cloud security. ISO 27001, noted for its systematic approach to information security management, can be modified for cloud security by broadening its controls and principles to include cloud-based assets and services. The NIST Cybersecurity Framework, known for its versatility and adaptability, allows enterprises to adjust its features to fit their individual cloud security requirements, making it appropriate for the dynamic nature of cloud computing. CSA Security Guidance, on the other hand, offers a complete set of best practices for protecting cloud systems, covering a wide variety of subjects such as operations, risk management, governance, and compliance.

TABLE 7.1
Security Frameworks

Security Framework	Overview	Applicability to Cloud Security
ISO 27001 (Information Security Management System) [12]	Using the internationally acclaimed ISO 27001 standard, an information security management system (ISMS) may be created, deployed, maintained, and enhanced over time [13]. A systematic approach to security and risk management is emphasized.	By extending its controls and principles to cloud-based assets and services, ISO 27001 may be modified for cloud security [14].
NIST Cybersecurity Framework [14]	This framework provides a versatile and adaptive way to handle cybersecurity threats; it was developed by the US National Institute of Standards and Technology (NIST) [15].	The NIST framework is adaptable, so businesses may modify its features to meet their own cloud security requirements, making it a good fit for the ever-changing world of cloud computing [16].
CSA Security Guidance (Cloud Security Alliance) [14]	For protecting cloud environments, a thorough set of best practices is provided by the CSA Security Guidance [16]. It addresses a number of topics, including operations, risk management, governance, and compliance.	The unique security issues in the cloud are best tackled with the help of CSA's cloud-centric architecture [17].

Each of these security frameworks offers a methodical plan for dealing with security threats, but they have different foci and aren't all suitable for use in the cloud. Any business may choose the cloud security architecture that works best for its specific requirements.

7.3.2 Comparative Analysis of Security Frameworks

Standards like PCI-DSS, ITIL, and ISO-27001/27002 [6, 18] are used to handle and evaluate security risks and issues. Table 7.2 evaluates the merits and limitations of three main security frameworks: ISO 27001, the NIST Cybersecurity Framework, and the CSA Security Guidance, with an emphasis on their relevance and efficacy in solving cloud security issues. ISO 27001 is widely recognized and approved, emphasizing a risk-based strategy that can be applied to a variety of industries and circumstances; yet, it may be resource-intensive to implement and maintain. The NIST Cybersecurity Framework provides a comprehensive structure with adjustable features, encouraging customization based on organizational needs, albeit customization can be difficult for some. CSA Security Guidance stands out for its focus on cloud security, addressing unique difficulties and encouraging transparency from cloud providers; yet, it may require adaption to individual cloud environments and may not cover all developing cloud security challenges.

7.4 UNIQUE SECURITY CHALLENGES IN CLOUD COMPUTING

Cloud computing offers several benefits, but, it also introduces certain security concerns that enterprises must address. These difficulties must be overcome.

TABLE 7.2
Comparative Analysis of Security Frameworks

Aspect	ISO 27001	NIST Cybersecurity Framework [11]	CSA Security Guidance
Strengths	Globally recognized and accepted standard [12].	Comprehensive framework with five core functions [15].	Specialized in cloud security, addressing unique challenges.
	Adaptable to various industries and contexts.	Flexible and adaptable to different organizations [9].	Community-driven and continually updated guidance [19].
	Emphasizes risk-based approach [13].	Promotes customization based on risk profiles [12].	Promotes transparency from cloud providers.
Weaknesses	Resource-intensive to implement and maintain.	Customization can be challenging for some organizations [17].	Requires adaptation to specific cloud providers and services.
	General may need customization for cloud-specific concerns [6].	Might lack specificity for certain industries.	May not cover all emerging cloud security challenges [8].

7.4.1 Discussion of Security Challenges Unique to Cloud Environments

The management and use of IT resources by enterprises have been completely transformed by cloud computing. But this change has brought with it a new set of special security issues that need to be carefully considered. Table 7.3 includes many security concerns specific to cloud environments, along with descriptions of each challenge and possible mitigating measures. These issues include the shared responsibility paradigm, which requires a clear separation of obligations between cloud providers and clients in order to avoid misconfigurations and vulnerabilities. Data isolation is identified as a significant concern in multi-tenant setups, emphasizing the importance of strong security measures such as encryption and access controls. The hazards of multi-tenancy are also explored, emphasizing the significance of adequate isolation and regular monitoring to prevent any breaches. Compliance in the changing cloud environment necessitates a full awareness of legal obligations, as well as regular audits.

The majority of people believe that cloud providers are responsible for taking care of cloud security. On the other hand, throughout the last several years, an ever-increasing number of businesses have shifted their operations, data, and apps to the cloud [8]. One definition of a cloud safety risk is "something terrible that can happen to digital assets that are stored in the cloud" [12]. Certain assets may be included in this category, such as data, software, infrastructure, the confidence of customers, and the reputation of the firm [13]. A broad variety of industrial applications may benefit from scalability as well as frequent software and hardware updates thanks to cloud computing [10, 11, 15]. A number of particular obstacles stand in the way of the widespread use of cloud computing, including but not limited to complexity, compliance, security, dependency, privacy, control, and cost [14].

7.4.2 Strategies for Addressing These Challenges and Mitigating Risks

In order to effectively mitigate the wide array of cyber threats that affect cloud computing, a comprehensive strategy is necessary, integrating technological remedies, resilient regulations, and ongoing surveillance [11]. To strengthen access controls and prevent unauthorized account access, organizations must initially implement robust authentication mechanisms, such as multi-factor authentication [20]. Consistent vulnerability assessments and security audits aid in the detection and remediation of potential deficiencies within cloud infrastructure. Encrypting data while it is in transit and at rest is an effective measure to bolster data security. Organizations must give precedence to employee cybersecurity awareness training to mitigate the potential hazards that may arise from social engineering and phishing attacks. Fostering partnerships with reputable CSPs that rigorously enforce security protocols can contribute to the overall fortification of cloud environments' security [6]. The implementation of a comprehensive incident response plan and disaster recovery strategies is of the utmost importance to mitigate the consequences of cyber threats and enable an efficient and timely reaction to security incidents. Early threat detection and response are facilitated by the implementation of intrusion detection systems and continuous monitoring of cloud resources [15]. Moreover, organizations can maintain a competitive edge in the ever-changing realm of cyber risks in cloud computing by remaining updated on emergent threats and embracing a proactive and

TABLE 7.3
Security Challenges in Cloud Computing

Security Challenges Unique to Cloud Environments	Challenge	Discussion
1. Shared responsibility model	Security of the underlying infrastructure is the responsibility of cloud providers, who operate on a shared responsibility paradigm; security of data and applications is the responsibility of customers. Miscommunication and security holes could result from improper management of this paradigm [6].	Organizations must possess a comprehensive understanding of and precisely delineate their obligations within the framework of shared responsibility [12]. This includes identifying which security measures are the responsibility of the cloud provider and which fall on the customer. Misconfigurations or oversight in either party's responsibilities can lead to vulnerabilities.
2. Data isolation	In multi-tenant cloud environments, multiple customers share physical resources. Ensuring data isolation and preventing data leakage between tenants is a critical challenge [17].	Cloud providers implement strong isolation mechanisms, but customers should also take measures to secure their data [19]. This includes using encryption, robust access controls, and monitoring to prevent unauthorized access or data leakage.
3. Multi-tenancy	The multi-tenancy model can create risks where malicious activities or vulnerabilities in one tenant's environment may affect others, leading to potential security breaches [12].	Effective isolation of tenants, strong access controls, and continuous monitoring are key strategies to mitigate multi-tenancy risks. Intrusion detection systems and rapid incident response plans can help contain and address breaches promptly.
4. Compliance issues	Meeting regulatory compliance requirements in the cloud can be complex due to the dynamic nature of cloud environments and the need to adhere to specific industry standards [7].	Organizations must identify and understand the compliance requirements relevant to their industry and geography. Leveraging cloud-native compliance tools and conducting regular compliance audits are essential to maintaining adherence [14].
5. Data transfer and egress costs	Data transfer between on-premises systems and the cloud, as well as egress costs (data leaving the cloud), can impact both security and operational expenses [18].	Optimize data transfer by using efficient compression and deduplication techniques. Implement data transfer controls to prevent unauthorized data exfiltration [10]. Monitor and control egress costs through usage tracking and budgeting.
6. Vendor lock-in	Over-dependence on a single CSP can create vendor lock-in, making it challenging to migrate to a different provider or revert to on-premises infrastructure if needed [3].	Develop a multi-cloud or hybrid cloud strategy to reduce vendor lock-in risks. Use containerization and orchestration tools to ensure the portability of applications across cloud providers. Establish exit strategies and backup plans to mitigate potential vendor lock-in [12].

adaptable cybersecurity stance. By taking a proactive and planned approach, cloud computing security challenges of this kind may be addressed:

1. Clearly Identify Duties and Roles
   - Create and uphold a clear shared responsibility agreement that specifies the obligations of the client and the cloud provider. Make sure the security controls are allocated correctly.
2. Put Strict Data Security Measures in Place
   - To ensure secrecy, encrypt critical information both in transit and at rest.
   - To limit access to just those who are permitted, implement strong access controls and authentication procedures [12].
   - To find and fix vulnerabilities, audit and evaluate data security procedures on a regular basis.
3. Improve Incident Response and Monitoring
   - To identify and address security issues quickly, put intrusion detection and continuous monitoring technologies into place.
   - Create and evaluate incident response strategies to quickly control and mitigate security breaches [10].
4. Solutions for Cloud-Native Compliance
   - To make compliance management easier, make use of cloud provider-native compliance tools and services [18].
   - Conduct compliance evaluations on a regular basis to make sure that industry-specific rules and guidelines are being followed.
5. Integrity-Agnostic Products
   - To reduce vendor lock-in, create solutions and architectures that aren't dependent on any one cloud provider [16].
   - For more portability, make use of orchestration and containerization technologies.

Organizations may optimize the security of their cloud systems while using the advantages of cloud computing by addressing these challenges and putting effective policies into place.

7.5 PROPOSED FRAMEWORK

Cloud computing has been plagued by security issues since its inception. As a consequence of new technologies and cloud architectures, researchers still need to be aware of certain new security threats connected to cyber-attacks, multi-tenancy, and other cyber-attacks. Data assets in a cloud computing setting could be stored in several locations or different forms. Information assets must be classified according to their level of importance to effectively handle security problems. Consequently, security might be maintained with minimal effort and expense. Since different companies may place different values on different pieces of information, it could be challenging to classify data sent across many users and organizations. To ensure that safeguards against data loss or manipulation are in place, CSPs must overcome difficulties. Forensic analysis uncovered more information, and network monitoring and recording became more easier, on physically connected devices, data from the IoT, and networks. Two sources: [14] and [9]. Secure data and application maintenance, protected interfaces, data acquired only by authorized persons, and data accessibility when required make a data breach or hacking incident controlled. CSPs need to have safeguards in place to deal with these issues. Cloud providers provide mirroring as a subscription service; however, the added bandwidth consumption drives up the price. Providers of cloud services sometimes find it challenging to meet the demands of all cloud customers without increasing their prices. A plethora of security holes may be associated with the various application frameworks and platforms, references [17] and [19]. The framework for safe cloud computing is shown in Figure 7.3. It is based on a security model that specifies the characteristics of each component and applies the required security technologies for implementation between components in cloud computing.

- *Cloud Service Provider*: A CSP provides cloud computing services and resources to organizations and consumers online. CSPs run data centers and provide Infrastructure as a Service (IaaS), Platform as a Service (PaaS), and Software as a Service (SaaS) [12]. Cloud computing hardware, networking, and software infrastructure are managed by them. Amazon Web Services (AWS), Microsoft Azure, and Google Cloud Platform (GCP) provide scalable, on-demand computing power, storage, and other IT services without the need to invest in and maintain physical infrastructure [20]. The cloud computing ecosystem relies on CSPs to help enterprises decrease expenses, boost flexibility, and scale [6].

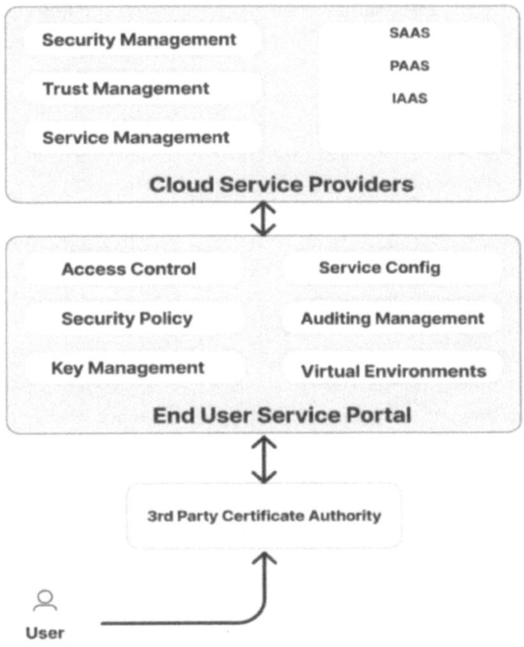

FIGURE 7.3 A framework for security in cloud computing.

- *Last User Service Portal*: End users, such as employees or customers, may access and control IT services, applications, and resources supplied by an organization or CSP via an end-user service portal [9]. These portals streamline IT service consumption and simplify the user experience [16]. End-user service portals typically offer user authentication and authorization, self-service password resets and software installations, cloud-based applications and resources, and a dashboard or interface for IT environment monitoring and management. These portals boost productivity, simplify access to vital services, and help firms optimize IT consumption via reporting and analytics [18].

Figure 7.3 shows the secure cloud computing framework. It is built on a security model that describes each part and uses the right security technologies to make it work. The term "secure cloud computing framework" is used in the context of cloud computing to describe a set of rules, best practices, and technological solutions created to safeguard the confidentiality of data and applications stored in cloud-based environments. Cloud computing has several advantages, including its scalability, cost-effectiveness, and flexibility; yet it also presents new security risks, which is why a framework is required to address these concerns. The detailed description of cloud security frameworks is given in Table 7.4.

TABLE 7.4
Description of Cloud Security Framework

Component	Description	Key Functions
Security management	Security management defines, implements, and maintains the security posture of cloud-based systems and services [6]. It includes policies, procedures, and tools for threat detection, risk assessment, and incident response.	• Security policy development and enforcement. • Security monitoring and incident management. • Vulnerability assessment and patch management.
Trust management	Cloud trust management establishes and manages trust connections between entities. Given the complexity of cloud service composition dynamics, trust, and access control frameworks should include delegation primitives [20].	• Identity and access management. • Authentication and authorization mechanisms. • Trust negotiation and attestation.
Service management	Service management involves provisioning, monitoring, and maintaining cloud services [18]. It ensures availability, performance, and reliability while adhering to security standards.	• Service provisioning and orchestration. • Service-level agreements (SLA) monitoring. • Incident response and service recovery.
Access control	Access control regulates access to cloud resources through policies and mechanisms [11]. It encompasses authentication, authorization, and privilege management.	• Role-based access control (RBAC). • Access policy enforcement. • Access auditing and logging.

(Continued)

TABLE 7.4 (*Continued*)
Description of Cloud Security Framework

Component	Description	Key Functions
Service configuration	Service configuration manages the secure configuration settings of cloud services to ensure they are optimized and protected [17].	• Configuration management. • Secure parameter settings. • Configuration drift detection.
Security policy	Security policy defines rules and standards governing security practices within the cloud environment, ensuring consistent and compliant security measures [14].	• Policy creation and enforcement. • Continuous policy assessment. • Policy documentation and communication.
Auditing management	Auditing management handles logging and auditing activities within the cloud environment, providing a trail of actions and events for security analysis and compliance purposes.	• Log collection and storage. • Log analysis and anomaly detection. • Compliance auditing and reporting.
Key management	Key management is responsible for generating, storing, and distributing cryptographic keys for encryption and authentication, ensuring data confidentiality and integrity [4].	• Key generation and rotation. • Key storage and protection. • Key distribution and revocation.
Virtual environments	Virtual environments create and manage isolated and secure virtual spaces within the cloud infrastructure, maintaining security boundaries for multi-tenancy.	• Virtualization technology implementation. • Network segmentation and isolation. • Resource allocation and isolation.
3rd party certificate authority	A 3rd party certificate authority (CA) issues digital certificates for secure communication and authentication, ensuring the authenticity and trustworthiness of entities.	• Certificate issuance and management. • Certificate revocation and renewal. • Trust chain validation.
User	Users are individuals or entities accessing cloud resources and services [18]. They play a crucial role in adhering to security policies and practices to maintain a secure cloud environment.	• User authentication and credential management. • Adherence to security policies and practices. • Reporting security incidents and concerns.

7.6 DISCUSSION

The study of cyber vulnerabilities in cloud computing settings reveals an intricate terrain of dangers that need meticulous deliberation. The many types of risks highlighted, including classic issues like data breaches and identity theft, as well as more advanced difficulties like denial-of-service assaults and insider threats, highlight the need for a comprehensive cybersecurity strategy. The chapter highlights that the increasing use of cloud services intensifies the potential influence of these risks on the confidentiality, integrity, and availability of data. The results have ramifications for the wider field of modern IT environments, emphasizing the need for enterprises to prioritize flexible cybersecurity solutions that are in line with the recognized threat vectors. An analysis of widely used security frameworks reveals essential instruments for reducing cyber threats in cloud systems. Encryption plays a crucial role in ensuring the security of data, whether it is being sent or stored, while strong access controls are essential in preventing unwanted access. In order to secure the security of cloud infrastructures, it is important to take a proactive approach, which involves using intrusion detection systems and anomaly detection methods for early threat identification. Our conversation is enhanced by including real-world case studies and industry best practices, which provide concrete examples of successful initiatives. Decision-makers are advised to use these insights to make strategic investments, strengthen their defenses, and effectively deal with the changing nature of cyber threats with resilience and preparation, as cloud computing continues to influence the IT industry.

This chapter emphasizes the need for enterprises to modify their cybersecurity protocols in response to the constantly evolving threat landscape, as cloud computing continues to have a significant impact on modern IT settings. Decision-makers should prioritize expenditures that fit with the identified threat vectors, security frameworks, and best practices to strengthen their cloud infrastructure. Although the research offers significant insights, it is essential to recognize its limits. Continuous research is required to investigate particular methods of cyber-attacks, examine new technologies, and evaluate the long-term efficacy of defenses due to the ever-changing nature of these threats. To maintain the security of data and services in cloud computing environments, enterprises must adopt a proactive and well-informed approach to cybersecurity.

7.7 CONCLUSION

This research on cyber threats and security frameworks in Cloud Computing has yielded essential cloud security insights. Cyber threats, main security frameworks, unique challenges, and practical security instruments have been discussed. Our comprehensive framework provides recommendations for enhancing cloud security. This research is beneficial for practitioners and organizations seeking to improve security, manage risks, and ensure compliance. To advance cloud security practices, we encourage additional research into advanced threat detection, quantitative risk assessment, privacy in the cloud, and emergent security technologies. To develop the subject of cloud security, future work must concentrate on many crucial areas. These areas include the development of advanced threat detection mechanisms that leverage artificial intelligence and machine learning, the refinement of quantitative risk assessment models for precise risk quantification, the advancement of privacy-preserving techniques to make sure that private cloud data is safe, the exploration and evaluation of emerging cloud-native security technologies, addressing the unique security challenges posed by hybrid and multi-cloud deployments, and the development of advanced threat detection mechanisms that leverage artificial intelligence and machine learning. Addressing these issues will be critical in maintaining the security and dependability of cloud computing systems.

REFERENCES

1. Jensen, Meiko, Jörg Schwenk, Nils Gruschka, and Luigi Lo Iacono. "On technical security issues in cloud computing." In *2009 IEEE International Conference on Cloud Computing*, pp. 109–116. IEEE, 2009.
2. Dikaiakos, Marios D., Dimitrios Katsaros, Pankaj Mehra, George Pallis, and Athena Vakali. "Cloud computing: Distributed internet computing for IT and scientific research." *IEEE Internet Computing* 13, no. 5 (2009): 10–13.
3. Morad, Steve. "Amazon virtual private cloud connectivity options." *Amazon Web Services Documentation*. 2014. https://d1.awsstatic.com/International/fr_FR/whitepapers/aws-amazon-vpc-connectivity-options.pdf
4. Jensen, Meiko, Nils Gruschka, and Norbert Luttenberger. "The impact of flooding attacks on network-based services." In *2008 Third International Conference on Availability, Reliability and Security*, pp. 509–513. IEEE, 2008.
5. Regalado, Antonio. Who coined cloud computing: MIT Technology Review. 2011.
6. Kuyoro S. O., Ibikunle F., and Awodele O.. "Cloud computing security issues and challenges." *International Journal of Computer Networks (IJCN)* 3, no. 5 (2011): 247–255.
7. Fox, Armando, Rean Griffith, Anthony Joseph, Randy Katz, Andrew Konwinski, Gunho Lee, David Patterson, Ariel Rabkin, and Ion Stoica. *Above the Clouds: A Berkeley View of Cloud Computing*. Department of Electrical Engineering and Computer Sciences, University of California, Berkeley, Rep. UCB/EECS-2009-28. 2009.
8. Javed, Abdul Rehman, Muhammad Usman, Saif Ur Rehman, MohibUllah Khan, and Mohammad SayadHaghighi. "Anomaly detection in automated vehicles using multistage attention-based convolutional neural network." *IEEE Transactions on Intelligent Transportation Systems* 22, no. 7 (2020): 4291–4300.
9. Javed, Abdul Rehman, and Zunera Jalil. "Byte-level object identification for forensic investigation of digital images." In *2020 International Conference on Cyber Warfare and Security (ICCWS)*, pp. 1–4. IEEE, 2020.
10. Baker, Thar, Michael Mackay, Martin Randles, and Azzelarabe Taleb-Bendiab. "Intention-oriented programming support for runtime adaptive autonomic cloud-based applications." *Computers & Electrical Engineering* 39, no. 7 (2013): 2400–2412.
11. Al-Khafajiy, Mohammed, Thar Baker, Muhammad Asim, Zehua Guo, Rajiv Ranjan, Antonella Longo, Deepak Puthal, and Mark Taylor. "COMITMENT: A fog computing trust management approach." *Journal of Parallel and Distributed Computing* 137 (2020): 1–16.
12. Ahmad, Waqas, Aamir Rasool, Abdul Rehman Javed, Thar Baker, and Zunera Jalil. "Cyber security in IoT-based cloud computing: A comprehensive survey." *Electronics* 11, no. 1 (2021): 16.
13. Iwendi, Celestine, Zunera Jalil, Abdul Rehman Javed, Thippa Reddy, Rajesh Kaluri, Gautam Srivastava, and Ohyun Jo. "Keysplitwatermark: Zero watermarking algorithm for software protection against cyber-attacks." *IEEE Access* 8 (2020): 72650–72660.
14. Hina, Maryam, Mohsin Ali, Abdul Rehman Javed, Fahad Ghabban, Liaqat Ali Khan, and Zunera Jalil. "Sefaced: Semantic-based forensic analysis and classification of e-mail data using deep learning." *IEEE Access* 9 (2021): 98398–98411.
15. Shabbir, Maryam, Ayesha Shabbir, Celestine Iwendi, Abdul Rehman Javed, Muhammad Rizwan, Norbert Herencsar, and Jerry Chun-Wei Lin. "Enhancing security of health information using modular encryption standard in mobile cloud computing." *IEEE Access* 9 (2021): 8820–8834.

16. Xia, Tian, Hironori Washizaki, Yoshiaki Fukazawa, Haruhiko Kaiya, Shinpei Ogata, Eduardo B. Fernandez, Takehisa Kato et al. "CSPM: Metamodel for handling security and privacy knowledge in cloud service development." *International Journal of Systems and Software Security and Protection (IJSSSP)* 12, no. 2 (2021): 68–85.
17. Sehgal, Naresh Kumar, Pramod Chandra P. Bhatt, Naresh Kumar Sehgal, and Pramod Chandra P. Bhatt. "Cloud workload characterization." *Cloud Computing: Concepts and Practices* (2018): 61–83.
18. Popović, Krešimir, and Željko Hocenski. "Cloud computing security issues and challenges." In *The 33rd International Convention Mipro*, pp. 344–349. IEEE, 2010.
19. Xuan, Shichang, Wu Yang, Hui Dong, and Jiangchuan Zhang. "Performance evaluation model for application layer firewalls." *PLoS One* 11, no. 11 (2016): e0167280.
20. Shin, Dongwan, and G. Ahn. "Role-based privilege and trust management." *Computer Systems Science and Engineering* 20, no. 6 (2005): 401.

8 Cyber Threats in Cloud Computing

Unveiling Vulnerabilities and How to Strengthen the Defense

Faheem Ahmad and Mohammed Ishrat

8.1 INTRODUCTION

In simple terms, cloud computing is "Internet computing." Since the Internet is typically thought of as a collection of clouds, the term refers to the use of the Internet to deliver technology-enabled services to individuals and businesses. J. McCarthy proposed that "computer timesharing technology might lead to a future where computing power and even specific applications might be sold through a utility-type business model" In late 1960, the concept of cloud computing gained popularity [1]. With cloud computing, users may access resources via the internet at any time and from any location, without having to worry about maintaining the original resources' technical or physical state or managing them. It solves the issues of time zones and geographic locations by using automatic software integration, which may incorporate and customize applications without requiring users to do any work on their part [2]. Users can access information from any location with an Internet connection. Cloud computing resources are scalable and dynamic. Through the internet, resources are available from the cloud at any point in time and from any location in the world. Compared to other computing models, cloud computing is less expensive because it requires no maintenance. After all, the service provider maintains service availability and clients are relieved of resource machine maintenance and management issues. One important feature of cloud computing is scalability, which is made possible by server virtualization. This new wave of web-based computing eliminates the need for businesses to purchase and maintain their own internal IT solutions by using remote servers housed in incredibly safe and secure data centers for data management and storage. The structure of cloud computing is shown in Figure 8.1.

Following the setting up of a cloud, cloud computing deployment varies depending on the specifications and intended use. The main service models currently in use are:

- *Software as a Service (SaaS)*: Software is offered to customers as a service in accordance with their needs, allowing them to access services hosted on cloud servers. Clients can use SaaS to access cloud service provider's (CSP's) apps online, as they are hosted on cloud infrastructure. The thin

DOI: 10.1201/9781003409571-8

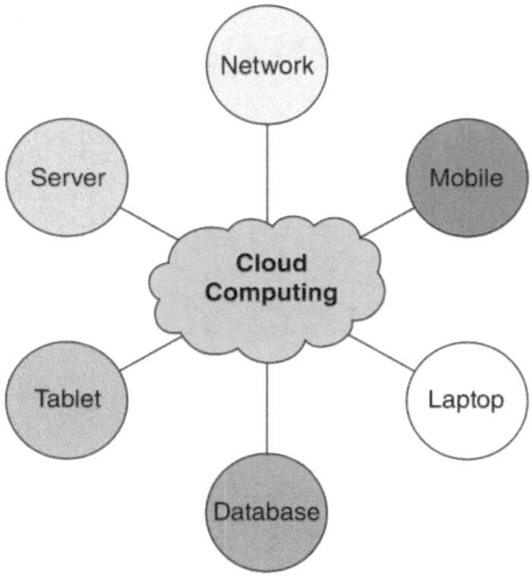

FIGURE 8.1 Cloud computing.

clients' interfaces allow users to use programs like web browsers. Software is only provided over the Internet using the SaaS approach. Instead of owning the program, customers pay for its consumption [3].

- *Platform as a Service (PaaS)*: Customer-owned apps needed a framework in order to function and be controlled. PaaS is used to supply the aforementioned services [4]. Clients that utilize PaaS are in charge of the applications that have been placed in the cloud; they do not have authority across the essential cloud infrastructure. It gives clients access to platforms so they can install other apps and build their own customized software on the cloud.
- *Infrastructure as a Service (IaaS)*: It allows users to manage operating systems (OSs), apps, storage, and network connectivity by renting computers, storage, and various other basic computing resources. The term IaaS describes the network, memory, processor, storage, and other computer capabilities that make up a CSP's physical infrastructure. The resources are provided via virtualized systems that can be accessed via the Internet. The types of the cloud models are shown in Table 8.1.

TABLE 8.1
Types of Cloud Service Models

Types of Cloud Service Models	Examples
SaaS	Google Drive, Dropbox, Amazon Web Service, Go to Meeting, Microsoft 365, Gmail
PaaS	OpenShift, OpenShift, Google App Engine and Windows Azure
IaaS	Microsoft Azure, Digital Oceans, and Cisco Metapod

8.2 LITERATURE REVIEW

Many studies on the security issues in the cloud have been conducted in the past few years. We will take a close look at cloud security in this article. We will also talk about difficulties and problems with cloud security. Table 8.2 lists earlier survey articles that addressed cloud security issues.

In addition to suggesting modern security frameworks to thwart attacks, it outlined all potential security issues with cloud computing and provided relevant solutions [10]. They also established security management in the cloud to lower risks and boost cloud trust. The concept, features, as well as framework of cloud computing, were discussed in this work [6]. While they encourage organizations and companies to utilize cloud computing technology in their operations, they do not advocate for its use in some sensitive applications, such as banking services. Addressing the security issues with vehicular cloud computing (VCC) while creating a roadmap for addressing them in the future, as well as talking about security breaches that compromise VCC security and offer fixes. Moreover, a thorough investigation into privacy concerns VCC security. The proposed framework of the blockchain was covered along with its features and cloud computing. Additionally, the effectiveness of blockchain technology in providing high security in the framework of cloud computing was demonstrated, and present blockchain-based applications for security in cloud computing were presented.

A detailed analysis of the policies and guidelines governing cybersecurity and data management in cloud computing can be found in the most recent paper. The management of cybersecurity controls to raise security and confidence within the cloud computing environment, in addition to addressing all aspects of cybersecurity as well as specialized operations in cloud computing to produce dependable and efficient results.

TABLE 8.2
Cloud Computing Objectives and Discussion

Objective	Related Discussion
Cloud Computing [5]	Examining the elements of cloud computing, as well as security and privacy issues, and locating current security fixes
Cloud Computing Security [6]	• Give a general overview of the data and features of cloud computing. • Outlining a few cloud-related issues and difficulties. • Suggestion for the cloud application's location.
Cloud Computing in Vehicular [7]	• Cloud computing security issues and their fixes for automobiles. • Research on cloud computing security and privacy for automobiles.
Cloud Computing and Blockchain [8]	• Talking about the features of the cloud and the blockchain. • Demonstrating how blockchain technology improves cloud computing security.
Cloud Computing and Cybersecurity [9]	• Talking About Cloud Computing Issues. • Create policies that provide a solid framework for thwarting cyberattacks in cloud computing. • Giving companies advice on how to manage these risks

8.3 KEY ASPECTS OF CLOUD COMPUTING

On-Demand Self-Service: Without the requirement for human assistance from the service provider, users are able to provision and manage computer resources as needed. When necessary, they can access, obtain, and improve service.

- *Broad Network Access*: A range of devices, such as computers, smartphones, and tablets, can access cloud services via the Internet. Clients' ought to have access to their apps, data, and cloud-based services via protocols and mechanisms that necessitate the availability of services to accommodate several types of thin or thick environments, like workstations, tablets, IoT devices, laptops, and mobile phones. Broad network access is commonly used to describe ubiquitous accessing networks.
- *Resource Pooling*: CSPs combine resources to cater to several clients, which results in cost savings and effective resource use. Many clients share the resources of the cloud in a multi-tenant environment. Customers are open and honest about the location of the data.
- *Metered Service*: Cloud computing is affordable and scalable since customers just pay for the resources they utilize. Resources are scaled up and down dynamically, and service usage is tracked and reported to CSPs and customers. By charging users according to the number of resources used, metering also helps with automatic resource optimization. The NIST has defined the five attributes of cloud computing which were previously mentioned. However, The Cloud Security Alliance (CSA) also recognized multi-tenancy as an essential feature.

8.4 SIGNIFICANCE OF CLOUD COMPUTING

An industry and corporation can be greatly impacted by the revolutionary technology known as cloud computing. Its potential to completely change how businesses use IT resources, provide services, and accomplish their objectives makes it significant. The importance of cloud computing is emphasized by the following key factors:

- *Scalability and Flexibility*: The cloud computing model offers businesses the flexibility and scalability to adjust their IT infrastructure in response to changing demand. Effectively managing diverse workloads requires this flexibility.
- *Accessibility*: Cloud services offer remote work, collaboration, and data access, as they can be used from any location with an internet connection.
- *Redundancy and Reliability*: Primary cloud carrier companies gift high availability and redundancy to decrease the opportunity of outages and records loss.
- *Security*: More safety is usually provided by using cloud carriers than even small organizations can acquire by way of closely investing in protection protocols and compliance certifications.

- *Innovation*: Businesses can speedily innovate and create new services and products way lots of services offered via the cloud, which includes the usage of Machine Learning, Artificial Intelligence (AI), as well as the Internet of Things (IoT).
- *Disaster Recovery*: Cloud offerings provide complete backup as well as catastrophe recuperation answers, making sure that information is safe and recoverable in the case of unforeseen activities.
- *Global Access*: Because cloud companies have data centers spread across a couple of regions, they can assist organizations in assembling facts sovereignty requirements while providing get right of entry to a global target audience.

8.5 THE SHARED RESPONSIBILITY MODEL

One of the main concepts in cloud computing is the Shared Responsibility Model, which outlines the security responsibilities that are shared by CSPs and cloud users (customers or businesses). Who is in charge of safeguarding various cloud service and infrastructure components is made apparent by this model. Depending on the cloud service model (IaaS, PaaS, or SaaS), various divisions of responsibilities may apply; however, general guidelines always apply. The divided responsibility model is depicted in Table 8.3.

8.5.1 Cloud Service Provider Responsibilities

- *Physical Infrastructure*: CSPs are in charge of maintaining the physical security of data centers, which includes monitoring, access control, and defense against external threats like unauthorized access and natural disasters.
- *Network Infrastructure*: To maintain the reliability and security of network services, CSPs oversee and safeguard the underlying network infrastructures which consist of switches, routers, and the internet connection.

TABLE 8.3
Shared Responsibility Model

<table>
<tr><td rowspan="3">Security in the Cloud (User's responsibility)</td><td colspan="4">Application Design, identity and access management</td></tr>
<tr><td colspan="4">Operating System, network, and firewall configuration</td></tr>
<tr><td>Data at rest(on-premises)</td><td>Data at rest(in Cloud)</td><td colspan="2">Data in transit</td></tr>
<tr><td rowspan="4">Security in the Cloud (Cloud provider's responsibility)</td><td colspan="4">Software</td></tr>
<tr><td>Compute</td><td>Storage</td><td>Database</td><td>Networking</td></tr>
<tr><td colspan="4">Hardware/Global Infrastructure</td></tr>
<tr><td>Regions</td><td>Availability Zones</td><td colspan="2">Edge Location</td></tr>
</table>

- *Hypervisor and Virtualization*: For IaaS offerings, CSPs are responsible for securing the hypervisor and virtualization layer, which abstracts and manages physical resources for virtual machines (VMs).
- *Host Operating System*: For IaaS, CSPs are responsible for securing the host OS running on physical servers
- *Physical Security*: Physical protection controls such as biometric controls, surveillance, and data center safety fall with CSP obligations.
- *Global Network Infrastructure*: CSPs manage the global network that connects their data centers to the Internet.

8.5.2 Cloud User Responsibilities

- *Data*: Cloud customers are liable for the safety of their data, which includes encryption, and data. This applies to all service types of models (IaaS, PaaS, SaaS).
- *Applications*: Cloud clients are in charge of protecting the apps and data they create or save on PaaS in addition to SaaS systems.
- *OS*: When using IaaS, cloud users are liable for securing the OS of the VMs they set up.
- *Configuration*: Cloud customers should responsibly configure and manipulate their cloud assets, which include access controls, firewall rules, etc.
- *Identity and Access Management (IAM)*: Managing access of the users, setting permissions, and implementing substantial authentication strategies are the responsibilities of the users.
- *Network Traffic Protection*: Cloud users are responsible for protecting data during transit and enforcing security measures like firewalls in addition to intrusion detection.
- *Application Security*: Ensuring the safety of applications developed by clients, inclusive of code vulnerabilities and runtime safety, is a user's responsibility.
- *Compliance and Governance*: Meeting industry compliance requirements and imposing governance policies are the responsibility of the users.

8.6 COMMON CLOUD SECURITY CHALLENGES

Cloud computing gives several benefits; however, it additionally introduces numerous security threats and demanding situations. To address these threats, groups have to implement a group of security practices, adopt robust protection technologies, and live up to date on the modern security developments in the cloud computing domain. Here are some common cloud computing threats that organizations are required to be aware of:

- *Data Breaches*: Any breach that exposes the end user or company's data kept on a cloud provider is taken into consideration as a cloud computing data breach [10]. Unauthorized hackers manage to get beyond the cloud issuer's or consumer's safety features and live hidden. After obtaining the customer's information, hackers sell it for income [11]. Breaches can arise due to poor access control, misconfigured settings, or vulnerabilities within the cloud company's infrastructure.

- *Insufficient IAM*: Poorly managed personal identities and access can lead to unauthorized users having access to resources. Weak authentication mechanisms, compromised credentials, and unsuitable role assignments contribute to this.
- *Insecure Interfaces and application programming interfaces (APIs)*: Cloud services are regularly accessed via APIs. Insecure or improperly implemented APIs may be exploited via attackers to advantage of unauthorized get entry to execute malicious moves or extract touchy statistics.
- *Data Loss*: Data stored in the cloud can be lost due to accidental deletion, hardware failures, or software bugs. Data loss can have severe consequences, especially if there are no reliable backup and recovery mechanisms in place.
- *Insecure Data Storage*: Inadequately encrypted or protected data that is at rest is vulnerable to theft. Without the right encryption and access controls, unauthorized users can also gain access to sensitive data.
- *Insecure Configuration Management*: Poorly configured cloud assets can cause vulnerabilities. Open ports, needless services, and weak security settings can be exploited by attackers.
- *Shared Resources Vulnerabilities*: Multitenancy, where a couple of users share an identical physical infrastructure, can lead to security risks. Vulnerabilities in one tenant's assets could doubtlessly affect others.
- *Lack of Visibility and Control*: As data and applications circulate to the cloud, companies might lose visibility and control over their assets. This can make it difficult to reveal threats and enforce security guidelines.
- *Insider Threats*: Malicious or negligent actions by employees or other authorized users can pose a significant risk to cloud security. Insider threats can involve data theft, unauthorized access, or intentional disruption of services.
- *Advanced Persistent Threats*: Attackers with advanced skills and a specific target that persistently evade organizations' defense are known as Advanced Persistent Threats or APTs. They often involve a combination of social engineering, malware, and long-term surveillance. The APT attacks were initially launched as a kind of cyber intervention against military institutions [12]. These attacks have grown in scope and are now directed toward state and industrial institutions [13]. APT attacks have also spread to the fields of education, finance, aviation, space exploration, energy supply, telecommunication, medicine, etc. To counter APTs in cloud environments, organizations should adopt a multi-layered security approach that combines proactive threat detection, continuous monitoring, access controls, strong authentication mechanisms, encryption, intrusion detection systems, etc. Collaborating with CSPs, staying informed about emerging threats, and conducting regular security assessments are crucial to mitigating the risk of APTs in the cloud. Figure 8.2 shows the multistep structure of APT.

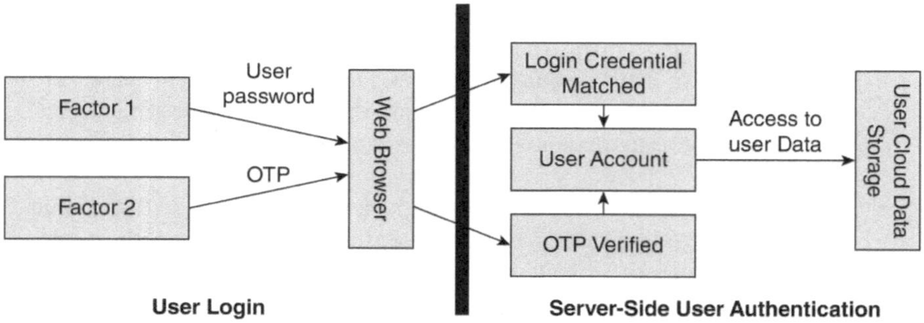

FIGURE 8.2 The description of the multistep structure of APTs.

- *Denial of Service (DoS) Attacks*: The DNS attacks block the availability of a network and its services by preventing authorized users from accessing resources available to them. Attackers initiate DoS attacks by flooding the network with requests in an attempt to overwhelm servers and use up bandwidth. This prevents legitimate users from connecting to the network, regardless of whether they are qualified to access it in the first place [14]. The Distributed DoS (DDoS) are attacks which are simultaneously launched by several sources [15]. Several techniques have been proposed to reduce DDoS attacks. Though [16] assert that SOS is acknowledged as the original solution that utilized overlay techniques, in order to protect against DoS attacks, the targeted network reroutes the packet received while concealing the precise location of the protected web server. As a preventive measure, SOS was suggested as a proactive strategy. In accordance with the view of [17], SOS prevents benign and anonymous users from accessing protected servers, and it also makes it difficult for users to access protected servers, by giving an attacker easy access to the majority of the tools required for a successful attack. This technique has a major disadvantage in that the threat is typically caused by attacks originating from counterfeit IP addresses, which allow DDoS to be launched against any internal firewall. The WebSOS is a proactive approach that was proposed by [18] and is the SOS's successor. It follows the same format as the SOS but makes use of CAPTCHA, which enhances the SOS's authentication procedure.
- *Loss of Governance*: Organizations might lose control over data and security practices if CSPs handle security in ways that conflict with an organization's policies or regulatory requirements.
- *Vendor Lock-In*: While not a security threat in the traditional sense, vendor lock-in can limit an organization's flexibility to move between cloud providers or back to on-premises infrastructure.
- *Data Residency and Compliance Issues*: Storing data in the cloud might raise concerns about where the data physically resides and whether it complies with specific industry regulations or legal requirements.

- *Supply Chain Attacks*: Attacks targeting third-party components, libraries, or dependencies that an organization's cloud services rely on can compromise the overall security of those services.

8.7 ZERO-DAY VULNERABILITIES AND CLOUD SERVICES

Zero-day vulnerabilities are security flaws or weaknesses in software, hardware, or firmware that are unknown to the vendor or developer. They are called "zero-day" because there are zero days of protection against them. Attackers can exploit these vulnerabilities before the vendor becomes aware of them or releases a patch. When it comes to cloud services, zero-day vulnerabilities present unique challenges and risks which are as follows:

8.7.1 Rapid Deployment and Proliferation

Cloud services can rapidly deploy and scale up infrastructure and applications. This implies that a zero-day vulnerability in a component has the potential to swiftly impact a large number of cloud instances [19].

One common feature of cloud environments is multi-tenancy, which allows multiple customers to share the same underlying infrastructure. A zero-day vulnerability could affect other customers who share the same resources if it compromises one customer's instance [20].

- *Limited Control*: Organizations have complete control and visibility over their security and infrastructure in traditional on-premises environments. On the other hand, cloud users have less control and insight into the fundamental cloud infrastructure, which makes it more difficult to identify and stop zero-day attacks. Clients using cloud services rely on the service providers to fix voids in the fundamental framework. Customers may have to patiently wait for a patch from a cloud provider if there is a zero-day vulnerability in the provider's infrastructure, which leaves them vulnerable.
- *Attack Surface Expansion*: Cloud environments are intricate, consisting of multiple interdependent parts and services. There is a chance that a zero-day vulnerability within a single element could be used to compromise others, increasing the attack surface. APIs, libraries, and other third-party components are frequently used by cloud services. Even in cases where the infrastructure of the cloud provider is secure, cloud services can be compromised by exploiting vulnerabilities in these dependencies.
- *Data Exposure and Loss*: Data breaches could result from the exploitation of a zero-day vulnerability inside a cloud provider, exposing private data kept in the cloud. Data loss may additionally stand up from zero-day vulnerabilities if hackers regulate or put off information from cloud offerings.

Implementation of reliable intrusion detection and safety tracking systems to find anomalous hobby and feasible zero-day assaults. Keep up with the vulnerabilities and patches that cloud carriers have released and install patches as soon as they are released to be available. To prevent exposure, encrypt sensitive facts in transit in addition to at rest inside cloud services. Furthermore, in the case of a zero-day vulnerability, enforce multi-component authentication along with robust access controls to avoid unauthorized access. Establish a well-described incident response method that outlines how to deal with security lapses and zero-day vulnerabilities.

8.8 STRATEGIES TO MINIMIZE ZERO-DAY VULNERABILITIES

Cloud companies need to discover and fasten 0-day vulnerabilities within their structures on the way to reduce this hazard. To lessen the publicity of such vulnerabilities, cloud customers have to additionally hold up with safety updates and patches supplied by way of their cloud company and make use of them as soon as they become available.

- *Cooperation to Resolve Cloud Zero-Day Vulnerabilities*: Many events, which include CSPs, safety specialists, governmental groups, and the bigger cybersecurity network, collaborate to combat zero-day vulnerabilities inside the cloud [21]. In order to guard cloud-based systems and the data that they hold, those tasks seek to detect, address, and disseminate statistics regarding zero-day vulnerabilities. Here are some salient facts about those joint endeavors:
- *Appropriate Disclosure*: In many instances, security researchers are important in finding zero-day vulnerabilities. They adhere to accountable disclosure suggestions, which dictate that earlier than making the vulnerability public, the impacted dealer or CSP needs to be notified in person.
- *Cooperation between CSPS and Researchers*: Researchers collaborate with the CSP to offer comprehensive records about zero-day vulnerabilities and their capacity impact as soon as they may be determined. After that, the CSP gets to collaborate in developing a mitigation plan.
- *Programs for Bug Bounty*: Numerous CSPs offer bug bounty applications that offer rewards to security researchers for reporting vulnerabilities [22]. These initiatives guide accountable disclosure and assist CSPs in locating and addressing vulnerabilities before hackers can take benefit of them.
- *Exchange of Information*: To cope with zero-day vulnerabilities, data sharing among CSPs, safety carriers, governmental organizations, and enterprise institutions is crucial. Initial identification and mitigation may be aided via changing data on potential threats and assault indicators [23].
- *Centers for Collaborative Research*: A few businesses and educational institutions set up cloud security research centers. These centers carry out studies on cloud vulnerabilities, disseminate results, and work with CSPs to create mitigation plans.
- *The Role of the Government*: Coordinating actions to deal with zero-day vulnerabilities is often the responsibility of government agencies. They may encourage cooperation among stakeholders and offer guidelines, rules, or regulations regarding cloud security.

- *Programs for Vulnerability Coordination and Disclosure*: Vulnerability disclosure and response activities are coordinated by groups like the Computer Emergency Response Team (CERT). They serve as liaisons between security experts and CSPs to guarantee that vulnerabilities are handled appropriately [24].
- *Industry Guidelines and Optimal Procedures*: Standards and best practices for cloud security are created by industry associations and standards bodies. By following these recommendations, organizations, and CSPs can lower the possibility of zero-day vulnerabilities by implementing security measures [25].
- *Consistent Security Updates*: In order to resolve known vulnerabilities, including zero days, CSPs are in charge of promptly releasing security updates and patches. To safeguard their cloud resources, cloud users need to make sure they apply these patches as soon as possible [26].
- *End-User Education*: Organizations and cloud users are essential to cloud security. They ought to teach their personnel and users about first-class practices for cloud security, which include the importance of making use of updates and practicing safe and secure computing.

8.9 EMERGING THREATS AND FUTURE CONSIDERATIONS

Emerging threats and future considerations in cloud security are of paramount importance as technology evolves. Staying ahead of these threats is crucial to maintaining the security and integrity of cloud environments. Here are some emerging threats and future considerations in cloud security:

- *Serverless and Edge Computing Security*: As serverless grow to computing advantage popularity, securing these distributed and dynamic architectures becomes a priority. Attackers can also target inclined serverless capabilities or side devices to compromise cloud assets [27].
- *Quantum Computing Threats*: The creation of quantum computing poses an ability hazard to encryption algorithms normally used in cloud safety. Organizations need to put together the submit-quantum cryptography technology by adopting quantum-resistant encryption methods.
- *AI and Machine Learning-Powered Attacks*: Attackers are increasingly more the use of AI and device studying to beautify their attack techniques [28]. AI-driven assaults may be more advanced and evasive, requiring superior AI-based security answers for protection.
- *Supply Chain Attacks*: Attacks targeting the supply chain, which include software program supply chain attacks, pose huge risks to cloud protection. Organizations must put into effect strict safety features to validate third-party components and dependencies. Attacks targeting the supply chain, such as software supply chain attacks, pose significant risks to cloud security. Organizations must implement strict security measures to validate and secure third-party components and dependencies.

- *Container and Kubernetes Security*: Container and Kubernetes adoption is growing, however misconfigurations and vulnerabilities in this technology can cause protection breaches. Organizations must recognize securing containerized packages and orchestrators.
- *Zero-Trust Security*: The concept of zero-believe protection, where belief is never assumed and verification is continuous, is gaining traction. Cloud environments want to move closer to 0-believe models to limit the assault floor and defend against insider threats.
- *Cloud-Native Security*: Cloud-local technologies like microservices, DevOps, and serverless require safety solutions mainly designed for these environments [29]. Cloud-local security tools and practices will retain to adapt.
- *Multi-Cloud and Hybrid Cloud Security*: Organizations frequently use a couple of cloud providers or an aggregate of on-premises and cloud environments. Securing these hybrid and multi-cloud architectures requires comprehensive protection strategies.
- *Data Privacy Regulations*: Data privacy rules like GDPR and CCPA preserve to adapt, putting strict necessities on how corporations take care of and defend personal records within the cloud. Compliance is a developing attention in cloud security.
- *Enhanced Authentication and Access Controls*: Beyond conventional username/password authentication, businesses will increasingly undertake stronger authentication methods, which include biometrics and continuous authentication, to shield get entry to cloud resources.
- *Cloud-Native Threat Detection*: Threat detection tools and practices designed especially for cloud-local environments become more state-of-the-art, permitting real-time chance detection and reaction [29].

8.10 MITIGATION STRATEGIES AND BEST PRACTICES TO STRENGTHEN THE DEFENSE

To safeguard information, applications, and infrastructure from constantly converting cyber threats, mitigation strategies and satisfactory practices for cloud computing environments are critical. Here are some key strategies and practices:

- *Shared Responsibility Model*: Understand and adhere to the shared responsibility model, which defines the division of protection obligations between the CSP and the client. Implement security features in line with your role within this version [30].
- *IAM*: Implement robust IAM strategies, consisting of least privilege (PoLP) and role-based Access Control (RBAC), in addition to strong authentication. Regularly overview and audit consumer permissions.
- *Data Encryption*: Secure facts both in transit and at rest by using robust encryption strategies. Utilize encryption key control practices to guard encryption keys.

- *Secure Configuration Management*: Ensure cloud assets are nicely configured and follow security exceptional practices. Implement protection organizations, firewall policies, and network segmentation to govern.
- *Regular Patching and Updates*: Stay knowledgeable of security updates and patches furnished by using your CSP and apply them promptly. This consists of updates for VMs, packing containers, and cloud offerings [31].
- *Threat Detection and Response*: To right away discover and deal with protection incidents, and computerized hazard detection systems in the area. To reduce the results of breaches, create an incident reaction plan.
- *Compliance and Governance*: Align your cloud security practices with industry-unique compliance guidelines and requirements. Regularly audit and record your cloud protection controls.
- *Employee Training and Awareness*: Educate personnel and cloud customers on security first-rate practices, social engineering threats, and how to perceive phishing tries.
- *Vendor and Third-Party Risk Management*: Assess the security practices of vendors and carrier providers who have entry to your cloud records or infrastructure. Ensure they meet protection standards.
- *Insurance for Cybersecurity*: To assist lessen the financial dangers associated with statistical loss and protection breaches, reflect on the consideration of purchasing cyber coverage.

Organizations can decrease the likelihood of incidents of security and facts breaches and build an amazing protection document in their IT cloud environments by means of placing those mitigation techniques and exceptional practices into practice. Evaluate and update your protection measures on an ordinary foundation to keep up with changing threats and cloud surroundings adjustments.

8.11 CONCLUSION

In the unexpectedly changing horizon of cloud computing, it is far crucial to recognize and counteract sophisticated cyber threats. The purpose of this chapter is to provide readers with an in-depth understanding of those sorts of threats and to offer recommendations on the way to improve cloud environments' security posture. Organizations can increase the security of cloud-based assets and preserve the integrity of their everyday data by means of retaining up with emerging risks and enforcing security features. As organizations depend increasingly upon cloud offerings to encourage innovation, reduce costs, and enhance scalability, in addition, they want to stand against the ever-intricate security threats linked to the cloud. In the end, in the latest digital landscape, cloud computing cyber threats pose a steady and changing task for organizations. Bolstering cloud computing's defense against cyber threats is a non-stop process that necessitates alertness, flexibility, and a dedication to the best safety standards. Through the implementation of the strategies and first-rate practices described right here, organizations may confidently navigate the cloud landscape and set up strong protection.

REFERENCES

1. R. Al Mogbil, M. Al Asqah and S. El Khediri (2020), IoT: Security challenges and issues of smart homes/cities. In *2020 International Conference on Computing and Information Technology (ICCIT-1441)* (pp. 1–6). IEEE.
2. A. Abdullah, R. Hamad, M. Abdulrahman, H. Moala and S. Elkhediri (2019), CyberSecurity: a review of Internet of Things (IoT) security issues, challenges and techniques. In 2019 2nd *International Conference on Computer Applications & Information Security (ICCAIS)* (pp. 1–6). IEEE
3. L. F. B. Soares, D. A. B. Fernandes, J. V. Gomes, M. M. Freire and P. R. M. Inácio (2020), "Cloud Security: State of the Art," in Security, Privacy and Trust in Cloud Systems, Berlin, Heidelberg: Springer Berlin Heidelberg, 2020, pp. 3–44.
4. F. Ahmad, S. Abidin, I. Qureshi and M. Ishrat (2023), Big Data and Its Role in Cybersecurity. In: Bhattacharya, A., Dutta, S., Dutta, P., Piuri, V. (eds), Innovations in Data Analytics. ICIDA 2022. *Advances in Intelligent Systems and Computing*, vol 1442. Springer, Singapore. doi: 10.1007/978-981-99-0550-8_10
5. H. Tabrizchi and M. Kuchaki Rafsanjani (2020), "A Survey on Security Challenges in Cloud Computing: Issues, Threats, and Solutions," *The Journal of Supercomputing*, vol. 76, no. 12, pp. 9493–9532, 2020.
6. L. Arockiam, S. Monikandan and G. Parthasarathy (2019), "Cloud Computing: A Survey," *International Journal of Computer and Communication Technology*, vol. 8, no.1, pp. 21–28, 2019.
7. H. Goumidi, S. Harous, Z. Aliouat and A. M. Gueroui (2021), "Lightweight secure Authentication and Key Distribution Scheme for Vehicular Cloud Computing," *Symmetry (Basel)*, vol. 13, no. 3, p. 484, 2021
8. G. Ashok, S. Shams, A. Shadab and S. Mohammed (2019), "Cloud ComputingSecurity Using Blockchain," *Journal of Emerging Technologies and Innovative Research*, vol. 6, no. 6, pp. 791–794, 2019.
9. N. Tissir, S. El Kafhali and N. Aboutabit (2021), "Cybersecurity Management in Cloud Computing: Semantic Literature Review and Conceptual Framework Proposal," *Journal of Reliable Intelligent Environments*, vol. 7, no. 2, pp. 69–84, 2021.
10. Y. Rahulamathavan, M. Rajarajan, O. F. Rana, M. S. Awan, P. Burnap and S. K. Das (2019), "Assessing Data Breach Risk in Cloud Systems", *2019 IEEE 7th International Conference on Cloud Computing Technology and Science (CloudCom)*, pp. 363–370, 2019.
11. C. Wang, S. T. K. Jan, H. Hu, D. Bossart and G. Wang (2018), "The Next Domino to Fall: Empirical Analysis of User Passwords across Online Services", *Proceedings of the Eighth ACM Conference on Data and Application Security and Privacy*, pp. 196–203, 2018.
12. P. Chen, L. Desmet and C. Huygens (2018), "A study on advanced persistent threats. Proc. Of the *IFIP International Conference on Communications and Multimedia Security* (2018)", pp. 63–72
13. D. McWhorter (2019), Mandiant exposes APT1 - one of China's cyber espionage units & releases 3,000 indicators (2019). *https://www.fireeye.com/blog/threat-research/2019/02/mandiant-exposes-apt1-chinas-cyber-espionage-units.html.*
14. W. Liu (2019), "Research on DoS Attack and Detection Programming", *2019 Third International Symposium on Intelligent Information Technology Application*, pp. 207–210, 2019.
15. B. Raju, P. Swama and M. Rao (2011), "Privacy and Security Issues of Cloud Computing," *International Journal of Engineering & Technology*, vol. 1, no. 2, pp. 128–136, 2011.
16. K. Lakshminarayanan, D. Adkins, A. Perrig and I. Stoica (2020), "Taming IP Packet Flooding Attacks," *ACM SIGCOMM Computer Communication Review*, vol. 34, no. 1, pp. 45–50, 2020.

17. W. Morein, A. Stavrou, D. Cook, A. Keromytis, V. Misra and D. Rubenstein (2020), "Using graphic during tests to counter automated DDoS attacks against web servers", *Proceedings of the 10th A CM conference on Computer and communication security - CCS* '20, pp. 8–19, 2020.
18. D. Cook, W. Morein, A. Keromytis, V. Misra and D. Rubensteint (2023), "WebSOS: Protecting Web Servers from DDoS Attacks", *The 11th IEEE International Conference on Networks. ICON 2023*, pp. 461–466, 2023.
19. K. Hashizume, D. G. Rosado, E. Fernndez-Medina and E. P. S. P. B. Fernandez (2020), "An Analysis of Security Issues for Cloud Computing," *Journal of Internet Services and Applications*, vol. 4, no. 1, pp. 1–13.
20. H. Tabrizchi and M. Kuchaki Rafsanjani (2020), "A Survey on Security Challenges in Cloud Computing: Issues, Threats, and Solutions," *International Journal of Supercomputing*, vol. 76, no. 12, pp. 9493–9532, 2020.
21. S. Saha, A. Vullikanti and M. Halappanavar, "FlipNet: Modeling Covert and Persistent Attacks on Networked Resources," *2017 IEEE 37th International Conference on Distributed Computing Systems (ICDCS)*, Atlanta, GA, USA, 2017, pp. 2444–2451, doi: 10.1109/ICDCS.2017.298.
22. T. Walshe and A. Simpson, "An Empirical Study of Bug Bounty Programs," *2020 IEEE 2nd International Workshop on Intelligent Bug Fixing (IBF)*, London, ON, Canada, 2020, pp. 35–44, doi: 10.1109/IBF50092.2020.9034828.
23. J. J. Thangaraj, "Threat Detection and Mitigation in Self-Organizing Wireless Communication Network," *2023 Second International Conference on Smart Technologies for Smart Nation (SmartTechCon)*, Singapore, Singapore, 2023, pp. 28–32, doi: 10.1109/SmartTechCon57526.2023.10391562.
24. I. Yurtseven and S. Bagriyanik, "A Review of Penetration Testing and Vulnerability Assessment in Cloud Environment," 2020 *Turkish National Software Engineering Symposium (UYMS)*, Istanbul, Turkey, 2020, pp. 1–6, doi: 10.1109/UYMS50627.2020.9247071
25. T. Moulahi, S. El Khediri, R. U. Khan and S. Zidi (2021), "A Fog Computing Data Reduce Level to Enhance the Cloud of Things Performance," *International Journal of Communication Systems*, vol. 34, no. 9, p. e4812.
26. H. Yong-Xiang, "A Study on the Security of Patch Management in a Cloud Computing Environment," 2018 *4th Annual International Conference on Network and Information Systems for Computers (ICNISC)*, Wuhan, China, 2018, pp. 278–282.
27. B. Li, S. Lv, Y. Zhang and M. Tian (2014), The Application of Trusted Computing Technology in the Cloud Security. In: Wong, W.E., Zhu, T. (eds) Computer Engineering and Networking. Lecture Notes in Electrical Engineering, vol 277. Springer, Cham. doi: 10.1007/978-3-319-01766-2_99
28. B. Brindavathi, A. Karrothu and C. Anilkumar, "An Analysis of AI-based SQL Injection (SQLi) Attack Detection," *2023 Second International Conference on Augmented Intelligence and Sustainable Systems (ICAISS)*, Trichy, India, 2023, pp. 31–35, doi: 10.1109/ICAISS58487.2023.10250505.
29. K. Che and S. Sheng, "Cloud Native Network Security Architecture Strategy under Zero Trust Scenario," *2023 IEEE 7th Information Technology and Mechatronics Engineering Conference (ITOEC)*, Chongqing, China, 2023, pp. 867–871.
30. B. F. Demissie and S. Ranise, "Assessing the Effectiveness of the Shared Responsibility Model for Cloud Databases: The Case of Google's Firebase," *2021 IEEE International Conference on Smart Data Services (SMDS)*, Chicago, IL, USA, 2021, pp. 121–131, doi: 10.1109/SMDS53860.2021.00026.
31. H. Yong-Xiang, "A Study on the Security of Patch Management in a Cloud Computing Environment," 2018 *4th Annual International Conference on Network and Information Systems for Computers (ICNISC)*, Wuhan, China, 2018, pp. 278–282, doi: 10.1109/ICNISC.2018.00063.

9 A Comprehensive Analysis of Vulnerabilities and Cybersecurity Issues for Digital Agriculture and Food Industry

Gausiya Yasmeen, Syed Adnan Afaq, and Saman Uzma

9.1 INTRODUCTION

In the 21st Century, the Web of Things (or Internet of Things (IoT)) and smart computing have fundamentally transformed how individuals live today. About ten years ago, data-driven services and IoT technology were unfathomable. They currently influence a wide range of industries, such as the smart home, automobile, and healthcare, and are all around us. The agribusiness or agronomics field, in particular, has adopted this high-tech involvement with open hands. The concept of cultivation and the practice of farming are shifting toward a new path, as a consequence of the rising usage of the Web of Things and Information and Communications Technology – ICT [1]. This notion is known as "Smart Farming." The technical advancement of smart farming in the agriculture sector is shown in Figure 9.1.

The United Nations (UN) predicts that by 2050, there will have been an increase in the global population of 9 billion, or one-third of the existing inhabitants. The organization of the UN for Food and Agriculture estimates that this population growth demands an upsurge in sustenance production of over 70%. The mistreatment of terrestrial, aquatic, and other natural possessions as well as other issues like increased competitiveness are also brought on by this population's rapid growth [1]. Due to these problems, it is vital to lessen how dependent our food system is on the environment. Subsequently, the development of precision farming or "smart farming" can greatly benefit from the usage of ICT and the growing IoT. UAVS: Unmanned Aircraft Vehicle Systems (drones), mechanoid (robots), Advanced Metering Infrastructure (AMI), advanced safety tools, high-tech facts for seed qualities, and smart soil conditions are just a few of the innovative new systems that provide farming quality management with previously unheard-of conveniences. The future networked systems combined for regulatory, monitoring, and mechanization will cause a considerable transformation in the substructures and facilities accessible in future farms. To sustain and improve operations, the level of competitiveness, and

DOI: 10.1201/9781003409571-9

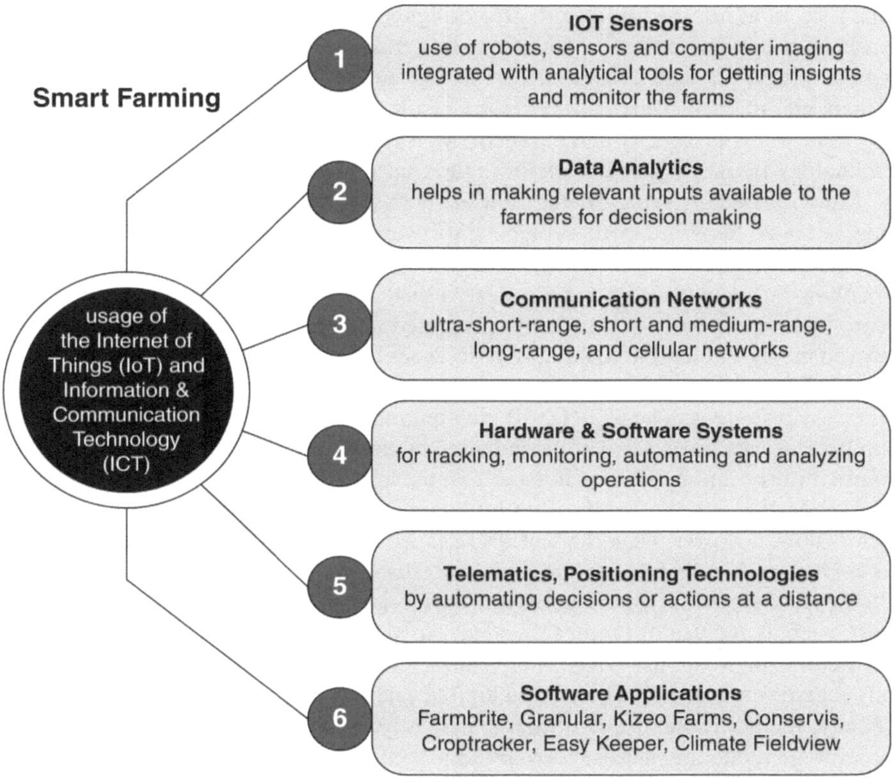

FIGURE 9.1 Technical advancement in the agriculture sector.

productivity, the agricultural industry will depend on information technology more and more.

The human population has expanded quickly over the last century. The boost in the nutritional value and amount of the world's food resources over the past 50 years has largely been attributed to technical developments in genetic engineering. Genomic engineering aids in growing seeds and floras which use lesser amounts of water and offer more nutrients to meet the needs of a growing population. The next technical advancement for the agriculture sector's sustainable food production is digital agriculture. Technology has enormous advantages, but it also has inherent hazards. As a result, the industry is being targeted more than ever because foreign rivals and hacktivists are after its intellectual property. Even if the majority of food and agriculture businesses do not yet invest in cybersecurity, the sector is starting to worry about the potential for cyber-agroterrorism. A well-planned terrorist assault could destroy any agriculture or Food Company's reputation as a reliable food supplier and erode public trust in the entire food supply chain or logistics network [2]. It is obvious that the industry's increasing digitization creates new potential for terrorists by giving them access to assault locations that were previously too far away or challenging to reach.

The risks of agroterrorism are too great to ignore, especially since cyberterrorism is a relatively low-cost endeavor with a huge potential for reward [2]. This indicates that a severe, well-planned cyberattack on an agricultural company might have serious human and financial repercussions. As a result, it's critical to raise awareness about the value of cybersecurity and the security risks associated with the extensive use of technology in the agricultural sector in the smart farming industry.

Malicious actors typically target less expensive and easier-to-access paths that may be weak, involving people, objects, software, procedures, or technology, under-protected by the user yet with highly serious repercussions. (J. Nikander) evaluated six dairy farms in Finland and discovered that a great deal of the networking hardware had been poorly substantially protected and that default credentials, which proved easily vulnerable to compromise, were being used. Additionally, the threat players have changed from amateurs to independent governments with essentially limitless resources. The COVID-19 incident made cyber-security fiasco one of the top ten risks according to a poll conducted by the World Economic Forum in 2022, while it ranks among the top five risks in the area [3]. The phrase "cybersecurity" is frequently used as a result of multiple assaults on digital infrastructure. Threat actors now, however, view the world's reliance on a reliable food logistics network as a chance to employ malware, such as ransomware, as leverage to accomplish their evil objectives. Logistics network cyber threats have been documented in the past, such as the use of target IT systems to aid cargo theft, drug sabotage, or the trafficking of weapons. Given their characteristics, similar dangers might affect any logistics network. The idea of a logistics network encompasses many facets of society, including the usage of mathematically modeled subsidy programs [4]. In order to generate the necessary awareness, some observations on the difficulties of smart farming are presented in this study, with a focus on security concerns and problems.

9.2 SMART FARMING

The practice of using numerous cybernated gadgets to surveil, evaluate, and regulate environmental factors that might have an impact on the production of food (crops, fruit, etc.) is known as digital agriculture (DigAg) or smart farming, sometimes known as smart agriculture or farming [4]. The development of smart low-power gadgets and widespread Internet access has permitted the digitalization of many labor-intensive occupations. The agriculture industry has been forced to react to the digital transition due to a skilled labor scarcity plus a rising demand for sustenance. In order to enhance agricultural operations and processes, smart farming makes use of ICT, particularly the IoT and associated big data analytics [4]. Modern sensors and systems are used to keep an eye on the environment, plants, water, soil humidity, crops, and plant diseases [2]. With the adoption of digital agriculture, more and better food would be readily available to the world's rising population. The estimated population of the globe in 2050 will be 9.32 billion, up from 6.60 billion in 2000. Therefore, it is essential to boost the productivity of the small amount of farmland. The development of digital and electronic crop monitoring, as well as those pertaining to soil moisture, weather conditions, pollination, irrigation systems, pests, and topsoil conditions, is the primary goal of smart farming in general. Smart farming makes use of mobile technology, smart networking, the adaptability and interoperability of agricultural operations, customer and supplier integration, and the adoption of cutting-edge business models [5]. With fewer human resources and interventions, digital tools like smartphones, different sensors, global positioning systems (GPS), mechanoid, and UAVS may be used to collect useful data, assess it, and make smart decisions that will enhance food production. Agriculture is essential to the survival of humans for the reason that it not only supplies food but also gives employment to large masses. The agricultural industry has to be modernized due to the increased demand for sustainable food production, a skills gap, and the effective use of finite environmental resources.

Cyber-physical system (CPS): CPSs are created on the pillar of smart systems which are the foundation of modern farming [1, 2]. A computer and communication network may monitor, synchronize, regulate, and integrate a CPS's action, empowering a group of networked agents to interact with the natural environment [2]. Such network agents can be broadly split into four layers as depicted in Figure 9.2. In order to give farmers real-time access to information and analysis on their crops, logistics, land, livestock, and sensors, the equipment can gather information such as pollination, soil moisture, and weather oscillations and transmit it through a cellular wireless network. Then actuators can carry out a previously programmed action [6]. By analyzing the data gathered and taking appropriate action to boost productivity or

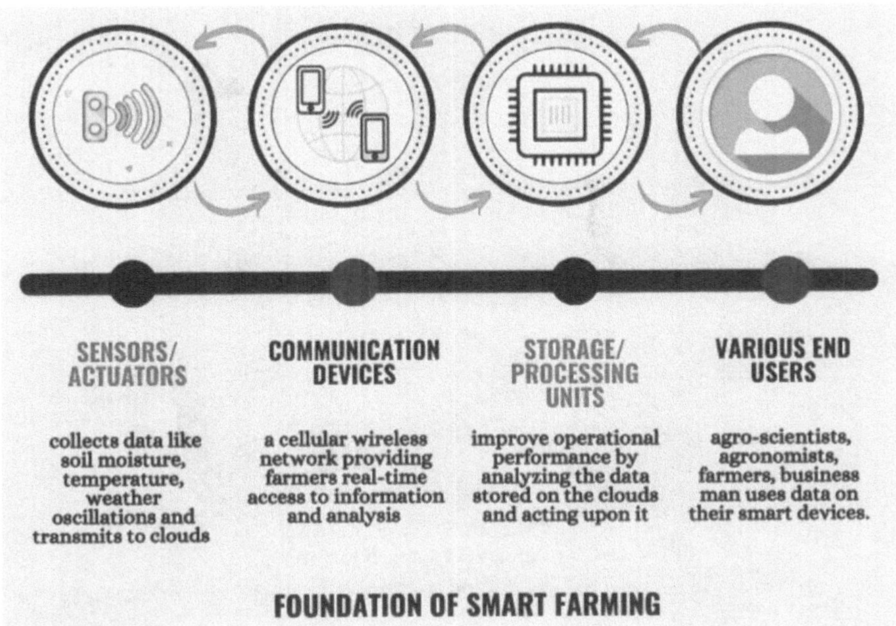

FIGURE 9.2 Four pillars of cyber-physical systems – based intelligent networks – are the foundation of smart farming.

streamline processes, the smart farm can enhance its operational performance. Even premium equipment will eventually be replaced by more advanced versions because of the growing invention and development of these technologies, which will drive down the cost of older equipment that would still benefit offline farms today. With Figure 9.3, let's explore potential applications of how digital agriculture IoT technology [4] might benefit farming and agriculture to better comprehend its benefits.

- *Irrigation Management System*: The irrigation system is one of the most essential components that any farm needs, as is widely known. Farmers or other virtual technicians can use and analyze data from remote sensors using a laptop, tablet, or smartphone. This enables them to define appropriate strategies for managing water resources, such as choosing the best distribution/irrigation approach and determining whether it should be directed or not, for how long, and with what kind of pressure [7].
- *Fertigation*: IoT can be used to implement approaches to remotely controlled fertigation. The soil is amended with fertilizers, soil conditioners, and other essential components during the fertilization process [8]. In this way, farmers may control the fertigation system from a distance, giving them the ability to actively alter the amount of fertilizer required as needed and keep an eye on the pH levels and fertilizer concentrations in the soil by employing remote sensors.

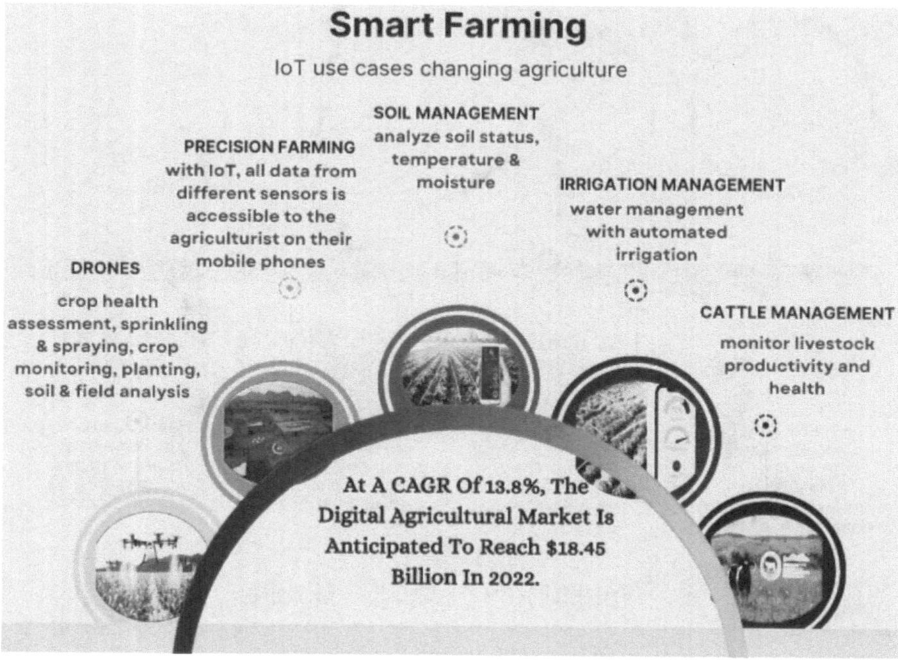

FIGURE 9.3 How IoT uses cases have changed the face of agriculture.

- *Monitoring of Livestock Maturity and Safety*: Farmers may follow stray animals with IoT-enabled devices that produce real-time livestock data using GPS location [6]. With the advancement of novel biometric sensors, these systems can now also give animals real-time biological data, including information on their body temperature, pulse, and even tissue resistivity. To learn more about the whereabouts, comfort, and health of their cattle, large farm owners may deploy wireless/mobile IoT applications [8]. Once sick animals are identified using the information acquired, proactive disease prevention measures can be taken. It can also be used to motivate people to take the proper steps to solve any problems in the quickest time possible and with minimal risk of damage. Since finding animals is simple and accurate thanks to IoT-based sensors, labor costs are also decreased. For high-value livestock as well as for the breeding of new animals, new clients could have the option to view the prospective animals before being acquired and follow their progress over time, keeping track of all the factors of their development.
- *Crop Management*: Farms will need to adapt and sustain the worldwide agricultural requirements as the world population rises and the climate becomes harsher. IoT solutions for smart farming may be created and implemented to increase output, decrease waste and expenses, and optimize resource use [7, 9]. Farmers will be better able to manage their resource use if they have access to a smartphone that can track the state of the soil where their most recent crop is being planted. UAVS in agriculture is becoming increasingly common and extremely useful, with significant IoT Smart Farming applications [9]. Remotely operated autonomous tractors designed to do tasks like drilling, sowing, and spraying are already being developed and used.
- *Crop Monitoring using UAVs (Drones)*: It is difficult and expensive to monitor a large field to identify problems with soil variance, fungus, and irrigation promptly. Tracking farm activities is now more manageable and considerably more accurate and precise, thanks to low-cost UAVS that are outfitted with various types of sensors, have limited prices, and are integrated with strong CPUs, GPS, and radio technology [10]. As a result, farmers may utilize time series animations to monitor the growth of their crops and problems related to soil variation, resulting in more effective crop management and an increase in the level of output and quality standards [11].
- *Food Logistics Network Management*: IoT sensors for managing the logistics network may be used across the product value chain to improve the quality of the final product. Physical fluxes and information-related parts of activities can be separated [12]. Physical proximity is no longer necessary for logistics network monitoring, which indicates that the physical products' journey or route from their source or point of origin to their destination is no longer reliant on the partners' locations for carrying out control and coordination operations [12]. Logistics networks may be remotely managed, planned, and optimized using new IoT-based systems.

The physical, edge, cloud, and communication or network layers [1, 4] make up the four levels of the architecture of digital agriculture, which are shown in Figure 9.4. The last layer stretches over all other levels and is what holds the other layers together.

- *Physical Layer*: The system's bottom layer is made up of actual, physical sensors and gateway hardware that are housed in a greenhouse or on farmstead structures. Such gadgets may include airborne UAVS, tractors that are self-directed, sensing tools for animals, or hubs that enable the connection between different smart objects or with the main central cloud [1–4]. These gadgets are used to accumulate real-time information about the climate conditions, the dampness level of the soil, and the temperature of the cattle's body. This information is then transmitted to distributed computing paradigms which are smart systems based on the cloud, for making decisions to support mechanization including other suggestions. For instance, information acquired by a field-based soil humidity sensing device analyzed at the edge or in the cloud may help predict the amount of water needed at the farm, expedite irrigation schedules, and offer end farmers a straightforward experience.
- *Edge Layer*: This layer is convenient for making decisions and local real-time calculations because it is close to ultimate prospects and gadgets. On the centralized cloud tier, both the network load and the computational

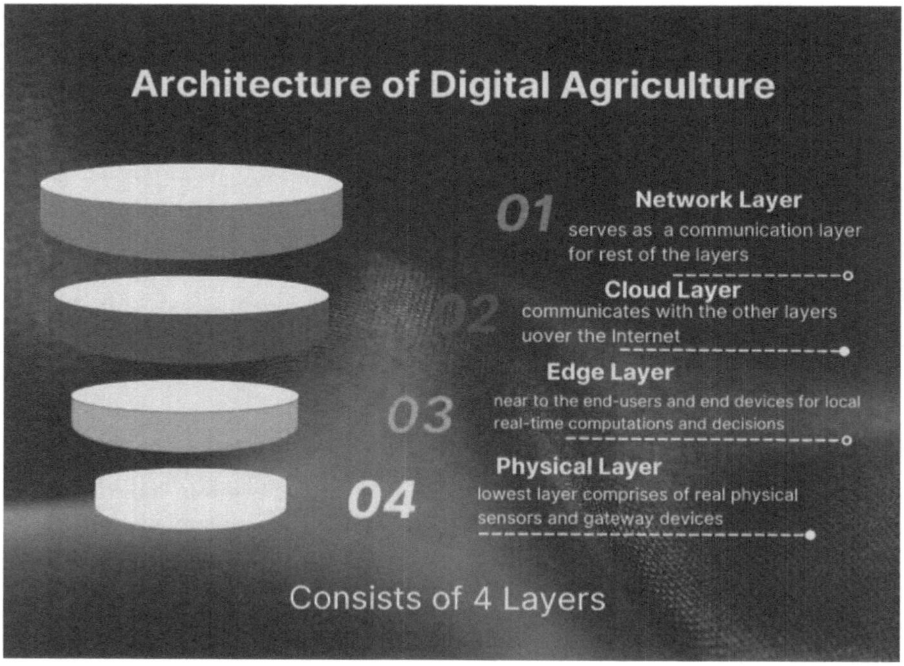

FIGURE 9.4 Four levels of the architecture of digital agriculture.

demand are reduced. The foundation of the edge computing layer is edge nodes. Every node acts as a gateway and performs a variety of tasks, namely, gathering data, security surveillance, and recognition, forecasting, and instantaneous verdict sustenance. Among the data capture services are data aggregation, filtering, encoding, and encryption of real-time data streams. Prediction services frequently have their roots in machine learning models that are built in the main cloud and executed on the edge layer. This information is utilized to forecast and categorize particular events involving plants or animals, such as predictions of crop yield, categorizations of the wellness of plants or animals, forecasts of the fertilizer and water requirements for a particular area of land to maximize yield, and estimations of soil erosion [1–4]. Real-time monitoring of abnormal events and their classification as malicious or benign may be accomplished by deploying security monitoring and detection technologies. Services including forecasting of devices fiasco and detection of anomalies fall under this category.

- *Cloud Layer*: The concepts of cloud computing and precision agriculture (PA) enable advancements to improve PA connection. In data centers, the cloud layer is often virtualized and uses the Internet to connect to other levels. The Platform as a Service (PaaS) architecture paradigm is typically followed by these cloud layer platforms, allowing customers to concentrate on executing applications and importing data. A data broker that gathers information being brought in from the edge layer and stores it in a Distributed File System (DFS) is administered and maintained by the PaaS provider. Analytical software uses this saved data to mine knowledge [1–4]. Insights are computed by this data analytics component and sent to the end user via a client program that runs on their computer. Several cloud-based technologies have been developed by well-known agricultural equipment producers, like John Deere and Farmers Business Network, to assist users in keeping track of different sensors and farm vehicles. Typically, these PaaS solutions are powered by well-known cloud computing infrastructures including Amazon Web Services and Google Cloud [1]. The construction of these farming PaaS systems utilizing these well-known cloud computing platforms also poses many security issues for the ecosystem of smart farming.
- *Network Communication Layer*: The word "connectivity" runs through foremost, perhaps all, of today's technologies. The concept of a network of smart gadgets is already a reality due to the increasing need for a boundary-less internet. The IoT is a concept that enables data sharing, monitoring, and control across linked objects. Numerous programs can analyze and utilize this data. The network layer in smart farming not only enables communication between the edge and physical layers but also gives those layers a way to communicate with the layer that is in the cloud. The network layer provides a means of communication to bind all other layers, whether it be for exchanging soil temperature via a peer-to-peer sensor communication system, sending farm monitoring data to

cloud data stores via high-speed mobile networks like 5G, or informing the cultivator regarding crop quality via a wireless ad-hoc topology [11]. A smart farming system's network layer is primarily responsible for two things. First of all, every layer of digital agriculture contains a different assortment of heterogeneous equipment. The network layer offers a safe and effective network stack that enables cross-layer, interoperable communication between wired, wireless, and mobile sub-networks [10]. Maintaining connectivity and hence raising availability is the layer's second duty. This layer is required for system-wide cyber communication across big data processing platforms used for analyzing collected data and individual sensors that gather information from the field.

The architecture of Smart Farming proposed by [1] shown in Figure 9.4 is based on actual scenarios of the smart farming world and an extensive literature survey. Examples of Smart Farming in Real-World: The weather stations that incorporate numerous smart farming sensors are arguably the most well-liked smart agricultural technology. As mentioned in the report released by the Ministry of Agriculture & Farmers Welfare posted on 20 Dec 2022, an endeavor aimed at the development of successful and productive agriculture technology is being carried out by the TIH Foundation of IOT and IOE, and it intends to create a comprehensive framework for agriculture technology (Agri Tech) to forecast crop production during the season [1, 8]. An IoT-based platform called EAgriS was created with this in mind. The TIH focuses on airborne robotics for soil component monitoring, drone-based imaging, and drone-based spraying as an IoT application in precision agriculture. The goal of TIH is to create a predictive data analysis model for intelligent decision-making based on environmental factors such as temperature, humidity, wind speed and direction, rainfall, soil parameters such as temperature, moisture content, the conductivity of electrical current, pH, NPK, and sulfur, as well as leaf moisture [8, 10]. The greatest agrarian information network in Latin America is called Agrosmart. In 2021, the business purchased BoosterArgo, South America's largest platform for farmers. Agrosmart is an interface that gathers and analyses data to produce insights that help 100,000 farmers overcome the effects of climate change and uncertainty and make data-driven decisions [6]. The most recent investment round for Agrosmart, a \$5.8 million series that concluded in 2019, brought in about \$9 million since its debut. Blackhills Farm, a 400-hectare property in New Zealand with more than 2000 cattle and 800 sheep, uses the SCADAfarm16 system, which enables the owners to remotely monitor water and energy use, the location of irrigators, measurements of soil moisture, and current meteorological conditions [2, 6]. Similar to this, the United States Department of Agriculture (USDA) created a 7,000-acre farm at Beltsville Area Research Centre to serve as a test site for smart farming technology. For a public-private program called Farmbeats, the farm was outfitted with a physical layer that included sensors, UAVs, and IoT-enabled farm equipment [11]. allMETEO, Pycno, Farmapp, Growlink, Arable, Semios, SCR by Allflex, Cowlar, CropX, Mothive, Doneseed, and SoilScout are a few available and frequently used products of IoT in agriculture.

9.3 CYBER THREATS AND VULNERABILITIES ISSUED IN DIGITAL AGRICULTURE

A new opportunity for adversaries to plot cyberattacks has arisen as a result of the use of cloud-based smart apps and sensor-based technology in agriculture. Currently, there are more and more opportunities for cyberattacks. New types of cyberterrorism assaults are being reported, and the practice of Information and Communication Technology and the Web of Things in agronomy has raised the sector's level of exposure. As a result, cyberattacks targeting the sector's production infrastructure and production equipment are growing more common in agriculture [13]. A system for irrigation, as an illustration, includes intelligent sensors and actuators, software, conventional networking hardware, and human interaction. Cybercriminals can breach one or more components of the agricultural application by taking advantage of flaws in the sophisticated systems that are frequently outsourced from many vendors and created for a variety of environments and applications [14]. While some of the vulnerabilities are unique to digital agriculture, others are comparable to threats faced by conventional technology or IoT networks. A Web of Sensors that regulates the production of crops containing genetically modified organisms (GMOs) is another example. It might be targeted, and the fundamentals of GMOs may be altered such that they are dangerous to humans [13, 15]. As a result, it's critical to be cognizant of and ready for any potential major safety hazards that Smart Farming may encounter. It is crucial to understand the main security risks that Smart Farming must address as well as the areas where the various stakeholders must concentrate their efforts to lessen their effects. Some dangers, however, are unique to digital agriculture. The foremost cybersecurity threats that are affecting digital agriculture are as follows:

9.3.1 Data Privacy and Security

In an intelligent farm, a massive figure of several sensors, gadgets, and equipment produces a considerable quantity of compound, dynamic, and spatial data. If such information is shared, whether by an insider or through unlawful means, there may be potential risks. All GPS systems use locational data, which might pose a serious security risk. The increased usage of GPS systems enhances the attackers' access to origin and position data specifics, allowing them to know precisely which field as well as which particular harvest might be attacked [15]. On a bigger scale, gathering crucial agricultural data on a certain nation could potentially be dangerous. In an environment in which farmlands and agronomists depend more and more on linked devices and services to increase productivity and product quality, the development of IoT introduces new obstacles. The loss of personal information is one of the concerns that worry farmers the most since it might be used against them when selling agricultural products. Such information might be released by a staff member (such as a disgruntled employee) to do harm or sell the knowledge for profit. For instance, a former Allen and Hoshall engineer had permission to use the organization's records for two years. The material this employee downloaded contained promotional information, plans for the project, and mechanical blueprints with a combined estimated value of $500,000. The proliferation of IoT will result in more privacy concerns [13, 15]. To ensure the efficient functioning of a digital farming ecosystem, maintaining data privacy and security is a critical prerequisite and one of the main objectives.

9.3.2 Authorization and Authentication Vulnerabilities

Social engineering, frequently referred to as the hacking of people, may be used to compromise even the most secure system because of human needs. For automated and compelling experiences, connected entities in digital agricultural applications, for instance, winged UAVs, self-directed tractors, and on-field sensors, communicate, interact, and give commands. According to [1], such communication can take place directly between machines or through a cloud- or edge-assisted network that can support IoT Communication Protocols, Message Queue Telemetry Transport (MQTT), Constrained Application Protocol (CoAP), and others. It's critical to confirm that the transmissions have originated from a trustworthy resource instead of a malicious adversary in both instances. Sensitive farm-related data such as information about the crop yield, fertilizer costs, or private information about cattle, related to data about their well-being, medical care, the decision to breed, and whereabouts, must have the owner of the field or another concerned party's consent before being transmitted to the cloud or a third-party application [16]. Furthermore, when buying cattle, the buyer may be given brief access to the animal's data to enable them to research the animal before making a purchase. In order for a farmer to be able to use an important section of agricultural machinery at the accurate hour for harvesting and planting, the OTA: Over-The-Air updates to the firmware for the equipment need to originate from a trustworthy source. Cross-cloud and multi-cloud trustworthy models [16] must be implemented when entities linked to different cloud providers communicate and access data remotely. The proper authorization is required before a farmer can turn on a machine-operated sprinkler for irrigation or obtain the current information from a device that measures the dampness in the soil. Competing farmers could overflow or dehydrate a farmer's field using their connections to the sensors. For a critical operation, such as using a self-propelled tractor on a field or a drone to spray pesticides, a multi-level authorization may be required. It is yet unclear whether the many access control models that have been developed and distributed for IoT [17]-like applications are suitable for adaptive smart farming.

9.3.3 Ransomware and Third-Party Apps

Gradually, ransomware is becoming the biggest threat to the world after the extensive use of IoT and ICT in various sectors, whether it is an individual or commercial enterprise. As said in the chapter, smart farming is changing the face of agriculture by using the latest technologies. With the advent of smart tools, the agriculture sector is also on the verge of being attacked by hackers. Attackers can encrypt farm and farmer data using any type of ransomware, leaving it unusable unless the ransom price payment has been made [18]. Data on the use of pesticides and GMOs can potentially be ruined by ransomware software. To address these issues, reliable data implementing a backup and recovery plan is necessary. Backups should be kept in a secure, separate area so that they are difficult for attackers to access from local networks. A breach of security or data block at a single interconnected business will be sufficient to damage the entire chain and possibly the economy of a country, according to the enormous scope of cyberattacks such as WannaCry27 ransomware and the most recent outbreak of ransomware in cities all over the US [19]. Edge cloud has grown to be more and more important. As edge layer typically requires mostly remote access, and here attackers get an access path into the network. Furthermore, while surfing Shodan (a search engine), for every device connected through IoT, it turns out to be upfront to treasure edge endpoint IP addresses. In addition to actively hijacking IoT gadgets or gateway terminals to get exposure to the smart farm network, there is a potential that an indirect assault could compromise unrelated parties. Smart farms, for instance, commonly use third-party agronomy analytics to analyze the collected data, which may be used in a range of research domains including crop physiology and genetics, agriculture economy, supply forecasts, and prediction of diseases [16]. Such organizations may have been given immediate access to edge-based smart farm data for real-time analytics. If an attacker phishes for them, hacks their systems, and injects harmful programs, they run the risk of having the data sent to these third parties being diverted to third-party servers for theft. Since the attackers are utilizing real, stolen third-party credentials, these kinds of crimes are especially challenging to identify [18].

9.3.4 Logistics Network

The usage of digital technologies and IoT makes it feasible to connect various organizational contexts, increasing the logistics network's efficiency and integration. One of the most reliable strategies to generate long-term employment and value addition in rural regions is to adopt an integrated strategy for the entire agriculture/food logistics network [12]. The agriculture industry likewise deals with goods that frequently have a relatively limited preservation time, are produced in large quantities under uncertain conditions, and, in some cases, the availability of goods can be unpredictable. IoT technology at every stage of the supply chain raises serious cybersecurity risks while a privacy compromise in the immediate delivery network system might have substantial cascading repercussions on the whole logistics network [14]. Over 250,000 clients of US Foods who buy goods will be negatively impacted if a cyberattack or ransomware assault damages the company's IT infrastructure, disrupts its computer-aided facility, and affects crucial business processes [19]. A key aspect in preserving the product's freshness is the temperature and circumstances in which it is transported. Delays, temperature changes, and other environmental conditions have a significant impact on agricultural products like fruits, vegetables, meats, and dairy as they move from farms through processing and distribution facilities to consumers. A successful agricultural business depends heavily on the availability as well as the effectiveness of the logistics network, and supporting it with adaptable and competitive IoT technologies can assist in optimizing the extremely complex logistics network for agriculture. The use of such systems, it must be noted, exposes potential security risks at every point in the logistics network, making it difficult to optimize the logistics network using technology. A lot of data is exposed due to security flaws in the supplier, making it vulnerable to phishing scams and unauthorized use of privileged credentials [12]. The most significant vulnerabilities, however, are those that appear at the top of the logistics network, giving them the ability to spread to the remainder of the organizational processes through its reliant systems. As suggested in [2] security awareness, access control via authentication mechanisms, and cryptographic procedures (using, for example, blockchain technology) are some of the security measures that can help to prevent logistics network exposure and assault. Additionally, for products used in animal feed, if an IT system monitoring the feed's contents is compromised, it could result in wide-ranging contamination in livestock which rapidly spreads to citizens.

9.3.5 Side-Channel Attacks

Side-channel attacks refer to the kind of cyberattack where more emphasis is drawn on the data extraction while accessing or communicating through the system rather than the loopholes present in its architecture or functionality. In other words, side-channel attacks are the ones that get their data from the way a system functions instead of using a system's implementation's vulnerabilities [20]. The stealth channel, which is used for communicating clandestinely to evade middle-of-the-conversation listeners or remove information, is a notion similar to the side channel. In order to obtain valuable confidential data from the intended device's internal operations and functioning, side-channel and clandestine attacks make use of the physical qualities of the software, hardware, or transmission media. As it is already well recognized that smart farming depends heavily on cutting-edge technology like IoT, it will inevitably inherit IoT vulnerabilities like side-channel attacks [20]. With these sorts of attacks, attackers may employ a range of routes. Competitors may employ two attack vectors in timing channel attacks, such as processing time and the timing patterns of cache hits and misses. Other potential attack vectors include system clock period variations and hardware glitches in the form of voltage fluctuations. Sound and acoustic channels, power usage patterns, or potential electromagnetic breaches are additional ways to launch an effective invasion [21]. Side-channel attacks can be further classified into physical and functional side-channel attacks. A quantifiable number that results from the implementation serves as the foundation for the physical categorization. Examples include electromagnetic emission, power output, user interaction, the timing of clocks, and network inference (wired/wireless) for thermal, acoustic, and optical signals [21]. The functional type is based on how the computing system is operating and how internally implemented functions could result in data leaks. Examples include memory implementation, CPU/GPU design, and cryptography implementation/coding in software and firmware.

9.3.6 Denial-Of-Service Vulnerability Attack

The interoperability of IoT devices and their web services is essential to smart farming. The more extensive, efficient, and legitimate the interconnection, as well as the connection's effectiveness and economy, the better the IoT's level of service quality. There are many nodes and groups in the IoT, which could lead to denial of service

(DoS) assaults [22]. The DoS aims to take down the system, rendering it worthless. This can be accomplished via a gadget that transmits signals with nefarious purposes and can interfere with active IoT devices or disable them. DoS assaults are frequently unpredictable and highly challenging to stop, thus the best protection against them is to avoid them. The most recent DoS incident was in 2016 when a malware named Mirai was deployed as a botnet to take control of vast networks of CCTVs and routers that were targeted toward online customers' remotely operated Linux-based devices [1]. Another instance is a significant DDoS attack launched against GitHub in 2018 caused an unprecedented spike in traffic that reached 1.35 terabits per second. The frequency, power, and sophistication of DDoS attacks are unavoidably escalating. As a result of utilizing IoT to enhance farming and agriculture, the idea of "Smart Farming" also heavily relies on cloud computing, which allows attackers to exploit system weaknesses through DoS attacks [23]. Any new high-tech farm may find the consequences to be quite expensive. These attacks may result in tangible harm (the need to replace and re-program servers and sensors, reconfigure networks, and restart systems), as well as operational and monetary damage (service interruption, the need for fresh instructional materials for farm operators, and the creation of new metering infrastructures). Such attacks have the potential to interfere with the normal operation of various modules within a single farm, and they also have the potential to interfere with lawful cyber services in other domains [24].

9.3.7 Other Vulnerabilities

Passing misleading information using any kind of technology is also a kind of cyberattack. Putting data integrity in danger is the goal of this attack. An intruder might spread incorrect information related to the smart farm, claiming that crops or cattle are infected with a disease. These made-up data reports adopt the format of an actual report that the targeted smart farm has made public. Therefore, proving that the reported news is false will require a lot of time, effort, and money. Radiofrequency networks like cellular or satellite ones are widely used by intelligent farming equipment. Smart farming equipment typically uses global navigation satellite systems (GNSS) to boost productivity with tools and techniques like path seeding, spray rates, auto steering, and planning [1]. The GNSS system, which combines real-time kinematics (RTK) technology and GPS, improves the precision of real-time position data. Attackers may use a large number of scattered low-power jammers to disrupt GNSS for malicious intentions, preventing smart farming equipment from working as intended across a wide area [24]. Data about the consumption of agricultural products, such as the purchase of root vegetables, fruits, and cattle, as well as details on agricultural equipment, might be stolen by malware. It can also recruit technologically advanced devices to create a botnet that would be used to execute malicious acts under the command of an attacker. Smart farm devices are not typically constructed with security in mind, and even when they are, users frequently skip the necessary setup procedures for a sufficient cybersecurity defense mechanism. In Table 9.1, various cyber security threats and vulnerabilities are presented which should be considered to safeguard the interests of farmers, agronomists, farms, and the food sector [23].

TABLE 9.1
Different Cyber Security Threats to Smart Farming and Food Sector

Threats	Layer	Type of Attack	Elucidation	Consequences on DigAg
Insider Data Leakage	User level	Data attacks	Fear leakage of confidential data by an insider	Farmers are concerned that sensitive information may be leaked since it may be used against them in the commodity market
Cloud Data Leakage	User level	Data attacks	Data is in distributed form at cloud data centers so data might be less secure	Data is stored on clouds which can be placed in any country increasing vulnerabilities
False Data Injection Attack	User level	Data attacks	Fabrication and falsification of data to harm the interest of others	False information about the soil's moisture content will cause overwatering, which will harm the crops
Misinformation Attack	User level	Data attacks	The aim is to endanger data integrity	Broadcast fabricated information about a smart farm that indicates a disease epidemic in crops or livestock
Radio Frequency (Fr) Jamming Attack	Network level	Networking and equipment attacks	Attackers may disable GNSS (global positioning satellite systems) in order to harm	Attackers may use several dispersed low-power jammers to disrupt GNSS for malicious purposes to disable smart farming equipment over a large area.
Denial of Service Attack	Network level	Networking and equipment attacks	Potential vulnerabilities can be used to launch a variety of attacks by sniffing out and gaining access to data	Utilizing the ability to interfere with a farm's usual operations to interfere with legitimate cyber services in other areas
Botnet	Network level	Networking and equipment attacks	Manipulating and infecting every networked device using a central "Botnet" of devices	A farm full of infected zombies IoT devices can be used to quickly infect numerous additional networks using a variety of methods
Side-Channel Attack	Physical level	Networking and equipment attacks	Attacks related to the implementation of the system like "timing channel attack"	Hardware glitching in the form of voltage variations and variations in system clock period during execution tasks, power consumption patterns, potentially leaked electromagnetic energy, or even sound and acoustic channels

(Continued)

TABLE 9.1 (*Continued*)
Different Cyber Security Threats to Smart Farming and Food Sector

Threats	Layer	Type of Attack	Elucidation	Consequences on DigAg
Supply Chain Attacks	Physical level	Supply chain attacks	A security breach in a just-in-time distribution system	The design and operation of DigiAg products and applications should consider security vulnerabilities caused by the obtaining of hardware, software, and services from numerous vendors (both internationally and domestically)
Malware Injection Attack	Cloud storage	Malware attack	Injects malware like ransomware into a connected smart device	Information regarding the procurement of fruits, vegetables, cattle, and other agricultural products, as well as data on farm machinery, can all be stolen by malware
Third-Party Attack	Cloud storage	Malware attack	Malicious script is injected into the application by an outsider	Smart farms frequently employ third-party agronomy analytics to examine the gathered data could be redirected to other servers for theft

9.4 SAFEGUARDING THE FUTURE OF THE DIGITAL AGRICULTURE

Numerous equipment and sensors are customized for use in agriculture even though they are not specifically designed for DigAg applications. The modification is typically focused on operation in a harsh, uncontrolled outside environment. Device security is given little consideration. Security is frequently given the lowest priority, similar to how other technologies are, as opposed to being integrated into the design process. For digital agriculture, improved vulnerability detection/prevention mechanisms have to be created. Blockchain technology and AI at the edge of computers would also be beneficial in reducing some of the current assaults. Edge-based IDS solutions for digital agriculture require a significant amount of labor to deploy. As there is now no explicit security framework in place, DigiAg needs a better and upgraded cyber-security framework that takes into account its needs and vulnerabilities [25]. DigAg privacy-preserving techniques are a developing field. To safeguard the data from the malevolent user in all aspects, including security of data, analysis of data, data utility, and system-wide efficiency, innovative confidentiality-preserving strategies must be created specifically for digital agriculture. The use of new privacy-preserving techniques would reduce PII, commercially sensitive data, and IP theft. For varied applications, there are different DigAg systems and IT requirements. The attack surface is widened by the integration of hardware and software from various manufacturers into one system [25]. Before integrating the equipment, a comprehensive cyber threat and vulnerability analysis should be carried out, taking into account side-channel assaults, which are challenging to study and often not addressed in cyber-security frameworks. In the modern world, everybody must have complete information related to cyber threats and security attacks. One hundred people were selected for a poll on "Cyber Awareness," which was piloted to gauge people's comprehension of cyber threats and security attacks [26]. A questionnaire of 15 questions was offered out of which an average of only 25.26% was able to give correct answers as compared to an average of 75.4% incorrect answers.

The outcome of the poll is presented here, in Figure 9.5, through which it can be concluded that the majority of the people have petite information related to this issue and many are unaware of potential cyber hazards [27. As a result, cybercriminals are grabbing the opportunity to fulfill their evil objectives, due to this void. By raising awareness, providing appropriate training, implementing security tools (firewalls, antivirus software, etc.), and making sure the end users are physically aware of anomalies, many of the risks may be avoided. This magnifies the impact of cybersecurity on the future of agriculture and the food industry, especially when taking into account Smart Farming. Everyone involved in the process must understand how

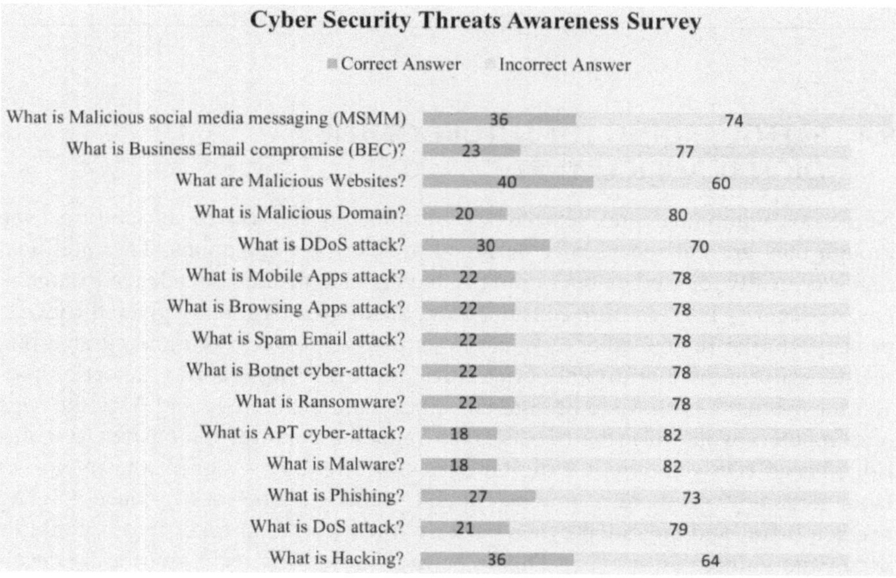

FIGURE 9.5 Result of a survey on "Cyber Security Threats Awareness".

smart farming overcomes new challenges, especially those relating to technology and cyber security. Currently, cybersecurity procedures are a problem. The need to protect all ICT systems will spur the creation of novel, smart agribusiness structures that are more robust and secure across their value chain.

9.5 CONCLUSION

Gradually technology is spreading its wings in the agriculture sector which is the backbone of every economy in the world. Precision or digital agriculture has uplifted the traditional format of farming and food logistics networks. IoT and ICT devices, as stated in the chapter, have established a foundational stone to modify the historically negative experiences of traditional agriculture and the food business, where one must rely on manual inspections and processes. The chapter mentions several smart systems that were created especially for farming and have increased productivity and efficiency. As food logistics networks and agriculture are advancing toward technology, the risk of cyberattacks is also growing. It has also been noted in this chapter that digital literacy is essential to combat such threats as cybercrimes and threats are not limited only to technology but also depend on how the user who's on the verge of such attacks is operating it. In developing nations like India where, according to the Economic Survey 2022–23, 47% of the population of India depends on agriculture for a living, while 65% of the population lives in rural areas, it becomes necessary to adopt smart farming with adequate data security and protection from cybercrimes.

REFERENCES

1. Gupta, M., Abdelsalam, M., Khorsandroo, S., & Mittal, S. (2020). Security and privacy in smart farming: Challenges and opportunities. *IEEE Access*, *8*. https://doi.org/10.1109/ACCESS.2020.2975142
2. Barreto, L., & Amaral, A. (2018). Smart Farming: Cyber Security Challenges. *9th International Conference on Intelligent Systems 2018: Theory, Research, and Innovation in Applications, IS 2018 - Proceedings*. https://doi.org/10.1109/IS.2018.8710531
3. Zahidi, S. (2018). The Global Risks Report 2022, 17th Edition. Available online: https://www3.weforum.org/docs/WEF_The_Global_Risks_Report_2022.pdf
4. Alahmadi, A. N., Rehman, S. U., Alhazmi, H. S., Glynn, D. G., Shoaib, H., & Solé, P. (2022). Cyber-security threats and side-channel attacks for digital agriculture. *Sensors*, *22*(9). https://doi.org/10.3390/s22093520
5. Yazdinejad, A., Zolfaghari, B., Azmoodeh, A., Dehghantanha, A., Karimipour, H., Fraser, E., Green, A. G., Russell, C., & Duncan, E. (2021). A review on security of smart farming and precision agriculture: Security aspects, attacks, threats, and countermeasures. *Applied Sciences (Switzerland)*, *11*(16). https://doi.org/10.3390/app11167518
6. Daum, T. (2018). ICT applications in agriculture. In *Encyclopedia of Food Security and Sustainability*. https://doi.org/10.1016/B978-0-08-100596-5.22591-2
7. Katta, S., Ramatenki, S., & Sammeta, H. (2022). Smart irrigation and crop security in agriculture using IoT. In *AI, Edge and IoT-Based Smart Agriculture*. Academic Press: Cambridge, MA.

8. Basso, B., & Antle, J. (2020). Digital agriculture to design sustainable agricultural systems. In *Nature Sustainability* (Vol. 3, Issue 4). https://doi.org/10.1038/s41893-020-0510-0
9. Shyamsunder, M., & Mohan, V. K. (2021). IoT-based soil quality monitoring for an efficient irrigation. *Alinteri Journal of Agriculture Sciences, 36*(2). DOI:10.47059/alinteri/V36I2/AJAS21110
10. Awasthi, A., Reddy, S. R. N., & Reddy, S. R. N. (2013). Monitoring for precision agriculture using wireless sensor network—A review. *Global Journal of Computer Science and Technology, 13*(7).
11. Fahmi, N., Prayitno, E., & Fitriani, S. (2019). Web of thing application for monitoring precision agriculture using wireless sensor network. *JURNAL INFOTEL, 11*(1). https://doi.org/10.20895/infotel.v11i1.421
12. Godfray, H. C. J., Beddington, J. R., Crute, I. R., Haddad, L., Lawrence, D., Muir, J. F., Pretty, J., Robinson, S., Thomas, S. M., & Toulmin, C. (2010). Food security: The challenge of feeding 9 billion people. In *Science* (Vol. 327, Issue 5967). https://doi.org/10.1126/science.1185383
13. Olcott, J. (2016). Input to the Commission on Enhancing National Cybersecurity: The Impact of Security Ratings on National Cybersecurity. Internet Security Alliance, Tech. Rep.,
14. Seselja, E. Cyber Attack Shuts Down Global Meat Processing Giant JBS. 2021. Available online: https://www.abc.net.au/news/2021-05-31/cyber-attack-shuts-down-global-meat-processing-giant-jbs/100178310
15. Rosline, G. J., Rani, P., & Gnana Rajesh, D. (2022). Comprehensive analysis on security threats prevalent in IoT-based smart farming systems. *Smart Innovation, Systems and Technologies, 243*. https://doi.org/10.1007/978-981-16-3675-2_13
16. Tudosa, I., Picariello, F., Balestrieri, E., de Vito, L., & Lamonaca, F. (2019). Hardware Security in IoT era: The Role of Measurements and Instrumentation. *2019 IEEE International Workshop on Metrology for Industry 4.0 and IoT, MetroInd 4.0 and IoT 2019 - Proceedings*. https://doi.org/10.1109/METROI4.2019.8792895
17. Hazrati, M., Dara, R., & Kaur, J. (2022). On-farm data security: Practical recommendations for securing farm data. *Frontiers in Sustainable Food Systems, 6*. https://doi.org/10.3389/fsufs.2022.884187
18. Chamarajnagar, R., & Ashok, A. (2019). Integrity Threat Identification for Distributed IoT in Precision Agriculture. *Annual IEEE Communications Society Conference on Sensor, Mesh, and Ad Hoc Communications and Networks Workshops, 2019-June*. https://doi.org/10.1109/SAHCN.2019.8824841
19. DHS. (2018). *US Department of Homeland Security: threats to precision agriculture*. Https://Www.Dhs.Gov/Sites/Default/Files/Publications/2018%20AEP_Threats_to_Precision_Agriculture.Pdf.
20. Devi, M., & Majumder, A. (2021). Side-channel attack in internet of things: A survey. In *Lecture Notes in Networks and Systems* (Vol. 137). https://doi.org/10.1007/978-981-15-6198-6_20
21. Randolph, M., & Diehl, W. (2020). Power side-channel attack analysis: A review of 20 years of study for the layman. In *Cryptography* (Vol. 4, Issue 2). https://doi.org/10.3390/cryptography4020015
22. Demestichas, K., Peppes, N., & Alexakis, T. (2020). Survey on security threats in agricultural IoT and smart farming. In *Sensors (Switzerland)* (Vol. 20, Issue 22). https://doi.org/10.3390/s20226458
23. Asif, M. R., Hasan, K. F., Islam, M. Z., & Khondoker, R. (2021). STRIDE-based Cyber Security Threat Modeling for IoT-enabled Precision Agriculture Systems. *2021 3rd International Conference on Sustainable Technologies for Industry 4.0, STI 2021*. https://doi.org/10.1109/STI53101.2021.9732597

24. Cupp, O. S., Walker, D. E., & Hillison, J. (2004). Agroterrorism in the U.S.: Key security challenge for the 21st century. In *Biosecurity and Bioterrorism : Biodefense Strategy, Practice, and Science* (Vol. 2, Issue 2). https://doi.org/10.1089/153871304323146397
25. Oerke, E. C., & Dehne, H. W. (2004). Safeguarding production – Losses in major crops and the role of crop protection. *Crop Protection*, *23*(4). https://doi.org/10.1016/j.cropro.2003.10.001
26. Alawida, M., Omolara, A. E., Abiodun, O. I., & Al-Rajab, M. (2022). A deeper look into cybersecurity issues in the wake of Covid-19: A survey. *Journal of King Saud University - Computer and Information Sciences*, *34*(10), 8176–8206. https://doi.org/10.1016/j.jksuci.2022.08.003
27. Nikander, J., Manninen, O., & Laajalahti, M. (2020). Requirements for cybersecurity in agricultural communication networks. *Computers and Electronics in Agriculture*, *179*. https://doi.org/10.1016/j.compag.2020.105776

10 Unveiling Critical Success Factors of Zero-Trust Cybersecurity Implementation in Food Industry from a Stage of Change Perspective

Pham Quang Huy and Vu Kien Phuc

10.1 INTRODUCTION

The fundamental focus of the diverse, intricate, and difficult food sector is the manufacturing of raw materials, semi-finished commodities, and final goods [1]. With consumers becoming more aware of what they consume, the food sector is becoming more competitive and dynamic. Modern consumers have high standards for food items, including, but not limited to, a wide range of flavors, minimal calorie count, affordability, and environmental friendliness. Food processing is growing more complex and harder in order to meet these consumer needs, requiring both changes to current food processing methods and the introduction of fresh, cutting-edge processing technology. To improve or replace outdated food processing methods and produce higher-quality, more consumer-preferred foods, a variety of innovative technologies have been researched and developed.

Industry 4.0, or smart industrial, was established in 2011 by the German federal government to aid in shaping the industrial sector's future. All physical assets must be fully digitalized as part of smart manufacturing, and partners from across the value chain must be integrated into these digital ecosystems. Through the use of digital technology, a network of stakeholders can work together in meaningful ways to achieve common goals and address common problems to create ecosystems. The proliferation of cyber threats is directly proportional to the rate of technological advancement. Ransomware, denial of service attacks, and malicious phishing attempts are the most common types of assaults carried out by hackers. Attacks are carried out to target every conceivable institution that is vulnerable, regardless of the size of the organization [2]. In addition to degrading production line efficiency or product quality, assaults on smart manufacturing systems could lead to economic collapse, injuries, or even human deaths—all of which would have far-reaching consequences for the country as a whole. As such, perspicuous criteria for data integrity and security are needed in

DOI: 10.1201/9781003409571-10

order to safeguard the firm and ensure digital trust [3]. Instead of becoming a fundamental component of smart manufacturing technology, cybersecurity is still seen as an add-on. Everyone in an organization, including contractors, should be cognizant of the significance of cybersecurity; it shouldn't be limited to the information technology department. Good cyber hygiene requires regular policy reviews. Since no two users are the same, a zero-trust security strategy is essential [2].

The goal of the zero-trust security strategy is to make it impossible for a company's internal and external users, apps, and infrastructures to trust one another. Instead of concentrating only on a certain technology, this methodology takes into account the full ecosystem of controls. Nevertheless, switching to implement zero trust might be difficult and time-consuming. When transitioning to a zero-trust ecosystem, the enterprise may face numerous obstacles and challenges. Despite the fact that the critical success factors (CSFs) of conventional security are no longer relevant for modern zero-trust cybersecurity, the body of existing literature implies that there has not been much academic research on CSFs [4]. Exploring the different elements influencing new technologies' industrial adoption is essential to maximizing their technological and economic potential as they are developed at an increasing rate.

Organizational adoption of zero-trust cybersecurity may be shifting. Therefore, categorizing organizations according to their stages is vital. Finding out how CSFs affect zero-trust cybersecurity implementation (ZTCI) and how to mitigate their effects is the driving force for this study. Considering this, the present research intends to fill a vacuum in readers' understanding by offering theoretical models that analyze the CSFs of ZTCI in the beverage and food business using the transtheoretical model (TTM) and the Capability, Opportunity, Motivation, and Behaviour (COM-B) model. In order to achieve this goal, the organizations are categorized into many stages according to TTM and COM-B: precontemplation, contemplation, preparation, action, and maintenance. This study developed three research questions to help direct the investigation and ensure that the research results could fill in certain gaps in the literature:

RQ1. What are the CSFs of ZTCI?
RQ2. How do these CSFs impact the implementation of ZTCI at different stages?
RQ3. Does ZTCI impact cybersecurity risk management (CRM)? How much does it impact CRM?

As for the research's theoretical importance, it fills a gap in the literature by conducting an empirical examination of the CSFs of ZTCI and, more especially, sheds light on the food and beverage sectors' little-known difficulties. This study uses a repeated cross-sectional design to address some of the limitations of previous studies that relied only on this type of research. As the most often used technique used during one-time data gathering, this type of data collecting has received widespread criticism from academics researching consumer behavior. This is due to the fact that collecting data just once provides a static picture of consumer sentiment and cannot capture the complex and ever-changing nature of a market. By empirically confirming the TTM and COM-B, this work specifically broadens the scope of ZTCI. The effects of CSFs on ZTCI at various phases are highlighted in this study. These results contribute to our understanding of ZTCI in the food and beverage industry by highlighting the importance of both organizational (OR) and environmental readiness (ER). It is worth mentioning that this study also confirms the impact of ZTCI on CRM. The results of this study add to the growing body of evidence showing that CRM is an effective indicator of ZTCI. As a result, this work enlarges the body of knowledge by supporting CRM as a fresh perspective on ZTCI research.

Since zero trust is a complex process with many moving parts, security leaders and professionals are either unsure of how to begin making it a reality or are worried about the necessary architectural and strategic changes [4]. Executives should be aware of the skills and expertise needed to implement data. Pertaining to practical significance, the research findings and insights obtained from this study can be useful to practitioners looking for ZTCI solutions before making decisions. The document gives managers in food and beverage businesses a comprehensive focus on OR management and operational hints, as well as a roadmap on the ZTCI. From a policymaking perspective, this research suggests that preliminary policy measures be taken to make decisions on ZTCI and CRM and to create future action plans. Additionally, the observations may provide policymakers with fresher perspectives to develop focused policies to improve and strengthen OR and ER, which can produce productive benefits over the long term.

The following sections make up the present manuscript and help to accomplish the study objectives stated above. The study's hypotheses are developed in Section 10.3, while the conceptual foundations are concisely summarized in Section 10.2. The methods and materials of the study are discussed in Section 10.4 based on these viewpoints. Section 10.6 discusses the consequences and potential future study paths, whereas Section 10.5 details the research analyses.

10.2 THEORETICAL UNDERSTANDING AND FOUNDATION

10.2.1 Transtheoretical Model

According to [5], adopting new habits should be seen as a progressive process to boost their likelihood of success and decrease resistance to existing ones. The TTM was selected for the current investigation due to its diversity in comparison to other theoretical approaches that try to influence behavior. To put it simply, in order to promote new behavior, diverse interventions should be targeted at different phases. The five stages of change that organizations go through—precontemplation (not contemplating change), contemplation (considering change), preparation (getting ready to change), action (trying a new behavior), and maintenance (maintaining the new habit)—formulate the core of the framework.

10.2.2 Conceptual Respects

Zero-trust cybersecurity. To better protect assets, data, systems, and people from threats that may evolve over time, zero-trust cybersecurity moves away from a focus on physical locations and toward a focus on data, according to [4]. According to zero-trust security, it is different from traditional perimeter-focused security in that it treats the organization's IT network as completely untrusted to the internet. One method for securing networks is the zero-trust security architecture, which starts with the premise that any device, user, or application could be malicious. Despite the fact that zero trust has not been defined by anyone, it is commonly known as a security model that acknowledges that firewalls do not necessarily protect all sensitive data and that trust must be established each time access control is used, without relying on assumptions made in the past.

CRM. Proceeding from the viewpoints put forth by [6], cybersecurity is all about protecting computing resources, data, and their confidentiality, integrity, and availability. It involves implementing rules, technologies, and strategies consistently to prevent attacks. Organizational cybersecurity, in a word, is the practice of safeguarding an organization's assets from potential dangers, both within and outside the company. Meanwhile, there are five key steps to risk management: identifying the risks, analyzing them, measuring them, dealing with them, and finally, monitoring and controlling them. Additionally, it stresses the importance of organizational coordination in promoting and controlling risk-related operations, as well as the interplay between organizations and their external environments. CRM is about much more than simply enhanced internal controls; it incorporates many other elements, such as strategy, information technology governance, managerial choices, employee actions, crisis resilience/business continuity, technological fixes, and real implementation and procedures.

10.3 SUBSTANTIATION OF RESEARCH HYPOTHESES

According to past studies, a variety of aspects, including the organizational aspect, have been found. Employee participation has a big impact on organizational aspects [7]. Several studies have reached the same conclusion that the availability of skilled personnel affects an organization's readiness to implement new technologies. On the other hand, growing competence and knowledge allow users of technology to make use of it more successfully. Because the transition to zero trust can expose vulnerabilities in older systems, apps, or infrastructure, it's possible that more resources will need to be meticulously controlled than first planned. Additionally, it might take longer and cost more money if these systems need to be redesigned or modified to enable zero trust. As a result, in order to integrate new technology, the corporation is required to have access to specialists and financial resources. As a result, organizations that invest more money in implementing digital technology are more likely to prosper [8]. Following the consideration of the points made earlier, the first hypothesis is proposed in this light.

Hypothesis 1 (H1)*. OR instigates a substantial and positive effect on ZTCI.*

The operating environment of an organization consists of all pertinent elements, including laws and regulations. The environmental context is the external social and technological support that has an impact on how advanced technology is implemented in companies. As of right now, no industry standard offers specific instructions on how to adopt zero trust in the organizational context. Due to this, it is difficult to determine whether the zero trust has been successfully applied in their businesses [9]. Additionally, several zero-trust platform components need to be standardized. Institutional readiness and ecosystem actor readiness to implement zero trust are indicators of environmental conditions in the current study. A flourishing digital economy cannot be achieved without strong legislative and policy frameworks pertaining to information technology [10]. Government regulations are essential for the widespread use and adoption of digital technologies. As a result, incentives and restrictions are part of government policy. Market constraints by governments can force the adoption of digital technology. A wide variety of actors outside of dyadic relationships are now increasingly realizing the manufacture and distribution of goods and services, thanks to technological developments. A client or employee can successfully navigate the ZTCI process thanks to the actor ecosystem readiness. Following the consideration of the points made earlier, the second hypothesis is proposed in this light.

Hypothesis 2 (H2)*. ER instigates a substantial and positive effect on ZTCI.*

Its prospective clients are aware that zero trust is a multi-tiered strategy that underlines the significance of always confirming all access to resources and never trusting anybody. As a result, the prospects of zero trust see it as an effective method for improving organizational network security. Zero trust's implementation secures access to the company's resources and increases security across the entire digital attack surface. Zero trust is emphasized as a paradigm resting on the tenets of no supposed trust, a risk-based approach to trust, and continual trust verification, building on the viewpoints of [11]. It specifically restricts each subject's ability to access the entire network. Zero trust raises a company's network visibility. As a result, it stimulates observation and the identification of suspicious traffic, in contrast to standard security procedures. The capacity to react swiftly to any hostile attack is enhanced by the possibility of moving less laterally. Zero trust also facilitates the transition to cloud computing and remote work for organizations. Ref. [12] suggests zero trust as a way to secure data stored in the cloud. Zero trust deployment in VPPs as a means to fortify security and shield users' personal information. Following the consideration of the points made earlier, the third hypothesis is proposed in this light.

Hypothesis 3 (H3)*. ZTCI instigates a substantial and positive effect on CRM.*

Figure 10.1 outlines the conceptual framework and underlying hypotheses, which are essential for fully understanding the cause-effect relationships among dimensions.

10.4 RESEARCH METHODOLOGY

10.4.1 Measured Variables Operationalization

The study uses an explanatory research methodology to look into the food and beverage companies in Vietnam. This strategy involves using quantitative techniques to look into a phenomenon that has not yet been adequately described. Quantitative analysis is done on the key success element for implementing zero-trust cybersecurity.

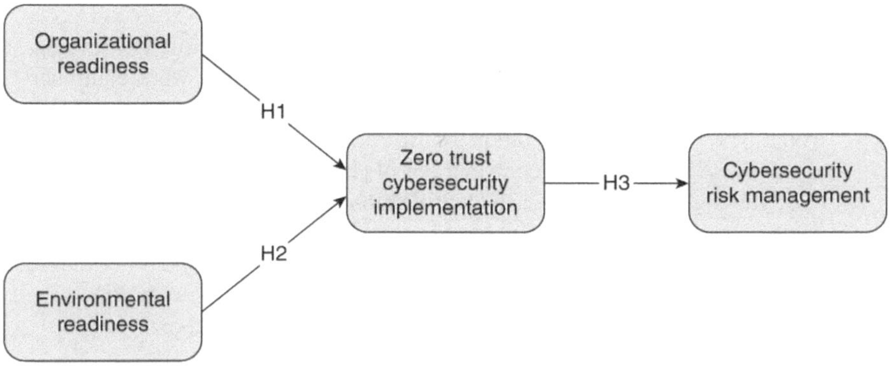

FIGURE 10.1 Hypothesized model.

To compile information for the suggested model's many paradigms, this study used a combination of a survey and a literature review. To meet the research setting and create a more focused questionnaire, this chapter modifies some measurements from the previous literature. The 5-point Likert scale is utilized in the current research. The questionnaire was properly created, including the order of the questions, the appearance, and the wording, to assure the representativeness of the participants. Five academics from academic institutions and five professionals from the accounting and information technology systems areas pre-tested the questionnaire to ensure its consistency and accuracy. Following a review of the instrument's face and content validity, minor changes were made to the arrangement and language of the constructions' elements. Thirty people participated in the pilot study; two people from each of the 15 companies were included. For each construct, the results produced a respectable level of Cronbach's alpha (>0.70). After the pilot test had been conducted for two weeks, the final survey was given:

ER. The first construct of ER is formulated by two second-order constructs, namely, institutional readiness and ecosystem actor readiness. More concretely, the criteria employed to evaluate institutional readiness are designed from those propounded by [13]. The criteria employed to evaluate ecosystem actor readiness are designed from those propounded by [14].

OR. The first construct of OR is formulated by two second-order constructs, namely, technology readiness, human resource readiness, and financial readiness. More concretely, the criteria employed to evaluate technology readiness are designed from those propounded by [13]. The criteria employed to evaluate financial readiness are designed from those propounded by [15]. The criteria employed to evaluate human resource readiness are designed from those propounded by [16].

ZTCI. The first construct of ZTCI featured a three-item scale that was modified based on findings from [17].

CRM. The first-order construct for defining CRM includes five second-order constructs—risk management environment, risk identification, risk evaluation, risk response, and risk monitoring and review—that were combined from the observations of [18, 19]. More concretely, the contributions of [18] as well as [19] formed the foundation for the criteria used to assess the risk management environment. The contributions of [19, 20] led to the development of the criteria used to evaluate risk identification. The measurement scales used to evaluate risk evaluation were based on the work of [19, 20]. While the contribution of the work of [19] established the cornerstone for the criteria used to assess the risk response, the obtained findings of [18, 19] constituted the foundation for the criteria used to assess risk monitoring and review.

10.4.2 Procedure for Sampling and Data Collection

The investigation population is paramount in the study to enhance reliability and validity. In this regard, the accountants and information technology staff members, managers of the food and beverage companies as well as their stakeholders were surveyed using a structured questionnaire. In order to collect information from the participants, researchers used two sampling methods: the convenience and the snowball. As a result, the information was gathered by conveniently contacting a small set of people. The qualified people were subsequently requested to extend invitations to others. In order to define and compare long-term perspectives on ZTCI, many rounds of data collecting and analysis were conducted. Each of the five data collection waves for this study lasted for seven months over the course of five years. This strategy might enable this research to get agreement on empirical findings in the collected data. Proposed model's correlation study between the time periods to ascertain the CSFs of ZTCI between each stage of change perspective. The cost of data collection and the requirement for enough statistical power to support the conceptual framework were taken into consideration while determining the sample size for this investigation.

Expanding on the views expressed by [21], the unknown population calls for a scale with 5–10 times the number of elements. Consequently, it was determined that a sample size of 400 replies was adequate. Beginning in the middle of December 2020 and lasting through the middle of May 2021, the first wave of data gathering was conducted. The second round of data gathering started from the middle of July 2021 and lasting through the middle of December 2021. The telephone interview was employed in these first two grounds of data collection. The third round of data gathering started from the middle of February 2022 and lasting through the middle of July 2022. The fourth round of data gathering started from the middle of September 2022 and lasting through the middle of January 2023. The fifth round of data gathering started from the middle of March 2023 and lasting through the beginning of August 2023. The chapter-based survey was employed in the next three grounds of data collection. The valid responses for the first, second, third, fourth, and fifth waves of database collection periods were 623, 623, 783, 812, and 812, respectively, after the removal of responses that were deemed unsuitable due to a significant quantity of missing data and the straight-lining of the responses. In this study, AMOS version 29.0 was utilized to conduct an analysis of the statistical data that was gathered in five waves. The sociodemographic profile of individuals who participated in the survey is presented in Table 10.1.

TABLE 10.1
Demographic Characteristics of Survey Respondents

Demographic Profile	Model 1 Precontemplation		Model 2 Contemplation		Model 3 Preparation		Model 4 Action		Model 5 Maintenance	
	Frequency	Percentage	Frequency	Percentage	Frequency	Percentage	Frequency	Percentage	Frequency	Percentage
Gender										
Male	498	79.94	502	80.58	419	53.51	595	73.28	568	69.95
Female	125	20.06	121	19.42	364	46.49	217	26.72	244	30.05
Age										
25 to under 35	16	2.57	19	3.05	15	1.92	32	3.94	24	2.96
35 to under 45	339	54.41	351	56.34	383	48.91	498	61.33	527	64.90
45 to under 55	253	40.61	235	37.72	360	45.98	255	31.40	240	29.56
55 to over	15	2.41	18	2.89	25	3.19	27	3.33	21	2.59
Experience (Years)										
10 to below 20	62	9.95	65	10.43	54	6.90	87	10.71	85	10.47
20 to below 30	482	77.37	477	76.57	627	80.08	605	74.51	613	75.49
30 to over	79	12.68	81	13.00	102	13.03	120	14.78	114	14.04
Education										
Undergraduate	586	94.06	599	96.15	744	95.02	779	95.94	783	96.43
Postgraduate	37	5.94	24	3.85	39	4.98	33	4.06	29	3.57

10.5 RESULTS AND DISCUSSIONS

10.5.1 Measurement Model

The following fit indices were used to evaluate the model's fit, namely chi-square (χ^2)/degree of freedom (df), root mean square error of approximation (RMSEA), comparative fit index (CFI), Tucker-Lewis index (TLI), and goodness-of-fit index (GFI). The hypothetical model's GFIs were confirmed only when they met the following requirements. First, the χ^2/df must be lower than 3.0. Second, the CFI and TLI should be higher than 0.90, whilst GFI could fluctuate from 0.774 to 0.923. Third, the RMSEA must be less than 0.08. All of the recorded indices of the measurement model met the cut-off criteria suggested by previous studies.

The convergent validity of the measurement model was assessed using construct dependability, factor loadings, and average variances extracted (AVE). Alternatively, loadings for all variables should be more than 0.6 [21]. The AVE values were greater than the needed minimum value of 0.5 [21]. Finding out how consistent and error-free the items used to measure variables are at any one moment is what the reliability evaluation is all about. The constructs should be considered internally consistent if Cronbach's alpha is greater than 0.7 [21]. Composite reliability (CR) scores, which reveal the range of indications for the latent construct, were greater than the indicated minimum value of 0.7 [21]. Table 10.2 shows the details of these results and emphasizes the reliability, validity, and internal consistency of the measurement model.

Our discriminant validity test followed the Fornell-Larcker criterion. Accordingly, the discriminant test is valid if all other correlations in the table are less than the greatest correlation on each diagonal. To evaluate the discriminant validity of the square root of the AVE [22], recommend values that exceed the correlations between each pair of model constructs. It follows that the variables measured the study's constructs accurately and discriminately. Table 10.3 suggests discriminant validity was not an issue.

10.5.2 Structural Model Assessment

The following fit indices were used to evaluate the model's fit, namely χ^2/df, RMSEA, CFI, TLI, and GFI. The hypothetical model's GFIs were confirmed only when they met the following requirements. First, the χ^2/df must be lower than 3.0. Second, the CFI and TLI should be higher than 0.90, whilst GFI can fluctuate from 0.774 to 0.923. Third, the RMSEA must be less than 0.08. The measurement and structural models were substantiated to precisely suit the recorded data based on the results presented in Table 10.4, as all of the recorded indices appeared to meet the cut-off criteria suggested by earlier studies.

TABLE 10.2
Results Summary of Convergent Validity and Construct Reliability

	Model 1 Precontemplation				Model 2 Contemplation			
	Convergent Validity		Construct Reliability		Convergent Validity		Construct Reliability	
Construct	Factor Loadings Ranges	AVE	Cronbach's Alpha	Composite Reliability	Factor Loadings Ranges	AVE	Cronbach's Alpha	Composite Reliability
Environmental Readiness								
Institutional readiness	0.766–0.854	0.670	0.896	0.899	0.766–0.853	0.672	0.897	0.900
Ecosystem actor readiness	0.736–0.834	0.614	0.872	0.872	0.734–0.833	0.619	0.874	0.875
Organizational readiness								
Technology readiness	0.735–0.898	0.648	0.888	0.889	0.739–0.898	0.653	0.890	0.891
Financial readiness	0.800–0.848	0.665	0.897	0.897	0.800–0.853	0.667	0.898	0.898
Human resource readiness	0.701–0.770	0.520	0.819	0.819	0.702–0.771	0.522	0.820	0.820
Cybersecurity risk management								
Risk management environment	0.673–0.790	0.524	0.820	0.821	0.669–0.794	0.525	0.821	0.821
Risk identification	0.664–0.829	0.547	0.865	0.866	0.668–0.830	0.549	0.866	0.867
Risk evaluation	0.702–0.874	0.571	0.875	0.878	0.705–0.872	0.573	0.876	0.879
Risk response	0.638–0.783	0.529	0.856	0.857	0.638–0.779	0.529	0.856	0.857
Risk monitoring and review	0.688–0.782	0.532	0.857	0.858	0.686–0.791	0.534	0.858	0.859
Zero-trust cybersecurity implementation	0.785–0.874	0.675	0.868	0.869	0.781–0.873	0.675	0.868	0.869

(*Continued*)

TABLE 10.2 *(Continued)*
Results Summary of Convergent Validity and Construct Reliability

Construct	Model 3 Preparation				Model 4 Action			
	Convergent Validity		Construct Reliability		Convergent Validity		Construct Reliability	
	Factor Loadings Ranges	AVE	Cronbach's Alpha	Composite Reliability	Factor Loadings Ranges	AVE	Cronbach's Alpha	Composite Reliability
Environmental Readiness								
Institutional readiness	0.781–0.927	0.686	0.905	0.907	0.735–0.975	0.680	0.901	0.904
Ecosystem actor readiness	0.774–0.889	0.664	0.894	0.897	0.763–0.905	0.670	0.895	0.899
Organizational Readiness								
Technology readiness	0.739–0.886	0.628	0.878	0.88	0.727–0.812	0.594	0.86	0.862
Financial readiness	0.779–0.849	0.667	0.898	0.898	0.754–0.878	0.667	0.897	0.898
Human resource readiness	0.722–0.813	0.566	0.846	0.847	0.727–0.833	0.606	0.866	0.868
Cybersecurity Risk Management								
Risk management environment	0.725–0.824	0.596	0.862	0.863	0.733–0.889	0.655	0.892	0.892
Risk identification	0.731–0.813	0.565	0.874	0.875	0.723–0.846	0.602	0.888	0.892
Risk evaluation	0.728–0.871	0.603	0.892	0.893	0.743–0.916	0.639	0.906	0.908
Risk response	0.733–0.845	0.596	0.889	0.89	0.721–0.872	0.646	0.909	0.911
Risk monitoring and review	0.762–0.801	0.600	0.891	0.892	0.787–0.818	0.633	0.904	0.906
Zero-Trust Cybersecurity Implementation	0.810–0.877	0.711	0.887	0.889	0.786–0.900	0.732	0.895	0.900

(Continued)

TABLE 10.2 *(Continued)*
Results Summary of Convergent Validity and Construct Reliability

Construct	Model 5 Maintenance			
	Convergent Validity		Construct Reliability	
	Factor Loadings Ranges	AVE	Cronbach's Alpha	Composite Reliability
Environmental Readiness				
Institutional readiness	0.736–0.975	0.681	0.901	0.904
Ecosystem actor readiness	0.763–0.905	0.670	0.895	0.899
Organizational Readiness				
Technology readiness	0.731–0.808	0.592	0.860	0.861
Financial readiness	0.765–0.858	0.656	0.892	0.893
Human resource readiness	0.728–0.834	0.606	0.866	0.868
Cybersecurity Risk Management				
Risk management environment	0.733–0.890	0.655	0.892	0.892
Risk identification	0.722–0.847	0.602	0.888	0.892
Risk evaluation	0.742 -0.916	0.639	0.906	0.908
Risk response	0.721–0.873	0.646	0.909	0.911
Risk monitoring and review	0.787–0.817	0.633	0.904	0.906
Zero-trust cybersecurity implementation	0.792–0.906	0.732	0.895	0.899

TABLE 10.3
Results Summary of Discriminant Validity Using Fornell-Larcker Process

Model 1 Precontemplation

	RE	RMR	RR	RI	IR	FIR	EAR	TER	RME	HRR	ZTCI
RE	1										
RMR	0.221	1									
RR	0.376	0.222	1								
RI	0.191	0.204	0.259	1							
IR	0.798	0.225	0.335	0.193	1						
FIR	0.258	0.575	0.295	0.295	0.328	1					
EAR	0.266	0.220	0.299	0.131	0.254	0.230	1				
TER	0.325	0.270	0.372	0.310	0.287	0.287	0.307	1			
RME	0.357	0.207	0.223	0.242	0.308	0.218	0.177	0.314	1		
HRR	0.250	0.132	0.320	0.237	0.281	0.215	0.217	0.240	0.216	1	
ZTCI	0.546	0.294	0.509	0.310	0.503	0.285	0.329	0.409	0.182	0.345	1

Model 2 Contemplation

	RE	RI	FIR	IR	RR	RMR	TER	EAR	HRR	RME	ZTCI
RE	1										
RI	0.223	1									
FIR	0.374	0.222	1								
IR	0.188	0.203	0.260	1							
RR	0.799	0.225	0.333	0.193	1						
RMR	0.260	0.575	0.295	0.297	0.330	1					
TER	0.268	0.218	0.302	0.130	0.254	0.231	1				
EAR	0.324	0.274	0.374	0.312	0.282	0.293	0.306	1			
HRR	0.353	0.206	0.225	0.240	0.308	0.216	0.179	0.315	1		
RME	0.253	0.134	0.317	0.237	0.280	0.222	0.222	0.242	0.214	1	
ZTCI	0.543	0.296	0.510	0.312	0.507	0.286	0.328	0.406	0.183	0.351	1

(Continued)

TABLE 10.3 (*Continued*)
Results Summary of Discriminant Validity Using Fornell-Larcker Process

Model 3
Preparation

	RE	RI	FIR	IR	RR	RMR	TER	EAR	HRR	RME	ZTCI
RE	1										
RI	0.263	1									
FIR	0.398	0.326	1								
IR	0.246	0.418	0.227	1							
RR	0.191	0.310	0.205	0.220	1						
RMR	0.331	0.296	0.272	0.167	0.260	1					
TER	0.183	0.219	0.309	0.147	0.356	0.345	1				
EAR	0.196	0.232	0.197	0.190	0.314	0.389	0.313	1			
HRR	0.157	0.230	0.147	0.168	0.191	0.209	0.179	0.090	1		
RME	0.339	0.270	0.375	0.268	0.217	0.218	0.217	0.182	0.119	1	
ZTCI	0.338	0.319	0.329	0.220	0.407	0.583	0.365	0.374	0.232	0.252	1

Model 4
Action

	RR	RE	RMR	RI	IR	FIR	EAR	RME	HRR	TER	ZTCI
RR	1										
RE	0.154	1									
RMR	0.321	0.223	1								
RI	0.296	0.293	0.218	1							
IR	0.212	0.239	0.285	0.167	1						
FIR	0.203	0.229	0.225	0.156	0.331	1					
EAR	0.210	0.124	0.180	0.130	0.347	0.261	1				
RME	0.208	0.190	0.324	0.138	0.284	0.174	0.212	1			
HRR	0.349	0.325	0.245	0.360	0.181	0.126	0.161	0.298	1		
TER	0.102	0.172	0.193	0.140	0.411	0.502	0.276	0.232	0.160	1	
ZTCI	0.197	0.271	0.343	0.173	0.510	0.557	0.313	0.413	0.271	0.521	1

Model 5
Maintenance

	RR	RE	RMR	RI	IR	EAR	RME	FIR	HRR	TER	ZTCI
RR	1										
RE	0.154	1									
RMR	0.321	0.223	1								
RI	0.296	0.293	0.218	1							
IR	0.212	0.239	0.285	0.167	1						
EAR	0.210	0.124	0.180	0.130	0.347	1					
RME	0.208	0.190	0.324	0.138	0.284	0.212	1				
FIR	0.200	0.185	0.206	0.135	0.271	0.251	0.150	1			
HRR	0.349	0.325	0.245	0.360	0.181	0.161	0.298	0.098	1		
TER	0.101	0.170	0.192	0.140	0.413	0.277	0.231	0.405	0.159	1	
ZTCI	0.197	0.271	0.343	0.173	0.510	0.313	0.413	0.412	0.272	0.520	1

Note: IR, Institutional readiness; EAR, Ecosystem actor readiness; TER, Technology readiness; FIR, Financial readiness; HRR, Human resource readiness; RME, Risk management environment; RI, Risk identification; RE, Risk evaluation; RR, Risk response; RMR, Risk monitoring and review; ZTCI, Zero-trust cybersecurity implementation.

TABLE 10.4
The Outcomes of Measurement and Structural Model Analysis

Model Fitting Index	Model Fitting Values	Standard Values of Measurement Model	Standard Values of Structural Model	Result Judgment
		Model 1 ***Precontemplation***		
Chi-square/df	< 3	1.548	1.618	Satisfied
TLI	≥ 0.9	0.967	0.963	Satisfied
CFI	≥ 0.9	0.970	0.965	Satisfied
GFI	0.774 ≤ x ≤ 0.923	0.923	0.917	Satisfied
RMSEA	< 0.08	0.026	0.028	Satisfied
		Model 2 ***Contemplation***		
Chi-square/df	< 3	1.600	1.701	Satisfied
TLI	≥ 0.9	0.950	0.942	Satisfied
CFI	≥ 0.9	0.955	0.945	Satisfied
GFI	0.774 ≤ x ≤ 0.923	0.907	0.898	Satisfied
RMSEA	< 0.07	0.031	0.034	Satisfied
		Model 3 ***Preparation***		
Chi-square/df	< 3	1.587	1.687	Satisfied
TLI	≥ 0.9	0.951	0.942	Satisfied
CFI	≥ 0.9	0.955	0.946	Satisfied
GFI	0.774 ≤ x ≤ 0.923	0.905	0.898	Satisfied
RMSEA	< 0.07	0.031	0.033	Satisfied
		Model 4 ***Action***		
Chi-square/df	< 3	1.782	1.875	Satisfied
TLI	≥ 0.9	0.959	0.954	Satisfied
CFI	≥ 0.9	0.963	0.956	Satisfied
GFI	0.774 ≤ x ≤ 0.923	0.916	0.909	Satisfied
RMSEA	< 0.07	0.031	0.033	Satisfied
		Model 5 ***Maintenance***		
Chi-square/df	< 3	1.780	1.868	Satisfied
TLI	≥ 0.9	0.959	0.954	Satisfied
CFI	≥ 0.9	0.962	0.956	Satisfied
GFI	0.774 ≤ x ≤ 0.923	0.916	0.909	Satisfied
RMSEA	< 0.07	0.031	0.033	Satisfied

The standardized path coefficients quantify the magnitude and direction of the causative variable's impact on the outcome variable in the proposed model and are presented in Table 10.5.

Concerning the precontemplation stage. The impact of OR (H1: $\beta = 0.563$; $p = 0.090$) demonstrates an unsubstantial interrelation with ZTCI. In the same vein, the impact of ER (H2: $\beta = 0.368$; $p = 0.114$) underscores an unsubstantial interconnection with ZTCI. The interrelation between ZTCI and CRM (H3: $\beta = 0.366$; $p = 0.058$) is unsubstantial. Therefore, H1–H3 are rejected. This is due to the fact that the majority of survey participants are approaching a scenario requiring a change while unaware. Regarding ZTCI, they have no plans.

Regarding the contemplation stage. The impact of OR (H1: $\beta = 0.577$; $p = 0.064$) demonstrates an unsubstantial association with ZTCI, whereas the impact of ER (H2: $\beta = 0.384$; $p = 0.015$) underlines a significant interconnection with ZTCI. The interrelation between ZTCI and CRM (H3: $\beta = 0.462$; $p = 0.083$) is unsubstantial. Therefore, H1 and H3 are rejected and H2 is buttressed. In this stage, respondents in the study start to notice changes in circumstances, make an effort to seek out additional information, and consider altering their conduct. While they would like to conduct the ZTCI, they do not take any concrete action.

Concerning to the preparation stage. The impact of OR (H1: $\beta = 0.583$; $p = 0.007$) demonstrates a substantial interrelation with ZTCI, and the impact of ER (H2: $\beta = 0.389$; $p = 0.014$) underlines a strictly substantial interconnection with ZTCI. The interrelation between ZTCI and CRM (H3: $\beta = 0.461$; $p = 0.012$) is staunchly reinforced. Therefore, H1–H3 are buttressed. In this stage, survey respondents make an effort to act even when they have not yet met the criteria for effective action. They are preparing for ZTCI.

In terms of the action stage. The impact of OR (H1: $\beta = 0.561$; $p = 0.000$) demonstrates a substantial interrelation with ZTCI, and the impact of ER (H2: $\beta = 0.389$; $p = 0.009$) underlines a strictly substantial interconnection with ZTCI. The interrelation between ZTCI and CRM (H3: $\beta = 0.500$; $p = 0.000$) is staunchly undergirded. Therefore, H1–H3 are buttressed. In this stage, survey respondents start taking action and look for assistance to improve the effectiveness of their actions. They start the ZTCI.

Pertaining to the maintenance stage. The impact of OR (H1: $\beta = 0.525$; $p = 0.000$) demonstrates a markedly substantial interrelation with ZTCI, and the impact of ER (H2: $\beta = 0.414$; $p = 0.006$) underlines a strictly substantial interconnection with ZTCI. The interrelation between ZTCI and CRM (H3: $\beta = 0.504$; $p = 0.000$) is staunchly undergirded. Therefore, H1–H3 are buttressed. In this stage, respondents to the poll begin to integrate their modifications. They intend to carry on with ZTCI in the future.

TABLE 10.5
Structural Coefficients (β) of the Hypothesized Model

Model 1 Precontemplation

Hypothesis No.	Hypothesized Path	Standardized	S.E.	C.R.	P-value	Inference
H1	OR → ZTCI	0.563	0.040	1.834	0.090	Rejected
H2	ER → ZTCI	0.368	0.021	1.903	0.114	Rejected
H3	ZTCI → CRM	0.366	0.242	1.472	0.058	Rejected

Model 2 Contemplation

Hypothesis No.	Hypothesized Path	Standardized	S.E.	C.R.	P-Value	Inference
H1	OR → ZTCI	0.577	0.076	1.764	0.064	Rejected
H2	ER → ZTCI	0.384	0.055	2.479	0.015	Buttressed
H3	ZTCI → CRM	0.462	0.261	1.557	0.083	Rejected

Model 3 Preparation

Hypothesis No.	Hypothesized Path	Standardized	S.E.	C.R.	P-value	Inference
H1	OR → ZTCI	0.583	0.079	2.707	0.007	Buttressed
H2	ER → ZTCI	0.389	0.058	2.461	0.014	Buttressed
H3	ZTCI → CRM	0.461	0.259	2.506	0.012	Buttressed

Model 4 Action

Hypothesis No.	Hypothesized Path	Standardized	S.E.	C.R.	P-Value	Inference
H1	OR → ZTCI	0.561	0.025	6.801	0.000	Buttressed
H2	ER → ZTCI	0.389	0.022	2.625	0.009	Buttressed
H3	ZTCI → CRM	0.500	0.378	5.502	0.000	Buttressed

Model 5 Maintenance

Hypothesis No.	Hypothesized Path	Standardized	S.E.	C.R.	P-Value	Inference
H1	OR → ZTCI	0.525	0.034	4.501	0.000	Buttressed
H2	ER → ZTCI	0.414	0.022	2.727	0.006	Buttressed
H3	ZTCI → CRM	0.504	0.378	5.573	0.000	Buttressed

10.6 CONSIDERATION AND AVENUES FOR FUTURE RESEARCH

10.6.1 Managerial and Policy Implications

Numerous managerial implications have been proposed since many important conclusions can be applied to the organization. First and foremost, founders and managers of food and beverage companies, especially information technology managers, should be very concerned about ZTCI's strategic role in those companies. ZTCI has the potential to reduce costs and revamp operations and competencies to enable novel flexible business paradigms. Second, food and beverage firms must accelerate ZTCI for sustainability. To do this, they must carefully evaluate and grasp ZTCI concepts, integrate and manage internal and external people, material, and financial resources, and design sensible ZTCI strategies. The senior management team must undertake ZTCI to continuously combine and reconstitute organizational talents and resources for the complex and unpredictable external environment. Improve existing capabilities by incremental changes and reformulate existing capacities through new operations. Helping food and beverage enterprises learn ZTCI implementation skills is important since financial readiness may affect their growth. Third, management teams should prioritize human aptitude development because it takes time and is essential to matching digital technology demands with available resources, especially human ones. Thus, food and beverage business owners and managers should adopt long-term goals, change their culture to be digital, relational, and creative, and give employee training. As essential components of organizational human resources, accountants have been encouraged to be proactive, stay up to date on new or improved skills and knowledge, and expand their competencies over a wide range of tasks by participating in appropriate training. Fourth, senior managers who make ZTCI decisions must have professional digital understanding, digital intelligence, and internal and external knowledge. This is because ZTCI would have long-term effects on the food and beverage sectors. Managers are encouraged to continue their education and receive job-specific training to improve their skills. Senior management must navigate multiple cultures and organizational frameworks. They must also balance short-term productivity with long-term adaptation and adapt their leadership style to diverse managers and their styles.

Government and universities delivering information technology training programs should be periodically and cautiously examined to ensure they are suitable for the food and beverage industry. Managers of food and beverage companies should also create a structure to encourage staff training. Government assistance helped the beverage and food industries progress toward ZTCI. Thus, the government should highlight ZTCI's long-term benefits for most social institutions. These benefits should be publicly promoted to persuade food and beverage companies to embrace ZTCI's potential. The government was asked to implement large-scale transformation policies to improve the food and beverage industries' digitalization capabilities, make sure these industries were aware of these policies, and set up service delivery and incentive channels. Identifying the populations most at risk of displacement and making plans to reintegrate them into the economy will help the government control and anticipate ZTCI's negative consequences.

10.6.2 Research Limitations and Agenda for Upcoming Studies

Caveats may open new research avenues. Vietnamese food and beverage companies are the only sample in this study. Follow-up research can include more firms and cultures. The conceptual model should be tested in countries with various technology infrastructures to promote universality. When applied to foreign food and beverage businesses, the findings must be carefully interpreted. Therefore, future studies should study the theorized link between other nations' food and beverage industries. The study's factors may be useful in different contexts. These extra elements may be intriguing for future empirical studies. Second, while the study's goals and objectives are satisfied and it uses quantitative research, a mixed-methods approach might yield better findings. The third constraint is the study's constrained dataset, despite the researchers' best efforts. The sample size is sufficient to establish the model's fit and proceed with statistical analysis, although larger samples could validate the model by providing more precise observations. Increasing the sample size would also illuminate variable relationships and support the results. Finally, individual-level data must illuminate organizational-level characteristics for this study to be valid. Surveys collect data efficiently and are a great starting point for further activities, notwithstanding their limited explanatory power. It is important to remember that survey data is collected individually and can be used to infer organization features using equivalence testing. The findings would be more reliable if all participants in all organizations provided feedback to boost validity and provide useful insight into the ZTCI's major success elements. Future studies should also gather data from diverse individuals.

ACKNOWLEDGMENT

This chapter was funded by the University of Economics Ho Chi Minh City (UEH) in Vietnam.

REFERENCES

1. M. E. Latino and M. Menegoli, "Cybersecurity in the food and beverage industry: A reference framework", *Computers in Industry*, vol. 141, pp. 1–16, 2022.
2. U. B. Chaudhry and A. K. M. Hydros, "Zero-trust-based security model against data breaches in the banking sector: A blockchain consensus algorithm", *IET Blockchain*, vol. 3, no. 2, pp. 98–115, 2023.
3. C. Buck, C. Olenberger, A. Schweizer, F. Völter and T. Eymann, "Never trust, always verify: A multivocal literature review on current knowledge and research gaps of zero-trust", *Computers & Security*, vol. 110, pp. 1–26, 2021.
4. W. Yeoh, M. Liu, M. Shore and F. Jiang, "Zero trust cybersecurity: Critical success factors and a maturity assessment framework", *Computers & Security*, vol. 133, pp. 1–13, 2023.
5. K. F. Yuen, Y. D. Wong, F. Ma and X. Wang, "The determinants of public acceptance of autonomous vehicles: An innovation diffusion perspective", *Journal of Cleaner Production*, vol. 270, pp. 1–44, 2020.
6. D. S. Berman, A. L. Buczak, J. S. Chavis and C. L. Corbett, "A survey of deep learning methods for cyber security", *Information*, vol. 10, no. 4, pp. 1–35, 2019.

7. M. N. Alraja, R. Imran, B. M. Khashab and M. Shah, "Sustainable green practices and SMEs sustainable performance in times of crisis (COVID-19 pandemic)", *Information Systems Frontiers*, vol. 24, pp. 1081–1105, 2022.
8. M. Nikopoulou, P. Kourouthanassis, G. Chasapi, A. Pateli and N. Mylonas, "Determinants of digital transformation in the hospitality industry: Technological, organizational, and environmental drivers", *Sustainability*, vol. 15, pp. 1–20, 2023.
9. B. Ali, S. Hijjawi, L. H. Campbell, M. A. Gregory and S. Li, "A maturity framework for zero-trust security in multiaccess edge computing", *Security and Communication Networks*, pp. 1–14, 2022.
10. A. Othman, A. Al Mutawaa, A. Al Tamimi and M. Al Mansouri, "Assessing the readiness of government and semi-government Institutions in Qatar for inclusive and sustainable ICT accessibility: Introducing the MARSAD tool", *Sustainability*, vol. 15, pp. 1–25, 2023.
11. A. Wylde, "Zero trust: never trust, always verify", In Proceeding 2021 International Conference on Cyber Situational Awareness, Data Analytics and Assessment (CyberSA). Dublin, Ireland, 12 July 2021, pp. 1–4, 2021.
12. S. Liu, Y. Zhuang, L. Huang and X. Zhou, "Exploiting LSB self-quantization for plaintext-related image encryption in the zero-trust cloud", *Journal of Information Security and Applications*, vol. 66, pp. 1–15, 2022.
13. Webster, A., and Gardner, J. (2019). Aligning technology and institutional readiness: the adoption of innovation. Technology Analysis & Strategic Management, 31(10), 1229–1241. https://doi.org/10.1080/09537325.2019.1601694.
14. Hein, A., Weking, J., Schreieck, M., Wiesche, M., Böhm, M., and Krcmar, H. (2019). Value co-creation practices in business-to-business platform ecosystems. Electronic Markets, 29, 503–518. https://doi.org/10.1007/s12525-019-00337-y.
15. Amarteifio, E. N. A., and Frimpong, S. (2019). Investment readiness and access to external finance among Ghanaian small and medium-size enterprises. International Journal of Entrepreneurship and Small Business, 37(2), 214. https://doi.org/10.1504/ijesb.2019.100104.
16. Vrchota, J., Maříková, M., Řehoř, P., Rolínek, L., and Toušek, R. (2019). Human Resources Readiness for Industry 4.0. Journal of Open Innovation: Technology, Market, and Complexity, 6(1), 1–20. https://doi.org/10.3390/joitmc6010003.
17. Yan, X., and Wang, H. (2020). Survey on Zero-Trust Network Security. In: Sun, X., Wang, J., Bertino, E. (eds) Artificial Intelligence and Security. ICAIS 2020. Communications in Computer and Information Science, vol 1252 (pp. 50–60). Springer, Singapore. https://doi.org/10.1007/978-981-15-8083-3_5.
18. B. Renault, J. Agumba and N. Ansary, "An exploratory factor analysis of risk management practices: A study among small and medium contractors in Gungormu", ActaStructilia, vol. 25, no. 1, pp. 1–39, 2018.
19. Goel, R., Kumar, A. and Haddow, J. (2020), "PRISM: a strategic decision framework for cybersecurity risk assessment", Information and Computer Security, Vol. 28 No. 4, pp. 591–625. https://doi.org/10.1108/ICS-11-2018-0131.
20. A. Karimi, N. Mousavi, S. Mousavi and S. Hosseini, "Risk assessment model selection in construction industry", *Expert Systems with Applications*, vol. 38, no. 2, pp. 9105–9111, 2011.
21. J. F. Hair, W. C. Black, B. J. Babin and R. E. Anderson, "*Multivariate Data Analysis*", 8th edn., Cengage, Hoboken, 2018.
22. C. Fornell and D. Larcker, "Evaluating structural equation models with unobservable variables and measurement error", *Journal of Marketing Research*, vol. 18, no. 1, pp. 39–50, 1981.

11 Understanding and Mitigating Cyber Threats in Smart Fruit Crops Safety

An Analysis of Vulnerability and Best Practices for Securing Agricultural Systems

Harish Chandra Verma and Mohammed Siddique

11.1 INTRODUCTION

The rapid development of technology and digitization, which has increased connectivity, productivity, and efficiency, has altered the agriculture/food industry. While these developments have enhanced the management and security of fruit orchards, they have also brought about new challenges in the form of cyber-attacks. Recent advancements in technology and digitization have increased productivity and efficiency in the global agricultural sector. Smart farming techniques, often known as precision agriculture [1, 2], are becoming more popular due to their ability to satisfy the world's growing demand for food. The significance of his new paradigm for conducting agriculture is indicated by the various smart farming use cases [3, 4] that are currently available globally. Utilizing big data analytics to offer helpful insights into the entire farming process has affected the entire food supply chain [5, 6]. One of the main worries that this shift to digital technology has given us is the safety and security of fruit harvests. As the globe grows increasingly connected, the agriculture industry has become a major target for cyber threats, necessitating a full awareness and mitigation of potential vulnerabilities. These cybersecurity factors are the main motivation behind the present study, in which authors examine the flaws in agricultural systems, particularly as they apply to fruit crops, and make recommendations for the most effective ways to strengthen them. This study aims to explore the urgent issue of cyber threats in fruit crop safety by evaluating the vulnerabilities that expose agricultural systems to hazards and offering best practices for securing these

 DOI: 10.1201/9781003409571-11

vital links in our food supply chain. Among the many benefits brought about by the swift uptake of digital technology in agriculture are precision farming, data-driven decision-making, and automated processes. But because of its greater reliance on digital technology and internet connectivity, it is now more open to cyber-attacks. Agriculture may become risky and unproductive as a result of potential attacks. For instance, abuses that can employ smart drones to overspray pesticides or obliterate all of the crops in a field, among other things, might lead to dangerous consumption and worsen the economy. Such concerted large-scale operations, sometimes known as agro-terrorism [7], can cause economic instability in a nation that depends heavily on agriculture.

Cyber risks to the security of fruit crops can come in a variety of shapes and sizes, including ransomware attacks, data breaches, and the obliteration of essential infrastructure. Hackers may target various elements of the agricultural ecosystem, including sensors, monitoring and surveillance technology, supply chain management software, and hardware. Successful cyber-attacks on fruit crops could end in catastrophe, losing farmers money, resulting in food shortages, and having a significant influence on food security. In order to address these new problems, agricultural stakeholders, governments, and researchers must collaborate to create efficient cybersecurity solutions. The shortcomings of fruit crop systems, such as insufficient security precautions, a lack of farmer awareness, and insufficient investments in cybersecurity infrastructure, will be examined in this inquiry. It will also highlight the best strategies and techniques that can be applied to properly protect agricultural systems.

In this chapter, the subject of cyber threats to fruit crop safety will be addressed, and crucial lessons about the possible risks faced by modern agriculture as well as the value of putting preventative security measures in place will be learned. Understanding the vulnerabilities and looking into workable solutions will help to create a more resilient and secure agricultural environment that upholds the safety of fruit crops and promotes the sustainability of food production for future generations.

11.2 LITERATURE REVIEW

The farming industry has become more sustainable and efficient due to its growing reliance on networked systems and technology. Fruit crops are now vulnerable to unprecedented cyber-attacks as a result of this digital revolution, putting both their security and the security of the global food supply at risk. In order to boost food production with fewer human resources and intervention, a variety of digital devices, including smartphones, various sensors, global position systems (GPSs), robotics, and drones, could be used to extract valuable data analysis and make successful decisions [8]. Cybersecurity concerns are raised by criminal acts that damage digital infrastructure, such as destroying or stealing data. According to [9], cybersecurity is a multidisciplinary field that includes biosecurity and cybersecurity. A growing number of agricultural systems are becoming connected to the internet as a result of the ongoing process of "digitalizing agriculture" [8]. Increased internet access makes it possible to increase productivity and enhances the farming system in other ways. Some of these systems, like the dairy industry's milking systems, are vital to the operation of the business. According to researchers in [10], these systems need to be operational at all times because any outage can quickly endanger the cattle. The agricultural industry has come to understand in recent years that internet-connected machinery exposes these systems to a variety of cyber threats [11–13]. An attack on the businesses that provide fertilizer and agricultural equipment could damage vital related machinery that is needed at a critical moment. It might alter the nutrient content of the fertilizers, which could badly harm the crops rather than help them [14]. Over the last several years, the agricultural industry has come to understand that when machinery is connected to the internet, it opens up to a multitude of cyber risks [11, 15]. Researchers in [16] collected 141 publications on agricultural cybersecurity on the Web of Science between 2008 and 2018, providing an overview. The bulk of the literature on agricultural cybersecurity that is now available focuses on external attacks on the agricultural IT infrastructure, sometimes even narrowing the focus to just the farm computers, or it addresses the issue from the perspective of the entire food supply chain. Among these are monitoring tasks, precision farming [17], automation, and crop management. The latter comprises monitoring or managing soil [18], weather, farms, pests, diseases, and irrigation and water quality [19, 20]. Thanks to sensors implanted in the ground to monitor blueberry irrigation, Chile has reduced the amount of water needed for cultivation by 70% by implementing smart irrigation [21]. Numerous agricultural goods are produced by farmers. Due to their small scale and low transaction volume, small farmers are excluded from the market by traditional e-commerce, which does not wish to service them. Transaction expenses can be significantly decreased and integrated into the market by using blockchain technology [22].

This chapter covers the following topics: the importance of cybersecurity for agriculture; agricultural systems' vulnerabilities; cyber threats that are particular to the safety of fruit crops; best practices for securing agricultural systems; and recommendations for safeguarding agricultural systems.

11.3 THE INCREASING IMPORTANCE OF CYBERSECURITY IN AGRICULTURE

The productivity and quality of fruit crops have increased as a result of the incorporation of digital technology in agriculture, including precision farming, intelligent irrigation, and remote monitoring. Agricultural systems are now more dependent on digital instruments, which have exposed them to cyber threats like data breaches, ransomware attacks, and interruptions of vital services. It is often known that as the volume and sophistication of cyber-attacks rise, so does the significance of cybersecurity. Effective cybersecurity is essential for defending against cyber-attacks against both individuals and companies. Identity theft, data breaches, and other online crimes can all be avoided with the use of cybersecurity.

11.4 VULNERABILITIES IN AGRICULTURAL SYSTEMS

There are several types of vulnerabilities in agriculture systems, and the possible causes of vulnerabilities in agricultural systems are as follows:

- *Out-Dated Infrastructure*: Many farms continue to run on out-dated or unlatched hardware and software, which leaves them open to known attacks.
- *Lack of Knowledge*: Farmers and agricultural professionals might not be fully aware of cyber dangers and the required security precautions to protect their systems.
- *Data Privacy Issues*: A lot of sensitive data, including customer information and information on crop yields, is gathered and stored by agricultural systems. Unauthorized access to sensitive data could have serious consequences. The ability of smart devices to communicate with one another is the most striking aspect of smart farming, producing an unprecedented volume of created data [21, 23]. The detection of insider information disclosure has always been a difficult issue because people often have valid access to the system, which makes it challenging to anticipate and recognize such attacks. There have been several studies on insider data leakage [24, 25], but none have focused on smart agricultural environments.
- *Supply Chain Risks*: Because of the agricultural supply chain's complexity, hackers have the chance to attack weak places in the network. The concept of "farm to plate" and the complete agricultural ecosystem involve some organizations that collaborate to just-in-time supply wholesome food to the ultimate consumer. In this supply chain topology, raw materials are produced at the farm [26]. The food business stores and processes these materials. The packaged processed food is then delivered to a store for distribution, where the final consumer purchases the processed foods. Since a security breach in the just-in-time distribution system could have major cascading implications on the whole supply chain, internet-of-things (IoT) technology adds potential cybersecurity concerns at each stage of the chain of supplies.

11.5 CYBER THREATS SPECIFIC TO FRUIT CROP SAFETY

Criminal organizations use malware and ransomware assaults to compromise data integrity and disrupt operations by infecting computer systems with malicious software. The food and agriculture industries may suffer greatly as a result of these attacks. Malware, for instance, has the potential to compromise vital infrastructure, such as food processing facilities or irrigation systems, by infecting control systems. Important data may become encrypted by ransomware attacks and become unreadable until a ransom is paid. Prominent instances comprise the NotPetya assault on the Danish shipping enterprise Maersk in 2017, which caused disturbances to worldwide supply networks; and JFC International, Loaves & Fishes, Home Chef, Harvest

Sherwood Food Distributors, etc. [27]. Some of the main cyber threats specific to fruit crop safety are discussed as follows:

- *Precision Farming Manipulation*: Cyber attackers might manipulate precision agricultural data, causing improper applications of fertilizer, herbicides, or water, which would harm crop health.
- *Manipulation of Automated Irrigation Systems*: Cyber threats may interfere with automated irrigation systems, leading to overwatering or underwatering, which affects the growth and output of fruit crops.
- *Pest and Disease Control Disruption*: Cyber-attacks on pest and disease monitoring systems can prevent prompt intervention, resulting in unchecked epidemics and agricultural losses.
- *Threats to Supply Chain of Food*: A cyber-attack on one vital component has the potential to cause havoc on the food industry as a whole, causing production, distribution, and availability of food to be disrupted. Consequently, strengthening the industry's resistance to cyber-attacks requires acknowledging and resolving these interconnections and interdependencies.

11.6 BEST PRACTICES FOR SECURING AGRICULTURAL SYSTEMS

The following cybersecurity measures are to be taken regularly to effectively safeguard digital agricultural systems (Figure 11.1):

- *Regular System Updates*: To reduce vulnerabilities, farmers should make sure that all software and hardware are current with the most recent security updates.
- *Users' Training on Cybersecurity*: The right cybersecurity awareness and training programs need to be developed to educate staff members about potential risks and safe practices. The best thing that food and agriculture enterprises can do to protect themselves from cyber-attacks is to provide workers with cybersecurity training. The social engineering element of most attacks, which is often an email, comes first. It might be challenging to tell modern phishing emails apart from trustworthy correspondence. By instructing personnel to be continually vigilant in detecting the detectable signs of a fraudulent email, users can function as a realistic first line of protection.
- *Secure Data Handling*: Access to sensitive data should be limited to those with a need-to-know basis, and sensitive data should be encrypted and stored securely.
- *Implementing Multi-Factor Authentication*: For crucial systems adds an extra layer of security and makes illegal access more difficult.
- *Network Segmentation*: By partitioning the agricultural network into sectors with restricted access, it can lessen the impact of a breach and contain possible

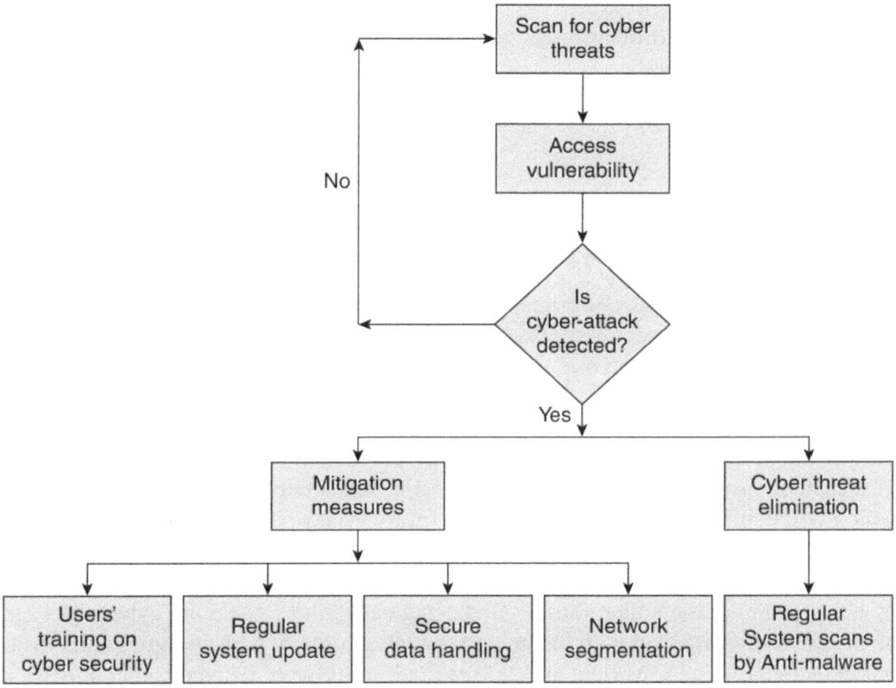

FIGURE 11.1 Schematic diagram showing steps of cybersecurity measures for the fruit/agriculture industry.

risks. The impact of a breach can be reduced and potential dangers contained by segmenting the agricultural network into areas with controlled access. IT managers in the agriculture and food sectors can increase security by dividing production networks into smaller segments and business networks from one another. Infrastructure for a corporation can be disconnected if suspicious conduct is discovered in a different part of the network because they are logically separated. As was already said, even segmented infrastructure can be compromised by newly discovered malware in a particular section of the network when updating programs, for instance. Segmenting, however, can prevent damaging infection from spreading across the entire firm.

- *Regular Backups*: Regularly protecting important data can lessen the effects of ransomware attacks and facilitate an efficient recovery. Attackers can limit the operations of food producers by denying them access to their business systems. Keeping a recent backup is the most practical mitigating approach to stopping ransomware thieves. Separate backup files should be maintained from original files to prevent hackers from stealing or encrypting backup copies as well as the original files [27]. The schematic diagram showing steps of security measures for the fruit/agriculture industry is depicted in Figure 11.1.

- *Collaborative Efforts*: To exchange threat intelligence and industry best practices, governments, agricultural groups, and technology suppliers should work together.
- *Third-Party Security Assessments*: Farms can use third-party experts to do security analyses of their systems in order to find and fix potential flaws.

11.7 RECOMMENDATIONS FOR SECURING AGRICULTURAL SYSTEMS

This chapter emphasizes the crucial subject of cyber hazards in fruit crop protection. If the agriculture sector adopts best practices and is aware of the threats, it may be able to increase its defenses against cyber-attacks. By combining data analytics and emerging technology, future cyber dangers can be forecasted and averted. Cooperation among stakeholders, lawmakers, and technological experts is essential for a resilient and secure agricultural system. Farmers may protect the productivity of their fruit crops by being proactive and being informed about new cyber hazards, which will contribute to sustainability and future food security.

The critical topic of cyber risks in fruit crop protection is highlighted in this book chapter. The agriculture industry may strengthen its defenses against cyber-attacks by being aware of the vulnerabilities and using best practices. Future cyber risks can be predicted and avoided with the help of the integration of data analytics and developing technologies. A resilient and safe agricultural system depends on cooperation between stakeholders, politicians, and technological specialists. Farmers may protect their fruit crop productivity by taking proactive measures and keeping up with new cyber risks, which will help to ensure future food security and sustainability. Understanding and minimizing cyber dangers is essential to maintaining the security of fruit crops and overall agricultural output as the agriculture sector continues to embrace digital transformation. Farmers can protect their systems and data by identifying vulnerabilities and putting best practices into practice. This will help to build a robust and secure agricultural ecosystem for the future. Sensors, IoT irrigation systems, milking robots, and drones are a few examples of digital agricultural equipment. Using devices with low-security standards, such as those that use unencrypted passwords, can make farm systems and networks more vulnerable. Before installing agricultural gadgets, it is advised that farmers inquire about the security measures from the technology providers to help mitigate this problem [28].

11.8 DISCUSSION

In this chapter, addressing both the technical and non-technical concerns related to digital agriculture and cybersecurity vulnerabilities has been addressed. Most farmers did not have regular backups of their data, a high-speed internet connection, or backup internet. In most farm networks, there is room for improvement in terms of design, implementation, and maintenance of the farm's local network as well as the larger cyber-physical environment. Farmers need to recognize the value of cybersecurity if they hope to increase it on their property. Farmers in India usually don't think much of the internet or cybersecurity. Automated "agro-bots" will monitor conditions and provide treatment using cutting-edge equipment designed to lower crop disease and boost yields. The intricate nature of the agri-food industry poses challenges in terms of product traceability within the vast supply chain. Blockchain technology provides a solution for this issue. Technology thus addresses issues related to food safety and quality that impact consumers, the government, etc. [22]. Blockchain's capacity to keep ownership records and tamper resistance can help solve serious issues with the current food system, such as food fraud, safety recalls, inefficient supply chains, and food traceability.

11.9 CONCLUSION

This chapter's conclusion emphasizes the urgent need for comprehensive cybersecurity solutions and sheds light on the vital subject of cyber hazards in fruit crop protection. By being aware of threats and putting best practices into practice, the agricultural industry can fortify its defenses, safeguard its fruit harvests, and uphold the requirements of world food security. Maintaining the security of fruit crops and total agricultural output is crucial as the agriculture sector continues to embrace digital transformation. Understanding and reducing cyber threats is also crucial. By detecting weaknesses and implementing best practices, farmers can secure their systems and data. This will contribute to the development of a strong and safe agricultural ecosystem in the future. Cooperation between stakeholders, continual awareness, and adaptive cybersecurity solutions are essential for remaining one step ahead of cyber threats in the ever-changing agricultural sector. In the quickly changing agricultural scene, collaboration among stakeholders, ongoing awareness, and adaptive cybersecurity measures are crucial for staying one step ahead of cyber threats. This chapter's conclusion emphasizes the urgent need for comprehensive cybersecurity solutions and sheds light on the vital subject of cyber hazards in fruit crop protection. By being aware of threats and putting best practices into practice, the agricultural industry can fortify its defenses, safeguard its fruit harvests, and uphold the requirements of world food security. The study paves the way for a secure and resilient agricultural future in which the benefits of technology are used while lowering the associated risks.

In conclusion, this chapter clarifies the pressing necessity for all-encompassing cybersecurity measures and highlights the significant issue of cyber risks in fruit crop safety. The agriculture sector can strengthen its defenses, protect its fruit crops, and uphold international food security requirements by being aware of vulnerabilities and implementing best practices. The study sets the way for an agricultural future that is safe and resilient, where the advantages of technology are utilized while reducing the dangers involved. Numerous user services have been made available as a result of the growth of communication and sensing-capable smart gadgets. At the same time, tasks have become easier and more effective for people to do. These systems are now vulnerable to cyber-attacks due to the widespread use of such internet-connected gadgets and data-driven apps across a variety of industries. This essay elaborates on unanswered research topics while discussing such cybersecurity issues in smart farming.

REFERENCES

1. M. Roser, "Future Population Growth," [Accessed: Sep. 12, 2023. Online], 2020. Available: https://ourworldindata.org/future-population-growth.
2. J. V. Stafford, "*Precision Agriculture '19*," Wageningen: Academic, 2019.
3. H. C. J. Godfray, J. R. I. Beddington, I. R. Crute, L. Haddad, D. Lawrence, J. F. Muir, J. Pretty, S. Robinson, S. M. Thomas, and C. Toulmin, "Food security: The challenge of feeding 9 billion people," *Science* 327, no. 5967, pp. 812–818, 2010.
4. D. Gollin, S. Parente, and R. Rogerson, "The role of agriculture in development," *American Economic Review* 92, no. 2, pp. 160–164, 2002.
5. A. Kamilaris, F. Gao, F. X. Prenafeta-Boldu, and M. I. Ali, "Agri-IoT: A semantic framework for Internet of Things-enabled smart farming applications," In: *Proc. IEEE 3rd World Forum Internet Things (WF-IoT)*, Dec. 2016, pp. 442–447, 2016.
6. D. Vasisht, Z. Kapetanovic, J. Won, X. Jin, R. Chandra, S. Sinha, A. Kapoor, M. Sudarshan, and S. Stratman, "Farmbeats: An IoT platform for data-driven agriculture," In: *Proc. 14th USENIX Symp. Netw. Syst. Design Implement (NSDI)*, pp. 515–529, 2017.
7. S. Ge. L. Wolfert, C. Verdouw, and M. J. Bogaardt, "Big data in smart farming—A review," *Agricultural Systems* 153, pp. 69–80, 2017.
8. S. Ksibi, F. Jaidi, and A. Bouhoula, "A comprehensive study of security and cyber-security management within e-health systems: Synthesis, analysis and a novel quantified approach," *Mobile Networks and Applications* 28, pp. 107–127, 2023. doi: 10.1007/s11036-022-02042-1.
9. S. Fountas, C. G. Sorensen, Z. Tsiropoulos, C. Cavalaris, V. Liakos, and T. Gemtos, "Farm machinery management information system," *Computers and Electronics in Agriculture* 110, pp. 131–138, 2015. doi:10.1016/j.compag.2014.11.011.
10. J. Nikander, M. Onni, and L. Mikko, "Requirements for cybersecurity in agricultural communication networks," *Computers and Electronics in Agriculture* 179, p. 105776, 2020. doi: 10.1016/j.compag.2020.105776.
11. L. Barreto, and A. Amaral, "Smart farming: Cyber security challenges," In: *Proceedings of International Conference on Intelligent Systems (IS)* (pp. 870–876). IEEE, 2018. https://ieeexplore.ieee.org/abstract/dS.ocument/8710531.
12. H. Chi, S. Welch, E. Vasserman, S. Kalaimannan, "A framework of cybersecurity approaches in precision agriculture," In: *Proceedings of the ICMLG2017 5th International Conference on Management Leadership and Governance. Reading* (pp. 90–95). UK: Acad. Conf. Publ. Int., 2017. https://search.proquest.com/docview/1897672343?pq-origsite=gscholar.
13. M. M. Jahn, W. L. Oemichen, and G. F. Treverton, "Cyber Risk and Security Implications in Smart Agriculture and Food Systems," *Technical Report;* College of Agriculture and Life Sciences, University of Wisconsin: Madison, 2019.
14. M. M. Jahn, "Cyber Risk and Security Implications in Smart Agriculture and Food Systems," Accessed: Oct. 05, 2023, 2019. [Online]. Available: https://jahnresearchgroup.webhosting.cals.wisc.edu/wpcontent/uploads/sites/223/2019/01/Agricultural-Cyber-Risk-and-Security.pdf.
15. J. West, "A prediction model framework for cyber-attacks to precision agriculture technologies," *Journal of Agricultural & Food Information* 19, no.4, pp. 307–330, 2018. https://www.tandfonline.com/doi/abs/10.1080/10496505.2017.1417859.
16. S. Nakhodchi, A. Dehghantanha, and H. Karimipour, "Privacy and Security in Smart and Precision Farming: A Bibliometric Analysis," In: *Handbook of Big Data Privacy*. Springer, Cham, pp. 305–318, 2020.
17. C. Hedley, and I. Yule, "A method for spatial prediction of daily soil water status for precise irrigation scheduling," *Agricultural Water Management* 96, no. 12, pp. 1737–1745, 2009. doi: 10.1016/j.agwat.2009.07.009.

18. S. Shamal, S. A. Alhwaimel, and A. M. Mouazen, "Application of an on-line sensor to map soil packing density for site specific cultivation," *Soil and Tillage Research* 162, pp. 78–86, 2016. doi: 10.1016/j.still.2016.04.016.
19. S. Katta, S. Ramatenki, and H. Sammeta, "Smart irrigation and crop security in agriculture using IoT," In: *AI, Edge and IoT-Based Smart Agriculture*. Academic Press, 2022. doi: 10.1016/B978-0-12-823694-9.00019-0.
20. A. Salam, "A path loss model for through the soil wireless communications in digital agriculture," In: *Proc. 2019 IEEE International Symposium on Antennas and Propagation (IEEE APS),* Atlanta, GA, USA, July 2019.
21. A. Alvino, and S. Marino, "Remote sensing for irrigation of horticultural crops," *Horticulture* 3, no. 2, p. 40, 2017.
22. M. Gupta, M. Abdelsalam, S. Khorsandroo, and S. Mittal, "Security and privacy in smartfarming: Challenges and opportunities," *IEEE Access* 8, pp. 34564–34584, 2020. doi: 10.1109/ACCESS.2020.2975142.
23. E.-C. Oerke, and H.-W. Dehne, "Safeguarding production losses in major crops and the role of crop protection," *Crop Protection* 23, no. 4, pp. 275–285, 2004.
24. O. S. Cupp, D. E. Walker, and J. Hillison, "Agro-terrorism in the U.S.: Key security challenge for the 21st century," *Biosecurity and Bioterrorism* 2, no. 2, pp. 97–105. doi: 10.1089/153871304323146397.
25. A. Walter, R. Finger, R. Huber, and N. Buchmann, "Opinion: Smart farming is key to developing sustainable agriculture," *Proceedings of the Natural Academy of Sciences of the United States of America* 114, no. 24, pp. 6148–6150, 2017.
26. M. M. Aung, and Y. S. Chang, "Traceability in a food supply chain: Safety and quality perspectives," *Food Control* 39, pp. 172–184, 2014.
27. H. C. Verma, S. Srivastava, T. Ahmed, and N. A. Usmani, "Cyber Threats in Agriculture and the Food Industry: An Indian Perspective," In: M. Husain, M. Faisal, H. Sadia, T. Ahmad, & S. Shukla (Eds.), *Advances in Cyberology and the Advent of the Next-Gen Information Revolution* (pp. 109–122). IGI Global, 2023. doi: 10.4018/978-1-6684-8133-2.ch006.
28. M. Hazrati, R. Dara, and J. Kaur, "On-farm data security: Practical recommendations for securing farm data," *Frontiers in Sustainable Food Systems* 6, p. 884187, 2022. doi: 10.3389/fsufs.2022.884187.

12 Cyber Resilience in a Digital World

Securing Healthcare, Agriculture, and Food Industry against Emerging Threats

Mohammad Ishrat, Wasim Khan, and Danish Ather

12.1 INTRODUCTION

As a symbol of an age of extraordinary technical development, the integration of digital systems is a driving factor in improvements across many different industries. The use of digital tools has greatly transformed several industries, including healthcare, agriculture, and food, by increasing productivity, decreasing operational costs, and better serving consumers [1]. Nevertheless, these crucial industries are now shrouded in mystery as a result of the fast transition to digitalization, which has revealed a dark realm full of complex cyber threats that are always changing. In this chapter, it has tried to explore the complex relationships between technical advancements and the terrifying dangers posed by malevolent cyber actions [2]. We are thrust into a future where cyber-attacks can have devastating real-world consequences as we explore this unexplored area and realize that the digital revolution has created a complicated network of dependency. Electronic medical records, telemedicine consultations, and connected medical devices have changed the medical industry's traditional view, characterized by the privacy of doctor-patient interactions and the reliability of written records. While healthcare systems are gaining importance, they also offer new opportunities to hackers. These criminals are exploiting vulnerabilities in connected medical devices to steal personal health information and potentially compromise critical medical procedures. Successful cyber-attacks in this area can lead to serious consequences, including loss of patient confidence in medical care and, in extreme cases, death [3]. Similarly, in agriculture, the combination of precision farming and Internet of Things (IoT) technology has enabled data-driven agriculture, good resource allocation, and increased yields. However, due to its dependence on the internet, agriculture is still vulnerable to communications that can be harmful to global production, consumption and supply chains, metals, and food. Protecting

DOI: 10.1201/9781003409571-12

agriculture from cyber threats is not only a financial issue but also important in terms of supporting the masses. While the food industry plays an important role in the world of food, it also faces unique digital challenges. Digitized products and improved business processes, including electronic production processes, make them vulnerable to cyber-attacks. Any damage to products, contamination of network equipment, or damage to consumer confidence as a result of an attack could have a significant impact on public health, financial security, and international economic relations.

This section is dedicated to a comprehensive review of cyber threats affecting major businesses. Our goal is to use qualitative research, case studies, and expert opinion to discover the weaknesses exploited by cybercriminals, identify vulnerabilities in each operation, and measure the outcome of a successful attack. The goal of sharing this information with policymakers, business leaders, and cybersecurity professionals is to provide them with the resources they need to strengthen their work, prevent and predict risks, mitigate their impact, and solve ongoing problems in the digital world. The goal is to perform a detailed analysis of this connection, aiming to uncover the intricate link between innovation capacity and the crucial need for restructuring. Let's venture into the digital landscape equipped with a profound understanding of the impending challenges, working collectively to safeguard the foundational elements of healthcare, agriculture, and the food industry from the persistent onslaught of cyber threats.

Various sectors, particularly healthcare, agriculture, and the food industry, have undergone significant transformations owing to the widespread incorporation of technology, making this study highly relevant. The authors' motivation comes from the notable increases in productivity, operational effectiveness, and consumer experiences made possible by digital solutions across a variety of industries [1]. The rapid integration of technology into these areas makes them more vulnerable to needs and evolving cyber threats, requiring a comprehensive investigation of these disadvantages. The authors' main task is to analyze the interaction between technology and the cyber risk phenomenon and conduct research to develop a deeper understanding. They aim to identify the evolution of cyber threats over time and the technological adaptations made to deal with evolving risks. Leading cybercriminals use a variety of techniques, and this study aims to identify these techniques, uncover specific malicious activity, and examine the consequences of a successful hack. Through extensive research, case studies, and insights from experts, our authors aim not only to shed light on the imminent dangers posed by cyber threats but also to equip stakeholders, including policymakers, industry leaders, and cybersecurity professionals, with the knowledge and understanding required to fortify defenses, mitigate risks, and navigate the digital realm with resilience and confidence [4]. This study seeks to offer a comprehensive and actionable analysis, illuminating the pressing need to secure the foundational pillars of healthcare, agriculture, and the food industry against the ever-evolving landscape of cyber threats. As shown in Figure 12.1, the various sources of cyber threats highlight the complexities of the digital threat landscape and their diverse origins.

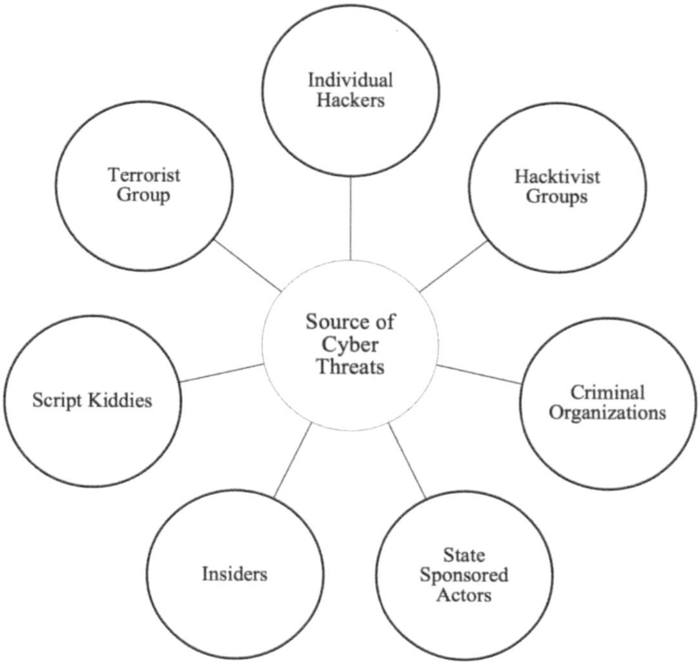

FIGURE 12.1 Source of cyber threat.

12.2 CYBER THREAT LANDSCAPE

Cyber threats refer to malicious activities conducted in the digital realm with the intent of compromising the confidentiality, integrity, or availability of computer systems, networks, and data [4]. These threats encompass a variety of techniques and attack vectors, each with its own characteristics and potential impact. The cyber threat landscape, depicted in Figure 12.2, categorizes different types of threats and their impacts on confidentiality, integrity, and availability. Here are some common types of cyber threats [5]:

12.3 LITERATURE REVIEW

Cybersecurity threats have become a critical issue for industries, such as healthcare, agriculture, and food industry. Recent research in healthcare has revealed the vulnerability of digital health systems and highlighted the need to keep medical records in electronic systems (electronic health records [EHRs]), medical devices, and connected hospital systems against cyber threats. Researchers are addressing the development of a cybersecurity framework by protecting patient information through advanced measures, such as vulnerability detection, encryption technology, and access control policies powered by artificial intelligence (AI) and machine learning (ML). In addition, new research explores the ethical and legal implications of cybersecurity

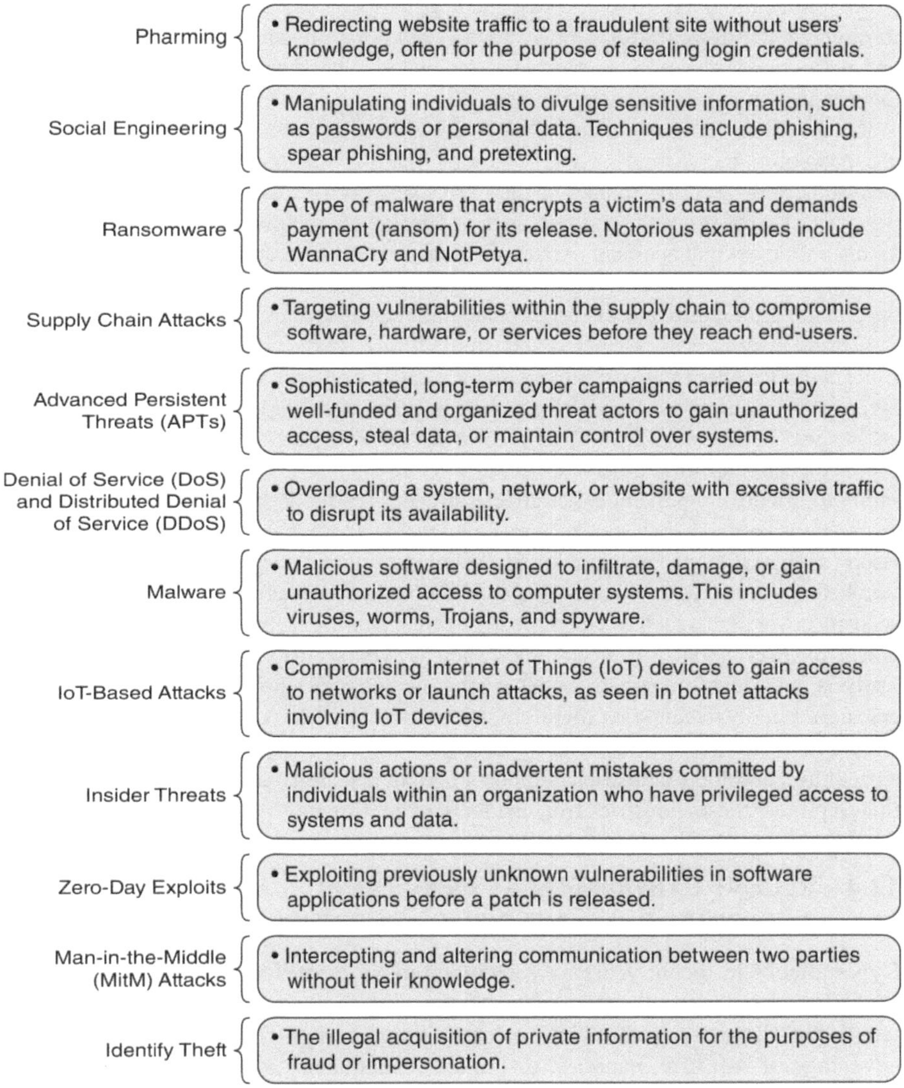

FIGURE 12.2 Cyber threat landscape.

vulnerabilities in healthcare, recommending strategies for risk assessment, compliance, and prevention, protecting patient information, and maintaining the integrity of the treatment process [6]. When it comes to agriculture, recent research confirms not only the evolution of digital technologies but also highlights the disadvantages it brings. Researchers are focusing on IoT devices, precision agriculture, and data farming methods to solve common cybersecurity problems [7]. The research highlights the risks posed by IoT devices in agriculture and suggests safeguards against automation, data integrity, and farm protection. Additionally, the study highlights the impact of cyber threats on global food security and highlights the need for preventive measures throughout the agricultural supply chain. Collaboration among academia, industry, and policymakers is seen as important to develop cybersecurity strategies tailored to the unique challenges facing agriculture [8].

In the food industry, recent data shows that food products and e-commerce are becoming increasingly digital, making them a major target for cyber-attacks. Researchers explore the impact of data breaches, product disruptions, and consumer privacy risks and propose solutions, such as Blockchain technology output, smart packaging, and smart management to increase transparency, traceability, and food safety [9]. The impact of cyber threats on reputation, consumer confidence, and regulatory compliance in the food industry is also significant and requires robust cybersecurity measures. Suggested responses are clearly stated. In the future, studies point to promises, such as Blockchain for chain transparency, AI-driven quality control, self-catering, smart packaging, and volunteer work to improve cybersecurity and efficiency in business-focused exchanges. Participants are encouraged to invest in digital transformation, prioritize product trust and transparency, adopt sustainable practices, and participate in collaboration to effectively prevent cyber-attacks. Recent academic research has explored the motivations behind cyber-attacks; it revealed the diversity of cybercrimes by examining issues, such as financial gain, espionage, sabotage, political targeting, hacking, information leakage, and cyber warfare [9–11]. Additionally, research tips on modern network security cover areas, such as cloud security, threat sharing, network security, biometrics, and crime prevention. Advanced topics, such as cybersecurity visualization, behavioral analysis, and quantum-secure cryptography, demonstrate the impact of cybersecurity research. Finally, recent data highlights the critical need for vigilance, innovation, and collaboration to overcome the many challenges posed by cyber threats in the healthcare, agriculture, and food industries. Support ongoing research, technology, and collaboration to protect these industries from evolving cyber threats.

12.4 RECENT CYBER-ATTACKS IN MEDICAL, AGRICULTURAL, AND FOOD INDUSTRIES

Cyber-attacks in recent years have included many crimes and caused digital damage. Their targets are people, organizations, and countries. Break into digital systems, steal data, disrupt business, or cause financial loss. These attacks often take advantage of software, network, or human inadequacies to cause serious harm to victims. Cybersecurity solutions continue to evolve to detect, prevent, and mitigate these attacks [12]. Recent cyber-attacks targeting the healthcare, agriculture, and food industries include the following.

12.4.1 Cyber Threats in the Healthcare Sector

- *Keyword Search and Data Filtering*: The basics of digital research include searching and filtering information using keywords. Searchers use these terms to find information, articles, and related information to help them in their searches. This approach allows people to make good decisions from big data and integrate the most relevant evidence [13–15].

- *File Signature Analysis*: To discover specific data, data signature analysis requires checking for certain patterns or signatures in the data, regardless of the details of the data. Data scraping uses this technique where searchers try to find missing or hidden data by browsing through the top and bottom pages [16, 17].
- *Hashing and Integrity Verification*: In digital forensics, hashing can ensure the integrity of digital evidence. To verify the consistency of the data during the search, researchers calculate the hash value and then compare the hash value [18].
- *Timeline Analysis*: The goal of real-time analysis is to reconstruct and reveal the nature of events using digital evidence such as timestamps and metadata. Time delays help researchers aggregate results by accounting for events caused by a user or system [19–21].
- *Live Memory Analysis*: Real-time memory scanning involves examining your computer's RAM (accessible memory) to identify operating system processes, network connections, and system performance problems. This approach helps identify artifacts stored in memory that cannot be detected with traditional data analysis techniques.
- *Data Decryption and Password Cracking*: If digital forensics experts come across encrypted data, they can use data decryption methods to access confidential data and messages. Recovering passwords to open encrypted files or password-protected files is another way to use password-cracking tools [22].
- *Steganography Detection*: Steganography is the practice of storing secrets in an inconspicuous environment. Technologists use steganography search tools and techniques to look for information hidden in other documents [23, 24].
- *Malware Analysis*: Malware analysis methods are used to investigate and assess potential malware infections or cyber-attacks. Analysis of malware lets investigators learn more about the threat's actions, capabilities, and effects on compromised machines [25].
- *Data Visualization*: In order to help investigators and stakeholders better comprehend the digital evidence at hand, data visualization techniques are used to present the data in a graphical format.
- *Forensic Reporting and Documentation*: To convey the results of a digital investigation clearly and simply, it is crucial to employ effective forensic reporting and documentation methodologies. Throughout an investigation, investigators should keep meticulous records of all their methods, techniques, and gathered evidence [26, 27].

12.4.2 Challenges Faced by the Healthcare Sector

Numerous obstacles threaten the healthcare industry's capacity to deliver high standards of care, protect patients, and run efficiently. The challenges faced by the healthcare sector are summarized in Figure 12.3, emphasizing the critical hurdles that impede operational and data security. Some of the most pressing problems in healthcare now are as follows.

Rising Healthcare Costs	Doctor visits, medicines, and hospital stays have been rising in cost for some time. Financial strain on individuals, families, and healthcare providers may limit community access.
Aging Population	Many nations are ageing. Due to an ageing population, medical, chronic sickness, and long-term care services are needed more.
Chronic Disease Management	General population illnesses like diabetes, cardiovascular disease, and obesity are rising. These disorders require consistent medical care, behavioral changes, and specialist communication.
Healthcare Access Disparities	Economic status, race/ethnicity, and location affect healthcare access and outcomes. Specificity makes reducing these inequities difficult.
Health Information Technology Adoption	Health information technology systems like electronic health records have improved data sharing and collaboration, but interoperability, data security, and usability remain challenges.
Data Privacy and Security	Cybercriminals must not access patients' personal data. Hospitals must safeguard sensitive patient data with strong cybersecurity.
Workforce Shortages	In many locations, doctors, nurses, and associated medical personnel are few. Due to this scarcity, healthcare providers may burn out and deliver poor service.
Medical Errors and Patient Safety	Misdiagnoses, as well as errors in the administration of medications, can have serious consequences for patients. Constant efforts at quality enhancement are necessary to guarantee patient safety.
Regulatory Compliance	The Health Insurance Portability and Accountability Act and other legislation govern patient data privacy and protection for healthcare providers. These rules are important but difficult to follow.
Public Health Emergencies	Pandemics, like the one caused by the COVID-19 virus, put a burden on healthcare facilities and personnel around the world. Coordination, resource allocation, and flexibility are essential for efficiently responding to such events.
Ethical Dilemmas	End-of-life care, budget distribution, and experimental treatments are among many ethically challenging healthcare decisions. Balance between patient independence, doing no harm, and fairness may be difficult.
Healthcare Innovation and Integration	Innovation in medical technology, treatments, and therapies must be balanced with evidence-based and cost-effectiveness. It may be challenging to integrate new medical technologies into healthcare systems.
Provider-Patient Communication	Accurate diagnoses, treatment adherence, and patient satisfaction depend on provider-patient communication. Language challenges, health literacy, and time limits can cause problems.
Mental Health Care	To minimize mental health stigma and increase treatment alternatives, much effort remains. Mental health requires medical, social, and educational interventions.

FIGURE 12.3 Challenges faced by the healthcare sector.

These challenges vary based on geographical location, healthcare system design, and cultural factors. The healthcare sector must continually adapt and develop innovative strategies to overcome these challenges and provide the best possible care to patients.

12.4.3 Implications for the Healthcare Sector

The challenges faced by the healthcare sector have far-reaching implications that affect patients, healthcare providers, policymakers, and society as a whole. These implications highlight the need for strategic planning, policy changes, and innovative solutions to address the complex issues within the sector [3, 28–30]. The implications of these challenges, as detailed in Figure 12.4, demonstrate the far-reaching consequences for stakeholders, including patients and providers. Following are some of the implications for the healthcare sector: figure 12.4.

The implications of healthcare challenges are multifaceted and interconnected. To address these implications effectively, stakeholders across the healthcare sector, including policymakers, healthcare organizations, professionals, and patients, need to collaborate on comprehensive solutions that prioritize patient well-being, equity, and quality of care [3].

12.4.4 Regulatory Landscape and Compliance

The regulatory landscape and compliance requirements in the healthcare sector are complex and dynamic, varying by country and region. Regulations are designed to ensure patient safety, maintain data privacy, and uphold ethical standards in healthcare practices. Following are some key aspects of the regulatory landscape and compliance in the healthcare sector [31, 32].

- *Health Insurance Portability and Accountability Act (HIPAA) (United States)*: HIPAA sets standards for the privacy and security of protected health information (PHI). Covered entities, including healthcare providers, health plans, and healthcare clearinghouses, are required to safeguard patient data and adhere to specific rules for its use and disclosure.
- *General Data Protection Regulation (GDPR) (European Union [EU])*: Personal data (including medical data) in the EU is subject to the Data Protection Law. When processing EU citizens' data, providers must comply with the GDPR's strict rules on data protection, consent, and breach.
- *Medical Device Regulations*: Medical devices and medications are licensed and monitored by regulatory agencies such as the Food and Drug Administration (FDA) in the United States and the European Medicines Agency (EMA) in Europe. Safety, effectiveness, and post-marketing surveillance must comply with regulations for medical facilities that use or distribute medical devices.
- *Clinical Trials Regulations*: Special regulations are in place to protect patients during the evaluation of new treatments and interventions. In order to gain acceptance and guarantee the reliability of trial data, adherence to protocols and reporting criteria is crucial.

Patient Health and Safety	Issues like medical errors and access inequities can harm patients. Patient well-being Healthcare includes improving care quality, reducing errors, and ensuring fair access to healthcare.
Healthcare Access and Equity	Healthcare access and outcomes can cause health disparities. Addressing these gaps is essential for healthcare equity and timely care for all.
Financial Sustainability	Healthcare systems might struggle with rising expenses and staff shortages. Healthcare systems must cut costs without losing quality to survive.
Quality of Care	Workforce shortages, weariness, and restricted resources threaten patient care. Only by improving healthcare delivery, adopting evidence-based approaches, and strengthening patient-centered treatment can high-quality healthcare be guaranteed.
Innovation and Technology	Patient care is threatened by staff fatigue and resource constraints. Quality healthcare can only be ensured by improving delivery, implementing evidence-based practices, and increasing patient centered treatment.
Workforce Well-being	Stressful conditions at work have been linked to higher rates of burnout, dissatisfaction with one's job, and employee turnover in the health care sector. Maintaining a competent and enthusiastic healthcare personnel depends on prioritizing their health and safety.
Public Health Preparedness	Effective techniques for preparation and reaction are essential in the face of public health emergencies. Protecting public health and reducing the severity of the effects of emergencies require fortified healthcare systems.
Policy and Regulatory Changes	Due to healthcare's problems, many legislative and regulatory adjustments have been proposed. These changes may impact population health, healthcare delivery, access, and funding.
Patient Empowerment and Education	Patient engagement and education are needed to treat chronic diseases and ensure patient safety. Empowering patients to manage their care improves outcomes.
Research and Evidence-Based Practice	Healthcare issues require constant research and evidence-based practices. By applying research, clinical practice can improve outcomes and patient satisfaction.
Global Health Implications	Many global healthcare issues affect multiple countries. Global cooperation and knowledge-sharing are needed to combat infectious disease outbreaks and health inequities.
Ethical Considerations	Healthcare professionals and governments must carefully evaluate and follow ethical norms when making resource allocation and end-of-life care decisions.

FIGURE 12.4 Implications for the healthcare sector.

- *Health Information Technology (HIT) Regulations*: HIT includes things like EHRs and telemedicine, but each country has its own set of rules for how to use them. Interoperability of data, consent from patients, and encrypted transmission of health information are all addressed by these rules [33].
- *Drug Approval and Regulation*: In countries like the United States and Europe, respectively, regulatory bodies like the FDA and the EMA manage the drug approval and regulation process. Before medications can be sold on the market, they must first pass rigorous safety, effectiveness, and quality tests.
- *Ethical and Research Regulations*: Regulations and guidelines, such as the Declaration of Helsinki, address ethical concerns in medical research. Research using human subjects requires ethical assessment and must protect the confidentiality of participants and prevent any harm to them.
- *Data Privacy and Security Regulations*: Data protection legislation, such as the GDPR in the EU and the Health Information Act in Canada, affect healthcare businesses in many nations.
- *Pharmacy and Dispensing Regulations*: Different jurisdictions have different rules concerning pharmacies and the distribution of medications. Medication prescribing, dispensing, and administration are guaranteed to be carried out following these rules.
- *Insurance and Reimbursement Regulations*: Insurance and payment rules for medical care differ from country to country. In order to get paid for their work, healthcare providers need to know how to bill, code, and document appropriately.
- *Telemedicine and Telehealth Regulations*: As the demand for remote medical assistance has grown, so too have the regulations governing telemedicine and telehealth services. License, consent, and quality of care regulations must be followed religiously.

There will always be new versions of healthcare regulations to follow. In order to comply with the ever-evolving healthcare regulations, healthcare providers must be informed, flexible, and resourceful [33, 34].

12.4.5 Mitigation Strategies and Best Practices

Proactive measures and best practices are needed to mitigate healthcare difficulties and hazards. These healthcare methods cover patient care, data security, workforce management, and regulatory compliance. Consider these mitigation methods and best practices:

- *Patient Safety and Quality of Care*: To ensure consistent and efficient patient care, enforce evidence-based clinical guidelines and processes. Create a safety-focused culture where healthcare workers can report near-misses and bad incidents without threat. Barcoding and electronic drug administration records can reduce pharmaceutical mistakes. Fund continual training and education for healthcare personnel to learn about best practices and new treatments.

- *Data Privacy and Security*: Complies with HIPAA and GDPR standards to protect patient information and privacy. Integrate firewalls, encryption, and intrusion detection systems into your network security strategy. Conduct regular security tests and penetration tests to detect and resolve vulnerabilities. Provide employees with cybersecurity training, including phishing detection and device security procedures.
- *Adopt HIT*: Establishing a communication health information system is essential to prevent patient information from being shared among healthcare providers. Updates and patches are necessary to maintain information security and address vulnerabilities in HIT systems. Training staff to know how to use HIT is critical to improving the quality of care and reducing errors.
- *Ensure Regulatory Compliance*: Regulatory compliance requires a compliance officer or team to monitor changes and ensure strict compliance. Compliance policies and procedures must be developed and implemented following regulatory requirements to comply with the law.
- *Emergency Preparedness and Response*: Effectively resolving public health crises, natural catastrophes, and other emergencies requires strong emergency response strategies. Regular exercises and simulations should assess these strategies' efficacy and suggest improvements. A well-coordinated response requires strong relationships with local and national health agencies.
- *Patient Engagement and Education*: Increase patient empowerment in healthcare decisions through education and shared decision-making. Communicate medical conditions, treatments, and prevention actions. Use telehealth and remote monitoring to improve patient involvement and follow-up care.
- *Ethical Decision-Making*: To solve difficult ethical problems and advise healthcare practitioners, ethics committees are essential. Open discourse and interdisciplinary teamwork can help solve ethical problems [35, 36]. Prioritize patient autonomy, beneficence, and fairness in decision-making.
- *Innovation and Research*: Encourage healthcare delivery, treatment, and technology research and innovation. When doing human subject research, follow ethical norms and get clearance. To ensure efficacy, prioritize evidence-based practices and extensively evaluate innovative initiatives.
- *Continuous Improvement*: Encourage learning from mistakes and adaptation in a culture of continuous improvement. Analyze data to improve patient outcomes, operational efficiency, and patient satisfaction. Encourage patient, family, and employee feedback to drive improvements [37, 38].

12.4.6 Future Trends and Recommendation

We must adapt these tactics to our healthcare organization's needs. For effective mitigation, stay current on healthcare regulations, technology, and best practices. Talking to specialists, attending industry conferences, and working with other healthcare professionals can help. Technology, demography, and patient expectations are changing healthcare. These themes and suggestions are emerging:

- *Telehealth and digital health*:
    - *Trend*: Telehealth and digital health initiatives will undoubtedly rise. These devices provide remote consultations, monitoring, and individualized health management [39].
    - *Recommendation*: Invest in user-friendly telehealth solutions and healthcare system interoperability. Prepare healthcare workers for virtual care.
- *AI and ML*:
    - *Trend*: AI and ML are utilized for medical diagnosis, treatment, and predictive analytics. They can improve clinical decision-making and administrative processes.
    - *Recommendation*: Explore AI applications that align with your organization's needs, such as diagnostic assistance or predictive modeling for patient outcomes. Invest in data governance and quality to ensure accurate AI insights [40].
- *Personalized medicine*:
    - *Trend*: Advances in genomics and molecular profiling enable personalized treatment plans tailored to an individual's genetic makeup and health history.
    - *Recommendation*: Collaborate with geneticists and specialists to implement personalized medicine approaches. Develop protocols for interpreting genetic data and integrating it into patient care [41].
- *Cybersecurity and data privacy*:
    - *Trend*: As healthcare becomes more digitally connected, the importance of cybersecurity and data privacy continues to rise.
    - *Recommendation*: Implement robust cybersecurity measures, conduct regular security assessments, and educate staff about cyber threats. Prioritize data privacy to maintain patient trust [1].
- *Workforce well-being and technology integration*:
    - *Trend*: Healthcare professionals are facing burnout and stress. Technology can play a role in alleviating administrative burdens and improving work-life balance.
    - *Recommendation*: Provide training on technology tools that enhance workflow efficiency. Implement EHR systems that prioritize usability and reduce documentation burdens [1].
- *Patient-centered care and shared decision-making*:
    - *Trend*: Patients are demanding more involvement in their healthcare decisions. Patient-centered care focuses on collaboration and shared decision-making.
    - *Recommendation*: Enhance patient education materials, involve patients in treatment decisions, and create avenues for patient feedback to improve care processes.

- *Regulatory compliance and ethical considerations*: Staying agile and adaptable to changing trends is essential for healthcare organizations. Regularly assess your organization's strengths, weaknesses, opportunities, and threats (SWOT analysis) to identify areas for improvement and innovation. Collaborate with other healthcare professionals, invest in staff development, and embrace a culture of continuous learning to navigate the future successfully [1, 42].

In summary, the healthcare sector faces various challenges and opportunities amidst evolving trends. Addressing rising costs, access disparities, and cybersecurity risks is vital for better care. Integrating innovative technologies like telehealth and AI can revolutionize patient treatment, while patient-centered practices and ethical decision-making foster trust. Regulatory compliance, data privacy, and workforce well-being are key priorities. The sector's mission of quality care, patient safety, and engagement remains central to thriving amidst change. Collaboration, innovation, and continuous improvement will drive better health outcomes globally. Figure 12.5 highlights the critical and multifaceted challenges in maintaining healthcare sector resilience amid evolving threats.

FIGURE 12.5 Vital challenges in the healthcare sector.

12.5 CYBER THREATS IN AGRICULTURE

The intersection of agriculture and the digital world promises innovative solutions that can revolutionize crop management, supply chains, and global food security. As technology becomes more deeply integrated into agricultural practices, the sector becomes vulnerable to cyber threats. This chapter aims to shed light on the problems caused by cyber threats and their impact on agriculture. Agriculture faces numerous cyber threats due to the increasing integration of technology. From data breaches to ransomware attacks to IoT vulnerabilities, vulnerabilities created by the digitalization of agriculture can disrupt operations, impair data integrity, and impact critical nutrition [43].

- *Data Breaches*: Data breaches are common in agriculture due to the collection and storage of sensitive data, such as crop data, livestock data, and financial transactions. Unauthorized access to information and agricultural data will raise concerns about privacy and financial loss.
- *Ransomware status*: Ransomware is a type of malware that encrypts files or systems and demands payment to decrypt them. Its impact on agriculture, from agriculture to supply chain management, has had a profound impact on the economy. A successful ransomware attack can cause operational disruptions, loss of revenue, and severe disruption to the entire farm.
- *Vulnerabilities in IoT devices*: The increasing use of IoT devices in agriculture is improving monitoring and automation processes. However, insufficient security measures will make these electronic devices vulnerable to attacks. Unauthorized access to the IoT from malicious sources can cause financial harm to farms by controlling data or disrupting operations.
- *Supply Chain Disruptions*: Since agricultural products are interconnected, cybercriminals can exploit this vulnerability in a variety of ways. Delays in deliveries due to disruptions in the economy can have a significant impact on food supplies and prices.
- *Infrastructure Compromises*: To support the digital transformation of agriculture, infrastructures, such as cloud services, sensors, and remote-control systems, need to be connected to the internet. Cyber-attacks on these systems can harm the functioning and health of crops by interfering with real-time monitoring, automatic irrigation, and equipment management.
- *Insider Threats*: Farming can also be impacted by insiders, such as employees, contractors, or others with authorized access. Any action taken by an insider at any time, whether intentionally or unintentionally, that compromises information security, disrupts operations, or exposes vulnerabilities to outside attackers is considered an insider threat [43, 44].

12.5.1 IMPLICATIONS FOR AGRICULTURE

The impact of cyber threats in agriculture extends beyond immediate concerns to the entire sector. This chapter presents the many benefits of cyber threats and examines their serious impact on agriculture, farmers' health, and global food security.

- *Operational Disruption*: Cyber threats have the potential to impact important agricultural activities, from planting and growing to harvesting and distribution. Ransomware attacks or compromised IoT devices can disrupt critical systems, causing delays, loss of productivity, and disruptions throughout the supply chain.
- *Data Integrity Compromise*: The integrity of agricultural data is important for informed decision-making. Cyber-attacks on data can lead to misinformation about the health of crops, soil, and weather conditions. Wrong decisions can affect productivity and cause crop damage [45].
- *Financial Losses*: While the direct financial loss is obvious, the impact can be huge. The cost of remediating cyber issues, investing in cybersecurity measures, and managing reputational damage places additional burdens on the agriculture industry. Small-scale farmers are particularly likely to be negatively affected due to limited resources.
- *Food Supply Chain Disruptions*: The connectivity of agricultural equipment increases the impact of cyber threats. Disruptions in a connection can cause disruptions that impact downstream processes, suppliers, retailers, and consumers. Food availability, price, and quality will be affected.
- *Farmer Livelihoods*: The emergence of cyber threats is affecting the lives of farmers, whose work is the basis of agriculture. Financial loss, business interruption, and data breach can harm their health as well as their business and health.
- *Global Food Security*: The far-reaching impact of cyber threats is not limited to a single farmer or field. Together, these factors can affect global food security by affecting the accessibility and affordability of staple crops. Malnutrition resulting from this condition can lead to food insecurity and weaken communities.
- *Broken trust:* In addition to lost property, broken trust can also have long-term consequences. Reduced consumer confidence in the safety and reliability of agricultural products can have a significant impact on businesspeople and business reputation [43–45].

The overall findings provide insight into the potential impact of cyber events on agricultural ecosystems. From crop failures to international transportation disruptions, these examples show how interconnected systems can expose local vulnerabilities. To effectively protect against these threats, you need to have a good approach to cybersecurity. The case study describes real-life examples showing the impact of cyber-attacks on agriculture. These examples show how individual events can have a large impact, affecting many actors in the ecosystem, from crop production to transportation disruptions. Understanding these consequences underscores the urgent need for cybersecurity measures. Agriculture must invest in defensive strategies to protect data, ensure operational efficiency, and promote a culture of cyber awareness. By implementing comprehensive cybersecurity measures, agriculture can improve its security, protect the health of farmers, and play an important role in ensuring global food security. In 2020, agriculture was the target of cyber-attacks targeting important sectors. Major food companies are facing cyber-attacks that disrupt their operations and cause delays in food deliveries.

12.5.2 Example of Cyber Threats in Agriculture

- *JBS Cyber-attack (May 2021)*: In May 2021, JBS, a global food company, fell victim to a ransomware attack that caused widespread damage in many countries. The strike led to the temporary closure of many factories and affected meat products. In the end, JBS chose to pay the hacker a ransom to restore its online systems.
- *Grain Cooperative Cyber-attack (2020)*: A significant American grain cooperative was the victim of a cyber-attack, which hampered its operations and caused grain shipments to be delayed. The cyber-attack brought attention to the susceptibility of vital agricultural infrastructure to attacks.
- *Criminals Targeting Agribusinesses*: There has been a rise in allegations of cybercrime aimed toward farms, agricultural equipment makers, and suppliers in recent years. These attacks may take the shape of ransomware, data breaches, or any other type of cyber exploitation.
- *Supply Chain Attacks*: Farmers, manufacturers, wholesalers, and retailers all play important roles in the agriculture industry's intricate supply networks. Any point along this chain could be attacked, causing chaos and a drop in food supplies.
- *IoT and Precision Agriculture Risks*: There are growing safety concerns regarding the interconnected systems used in modern agriculture as the sector embraces IoT devices and precision agriculture technologies. The security of agricultural operations and data is at risk if IoT devices are abused for their vulnerabilities.
- *Data theft and damage*: Cyber attackers can target the agriculture industry to steal confidential information, such as crop forecasts, ingredient pricing, and research data. Additionally, the destruction or manipulation of data can have a significant impact on agriculture and productivity.
- *Vulnerabilities in Farm Management Software*: Cybercriminals can exploit vulnerabilities in farm software used for farm management, data analysis, and monitoring.

Digitalization and globalization of agriculture have advantages and disadvantages. Nowadays, when technology plays an important role in agriculture, companies in the sector need to prioritize cybersecurity solutions to protect their infrastructure, information, and supply chains.

12.6 CYBER THREATS IN THE FOOD INDUSTRY

Digital search engines pose a unique challenge due to their functionality. The constant emergence of new materials, functions, and applications makes it difficult for researchers to keep up with the latest technological trends. Additionally, encryption and other data protection measures create problems in accessing and analyzing digital evidence. Analyzing big data also takes a lot of time and effort [46, 47]. Additionally, due to the complexity of digital systems and networks, it is very difficult to track the location and nature of security or data breaches. These industry impacts highlight the continued urgent need for forensic tools and techniques. The food industry continues to face serious challenges in 2021, with a US food importer falling victim. A ransomware attack involves malicious software encrypting files or systems and demanding payment in exchange for decryption, often causing significant operational and financial disruptions. The attack had a major impact on its equipment and operations. A few specific examples illustrate these problems.

- *Mondelez NotPetya Incident (2017)*: Like many other businesses, Mondelez International was hit by the NotPetya ransomware attack. It has been stated that Mondelez suffered huge financial losses as a result of the attack.
- *SolarWinds Supply Chain Attack (2020)*: The SolarWinds incident brought attention to the vulnerability of supply chain attacks in many industries, including the food industry, even though government and technological institutions were the principal targets. Potential risks to food production and distribution due to supply chain weaknesses exist.
- *Ransomware Targeting Food Distributors (2021)*: A ransomware assault in 2021 hampered the operations and supply chain of a US-based food distribution company. Attacks like these have the potential to impede food production and spread chaos throughout supply chains.
- *Food and Beverage Manufacturers' Vulnerabilities*: Manufacturers of food and drink, and the companies that provide them, are rapidly adopting digital methods. This might put them at risk for cyber-attacks like ransomware, data breaches, and interruptions to their supply chain.
- *E-commerce and Food Delivery Platforms*: Because of the popularity of online shopping and food delivery services, hackers now have more entry points than ever before. Security issues in the food industry: An attack on a food delivery platform could compromise customer information or disrupt critical services.
- *Data privacy concerns*: Increasing data breaches pose a threat to consumers' personal and financial information, especially at restaurants that collect and store customer data for online orders or loyalty programs.

It is important to recognize that cyber threats affecting the food industry can arise from a variety of factors, including economic growth, violence, surveillance, and response operations.

12.6.1 Implications of Cyber Threats in the Food Industry

The food industry faces significant risks from cyber threats affecting businesses and consumers. The benefits of cyber threats in the food industry include:

- *Data Breach and Customer Privacy*: Cyber-attacks can compromise the confidentiality of customer information to establish legitimacy, gain customers' trust, and gain the company's reputation. Customers affected by the breach may be subject to property damage or theft.

- *Supply Chain Disruption*: Strikes between suppliers or suppliers can disrupt the supply chain of products and services, causing production delays, shortages, and economic downturns that prevent the economy from meeting business needs.
- *Infection and related risks*: A cyber-attack on food processing can affect the integrity of food, create defective products, or cause serious harm by interacting with food.
- *Financial loss and fraud*: Cyber incidents can cause financial losses to businesses, including financial theft or fraud, which may impact customers due to inadequate payment methods.
- *Regulatory and Payments*: Violation of regulatory standards due to cyber threats can result in significant fines and sanctions from regulatory authorities for the financial stability of the food industry. Reputation and Customer Trust: Cyber incidents Customers are looking for other brands.
- *Intellectual Property Theft*: Valuable intellectual property, including designs and new technologies, can be stolen through cyber-attacks, destroying a company's competitive advantage and causing a major financial loss.
- *Operational Disruption*: Cyber-attacks can disrupt daily operations, impacting manufacturing processes, inventory management, and order fulfillment, causing outages.
- *Long-term damage to reputation*: Rebuilding trust and credibility after a cyber-attack is a long process. Despite high-security measures, some customers may still be concerned about the promotion of relevant names.
- *Competitive Disadvantages*: Companies affected by high-profile cyber incidents may face business problems, and customers choose competitors' products or services due to concerns about security and trust [48].

12.6.2 Examples of Cyber Threat in the Food Industry

In the food industry, point-of-sale (POS) system vulnerabilities are a significant cyber threat. In this scenario, an unauthorized person can access the POS system used by a restaurant, fast food, or grocery store [48]. The consequences of this situation are summarized as follows:

- *Attack Vectors*: Hackers can use a variety of methods, such as software vulnerabilities, employee phishing attempts, or leveraging technology as a whole to infect POS systems.
- *Data Crime*: Unauthorized access to POS can lead to the theft of customer information, including credit card numbers, names, and addresses. This stolen data can be used for theft or sold on illegal websites online.
- *Business Control*: Cyber attackers can control live businesses by modifying, altering, or issuing unauthorized refunds, causing financial loss.
- *Denial of Service (DoS) Attack*: Targeting POS for a DoS attack, the system may send a large number of requests to the server, which may cause system failure, resulting in downtime and financial loss.

- *Ransomware*: Cybercriminals can use ransomware to infect POS systems, encrypt sensitive information, and demand payment for decryption. Paying the ransom carries the risk of the company collapsing and suffering financial losses.
- *Non-compliance*: In some jurisdictions, failure to comply with data privacy laws may result in the company violating the law.
- *Loss of Customer Trust*: Brand pressure can damage customer trust and cause customers to turn to safe competitors. This can provide long-term financial benefits.

12.6.3 Future Trends and Recommendations

12.6.3.1 Future Trends

- *Blockchain for supply chain transparency*: The integration of Blockchain technology into the food supply chain holds great promise in increasing transparency and providing agricultural facts to the customer. The use of this technology has many advantages, especially the elimination of fake foods and increased consumer confidence.
- *AI-Supported Quality Control*: Based on AI and ML, quality control for food fraud can be provided. This automation helps identify any defects or contamination in the food to ensure the quality is always there.
- *Personalized foods*: Use analytics and AI algorithms to understand customers' eating patterns and preferences; this can help with personal and business personalized recommendations. This approach addresses individual needs directly, keeping customers happy and engaged.
- *Leverage smart packaging and labeling*: Adopting smart packaging systems, including QR codes and RFID tags, ensures customers receive instant updates on product base, ingredients, and shelf life. This transparency of information increases the customer's trust and confidence in the product.

These new technologies bring great benefits to the food industry, improving customer experience, increasing product quality, and making food better than ever.

12.6.3.2 Future Recommendations

Keeping these developments in mind, food industry stakeholders are advised to:

- The first step to maintaining competitiveness is to invest in digital transformation and widespread adoption of technology.
- Make supply chain trust and transparency a top priority, especially with customers.
- Sustainable and ethical sourcing techniques should be adopted.
- Partner with entrepreneurs and research institutions to innovate.
- Be vigilant about food safety and regulations.

Through proactive response to these trends and recommendations, the food sector can fulfill changing consumer wants and create a more sustainable and technologically advanced global food ecosystem.

12.7 MOTIVATIONS BEHIND CYBER-ATTACKS

Cyber-attacks in these industries can be motivated by a variety of factors. Attacks on computer networks can be driven by anything from monetary gain to political goals. In order to effectively build cybersecurity plans and respond to threats, it is essential to understand the reasons behind cyber-attacks. The following are some of the most typical causes of cyber-attacks:

- *Financial Gain*: Financial gain is often the impetus behind cyber-attacks. The goals of these intrusions include financial gain, personal data, or data with commercial value. One famous example of this motivation is ransomware assaults, in which the perpetrators demand payment to decrypt the victim's data.
- *Espionage and Intellectual Property Theft*: Organizations may be the target of state-sponsored actors and cybercriminals who are after intellectual property, trade secrets, research data, or confidential information. The exploitation of this stolen information for competitive advantage or economic espionage is not out of the question.
- *Disruption and Sabotage*: Critical services, infrastructure, or operations may be the target of a cyber-attack. In order to sow havoc, disrupt public services, or harm key systems, attackers may aim at industries like healthcare, energy, or transportation.
- *Political and Geopolitical Objectives*: There might be both political and geopolitical motivations behind state-sponsored cyber assaults. Cyber espionage, influence operations, and cyber-attacks can all be used by governments and allied parties to further their own strategic goals or interests on the international stage.
- *Data Breach and Identity Theft*: When committing identity theft, credit card fraud, or any number of other forms of cybercrime, hackers specifically target sensitive personal and financial information. Stolen information can be exploited for identity theft, financial fraud, and black-market trades.
- *Cyber Warfare and Nation-State Agendas*: It's possible that state-sponsored cyber-attacks are part of a country's larger military or geopolitical plan. Attacks might include anything from acquiring intelligence to conducting disruptive operations to destroying essential infrastructure.
- *Opportunistic Attacks*: The motivation behind certain cyber-attacks is unclear, and they instead take advantage of systems' flaws and weaknesses. Automated technologies could be used in these attacks to seek out targets in need of attack on the internet.
- *Revenge or Vendettas*: In order to exact revenge or settle scores, cybercriminals may target specific people, businesses, or government agencies.
- *Curiosity and Thrill-Seeking*: Some people or groups launch cyber-attacks because they are interested in the challenge of penetrating systems or networks for the sheer excitement of it.

These motivations illustrate the diverse and complex reasons behind cyber-attacks. Cybersecurity efforts need to address a wide range of potential motives to effectively protect systems, networks, and data from evolving threats.

12.8 DISCUSSION AND RESEARCH DIRECTIONS FOR DEVELOPING ADVANCED CYBERSECURITY SOLUTIONS

The study presented a comprehensive examination of cybersecurity threats prevalent in the healthcare, agriculture, and food sectors. Each industry faces distinct yet interconnected challenges in securing its digital infrastructures against evolving cyber threats.

- *Healthcare Sector*: In the healthcare domain, the study highlighted the increasing vulnerability of digital health systems to cyber threats. EHRs, interconnected medical devices, and telehealth platforms expose healthcare networks to heightened risks of data breaches and system vulnerabilities. The findings underscored the criticality of deploying advanced cybersecurity measures, leveraging AI-driven anomaly detection, encryption methods, and access controls to fortify healthcare systems. General discussions include ethical and legal considerations that address the need for risk assessment and regulatory compliance to protect patient information and maintain the integrity of the medical process.
- *Agriculture Sector*: Agricultural research shows the evolution and inconsistencies of technological technologies. It identifies risks associated with IoT devices and recommends countermeasures to ensure automation and data integrity in agriculture. Additionally, the study highlights the broader impacts of threats to global food security and highlights the need to build resilience across the agricultural supply chain. Collaboration among academia, industry, and policymakers is critical to developing cybersecurity strategies to address the unique challenges facing agriculture.
- *Food Industry*: An analysis of the food industry reveals that cyber threats targeting digitized food products, e-commerce platforms, and food delivery services are on the rise. This study highlights the massive impact of data breaches and related products and advocates for innovations such as Blockchain technology and smart packaging to improve screen rice transparency and tracking. Moreover, the study stressed the significance of robust cybersecurity measures in preserving brand reputation, consumer trust, and regulatory compliance within the food sector.

The research direction for advanced cybersecurity solutions includes critical domains, such as cloud security, threat intelligence and sharing, cyber insurance, human-machine teaming, multimodal biometrics, privacy impact assessments, deception technologies, emerging technology security, cybersecurity metrics, and ethical hacking. These studies are indicative of the ever-changing nature of the cybersecurity industry. Ongoing research in these areas can contribute to the development of state-of-the-art methods to combat new forms of cybercrime.

12.9 CONCLUSION

In the rapid evolution of digital photography, the connection between technology and cyber threats has become apparent. The integration of digital technology has brought significant advances to the medical, agricultural, and food sectors, increasing efficiency, productivity, and people's knowledge of used products. However, this digital transformation also exposes these important operations to various sanctions as well as various cyber threats. Our research on cyber threats in the healthcare industry shows that the shift from data collection to digital health systems is leading to changes in patient care, revealing vulnerabilities that can compromise critical medical procedures and patient privacy. The application of IoT technology in agriculture improves crop quality, but it also causes serious impacts that may affect global food security. The use of digital technology and automation in the food industry has increased efficiency. However, this technological change also makes businesses vulnerable to cyber threats, putting product integrity at risk and damaging consumers' self-confidence. Our extensive research into the cyber threat landscape reveals many strategies and incentives that lead to crime in the digital world. Capturing these incentives, from fundraising to target areas, plays an important role in creating an effective and efficient cybersecurity strategy. Anticipating changes in the cyber threat landscape presents both challenges and opportunities. The combination of AI and ML will make it easier to combat cyber-attacks. At the same time, flaws in the supply chain and IoT systems remain significant challenges. This report outlines a strategic plan that advocates for an AI-centered defense strategy to overcome these challenges while also advocating for international cooperation to improve cybersecurity and strengthen defense. As a result, while the digital world has great potential in many areas, such as health, agriculture, and food, it is important to be careful about cybersecurity. Implementing new solutions, encouraging collaboration, and ensuring strong security are critical to protecting these operations from evolving cyber threats. Integration is key to ensuring the success, stability, and repeatability of your digital environment.

REFERENCES

1. T. Javid, M. Faris, H. Beenish and M. Fahad, Cybersecurity and data privacy in the cloudlet for preliminary healthcare big data analytics, in 2020 International Conference on Computing and Information Technology (ICCIT-1441), 2020, pp. 1–4.
2. W. Khan, M. Ishrat, M. Haleem, A.N. Khan, M.K. Hasan and N.A. Farooqui, *An Extensive Study and Review on Dark Web Threats and Detection Techniques, Advances in Cyberology and the Advent of the Next-Gen Information Revolution.* IGI Global, 2023, pp. 202–219.
3. F.H. Semantha, S. Azam, K.C. Yeo and B. Shanmugam, *A systematic literature review on privacy by design in the healthcare sector*, Electronics (Basel) 9 (2020), pp. 452.
4. P. Malhotra, Y. Singh, P. Anand, D.K. Bangotra, P.K. Singh and W.-C. Hong, *Internet of things: Evolution, concerns and security challenges*, Sensors 21 (2021), pp. 1809.
5. I. Mukhopadhyay, Cyber threats landscape overview under the new normal, in *ICT Analysis and Applications*, In *ICT Analysis and Applications*, Springer Singapore, 2022, pp. 729–736.
6. M.A. Usmani, K.A. Usmani, A. Kaleem and M. Samiuddin, Cyber threat migration: perpetuating in the healthcare sector and agriculture and food industries, in *Advances in Cyberology and the Advent of the Next-Gen Information Revolution*, IGI Global, 2023, pp. 62–85.
7. D. Van Der Linden, O.A. Michalec and A. Zamansky, *Cybersecurity for smart farming: Socio-cultural context matters*, IEEE Technology and Society Magazine 39 (2020), pp. 28–35.

8. T. Drape, N. Magerkorth, A. Sen, J. Simpson, M. Seibel and R.S. Murch, et al., *Assessing the role of cyberbiosecurity in agriculture: A case study*, Frontiers in Bioengineering and Biotechnology 9 (2021), pp. 737927.
9. J.A. Chua and C. Pmp, *Cybersecurity in the healthcare industry*, Physician Leadership Journal 8, pp.229-31,(2021).
10. L.U. Khan, W. Saad, D. Niyato, Z. Han and C.S. Hong, *Digital-twin-enabled 6G: Vision, architectural trends, and future directions*, IEEE Communications Magazine 60 (2022), pp. 74–80.
11. Ylianttila M, and R. Kantola, *6G white paper: Research challenges for trust, security and privacy. arXiv preprint arXiv:2004.11665*, 2019.
12. A. Hassoun, A. Aït-Kaddour, A.M. Abu-Mahfouz, N.B. Rathod, F. Bader and F.J. Barba, et al., *The fourth industrial revolution in the food industry—Part I: Industry 4.0 technologies*, Critical Reviews in Food Science and Nutrition 63 (2023), pp. 6547–6563.
13. G. Pozsgai, G.L. Lövei, L. Vasseur, G. Gurr, P. Batáry and J. Korponai, et al., *Irreproducibility in searches of scientific literature: A comparative analysis*, Ecology and Evolution 11 (2021), pp. 14658–14668.
14. N.A. Adderley, *Graph-Based Temporal Analysis in Digital Forensics*, in scholar.afit.edu 2019.
15. W. Khan, *An exhaustive review on state-of-the-art techniques for anomaly detection on attributed networks*, Turkish Journal of Computer and Mathematics Education (TURCOMAT) 12 (2021), pp. 6707–6722.
16. E.A. Azeem, *The data carving – The art of retrieving deleted data as evidence*, International Journal for Electronic Crime Investigation 6 (2022), pp. 8.
17. R.R. Ali, K.M. Mohamad, S. Jamel and S.K.A. Khalid, *A review of digital forensics methods for JPEG file carving*, Journal of Theoretical and Applied Information Technology 96 (2018), pp. 5841–5856.
18. H.-J. Ke, J. Liu, S.-J. Wang and D. Goyal, Hash-algorithms output for digital evidence in computer forensics, in 2011 International Conference on Broadband and Wireless Computing, Communication and Applications, 2011, pp. 399–404.
19. S. Bhandari and V. Jusas, *An abstraction based approach for reconstruction of timeline in digital forensics*, Symmetry (Basel) 12 (2020), pp. 104.
20. M. Huber, M. Mulazzani, M. Leithner, S. Schrittwieser, G. Wondracek and E. Weippl, Social snapshots: Digital forensics for online social networks, in *Proceedings of the 27th Annual Computer Security Applications Conference*, 2011, pp. 113–122.
21. W. Khan, M. Haroon, A.N. Khan, M.K. Hasan, A. Khan and U.A. Mokhtar, et al., *DVAEGMM: Dual variational autoencoder with Gaussian mixture model for anomaly detection on attributed networks*, IEEE Access 10 (2022), pp. 91160–91176.
22. E.A. Vincze, *Challenges in digital forensics*, Police Practice and Research 17 (2016), pp. 183–194.
23. N. Alsaidi, M. Alshareef, A. Alsulami, M. Alsafri and A. Aljahdali, *Digital steganography in computer forensics*, International Journal of Computer Science and Information Security (IJCSIS) 18 (2020), pp. 54–61.
24. H. Kundra, W. Khan, M. Malik, K.P. Rane, R. Neware and V. Jain, *Quantum-inspired firefly algorithm integrated with cuckoo search for optimal path planning*, International Journal of Modern Physics C 33 (2022), pp. 2250018.
25. F. Aldauiji, O. Batarfi and M. Bayousef, *Utilizing cyber threat hunting techniques to find ransomware attacks: A survey of the state of the art*, IEEE Access 10 (2022), pp. 61695–61706.
26. G. Johansen, *Digital Forensics and Incident Response: Incident Response Techniques and Procedures to Respond to Modern Cyber Threats*, Packt Publishing Ltd, 2020.

27. A.A. Mughal, *A comprehensive study of practical techniques and methodologies in incident-based approaches for cyber forensics*, Tensorgate Journal of Sustainable Technology and Infrastructure for Developing Countries 2 (2019), pp. 1–18.
28. M. Sengupta, A. Roy, A. Ganguly, K. Baishya, S. Chakrabarti and I. Mukhopadhyay, *Challenges encountered by healthcare providers in COVID-19 times: An exploratory study*, Journal of Health Management 23 (2021), pp. 339–356.
29. S.U.D. Wani, N.A. Khan, G. Thakur, S.P. Gautam, M. Ali and P. Alam, et al., Utilization of artificial intelligence in disease prevention: Diagnosis, treatment, and implications for the healthcare workforce, Healthcare 10 (2022), pp. 608.
30. S. Markkandeyan, S. Gupta, G.V. Narayanan, M.J. Reddy, M.A. Al-Khasawneh and M. Ishrat, et al., *Deep learning based semantic segmentation approach for automatic detection of brain tumor*, International Journal of Computers Communications & Control 18 (2023). pp.1–16.
31. J.A. Diao, K.P. Venkatesh, M.M. Raza and J.C. Kvedar, *Multinational landscape of health app policy: Toward regulatory consensus on digital health*, NPJ Digital Medicine 5 (2022), pp. 61.
32. K. Begum, A. Dhamodhiran and N.M.S. Gokul, *Health care data privacy and compliance: Navigating regulatory landscape*, Central Asian Journal of Medical and Natural Science 4 (2023), pp. 596–608.
33. C. Anderson, R. Baskerville and M. Kaul, *Managing compliance with privacy regulations through translation guardrails: A health information exchange case study*, Information and Organization 33 (2023), pp. 100455.
34. D. Hawig, C. Zhou, S. Fuhrhop, A.S. Fialho and N. Ramachandran, *Designing a distributed ledger technology system for interoperable and general data protection regulation–compliant health data exchange: A use case in blood glucose data*, Journal of Medical Internet Research 21 (2019), pp. e13665.
35. J. Tully, J. Selzer, J.P. Phillips, P. O'Connor and C. Dameff, *Healthcare challenges in the era of cybersecurity*, Health Security 18 (2020), pp. 228–231.
36. R.M. Wells, *Identifying Trends Associated with Cyber-Crime in Healthcare Industries*, Northcentral University, 2021.
37. S. Nasiri, F. Sadoughi, M.H. Tadayon and A. Dehnad, *Security requirements of internet of things-based healthcare system: A survey study*, Acta Informatica Medica 27 (2019), pp. 253.
38. D. Lee and S.N. Yoon, *Application of artificial intelligence-based technologies in the healthcare industry: Opportunities and challenges*, International Journal of Environmental Research and Public Health 18 (2021), pp. 271.
39. Al Meslamani, A Z, *Technical and Regulatory Challenges of Digital Health Implementation in Developing Countries*, Taylor & Francis, 2023.
40. S. Abidin, M. Prasad Raghunath, P. Rajasekar, A. Kumar, D. Ghosal and M. Ishrat, Identification of disease based on symptoms by employing ML, in 2022 International Conference on Inventive Computation Technologies (ICICT), 2022, pp. 1357–1362.
41. M.N.O. Sadiku, O.I. Fagbohungbe and S.M. Musa, *Artificial intelligence in cyber security*, International Journal of Engineering Research and Advanced Technology 6 (2020), pp. 1–7.
42. N. Allahrakha, *Balancing cyber-security and privacy: Legal and ethical considerations in the digital age*, Legal Issues in the Digital Age 4 (2023), pp. 78–121.
43. K. Demestichas, N. Peppes and T. Alexakis, *Survey on security threats in agricultural IoT and smart farming*, Sensors 20 (2020), pp. 6458.
44. T.H.H. Aldhyani and H. Alkahtani, *Cyber security for detecting distributed denial of service attacks in agriculture 4.0: Deep learning model*, Mathematics 11 (2023), pp. 233.

45. A.N. Alahmadi, S.U. Rehman, H.S. Alhazmi, D.G. Glynn, H. Shoaib and P. Solé, *Cyber-security threats and side-channel attacks for digital agriculture*, Sensors 22 (2022), pp. 3520.
46. M. Stoyanova, Y. Nikoloudakis, S. Panagiotakis, E. Pallis and E.K. Markakis, *A survey on the internet of things (IoT) forensics: Challenges, approaches, and open issues*, IEEE Communications Surveys & Tutorials 22 (2020), pp. 1191–1221.
47. Z.A. Baig, P. Szewczyk, C. Valli, P. Rabadia, P. Hannay and M. Chernyshev, et al., *Future challenges for smart cities: Cyber-security and digital forensics*, Digital Investigation 22 (2017), pp. 3–13.
48. S. Sontowski, M. Gupta, S.S.L. Chukkapalli, M. Abdelsalam, S. Mittal and A. Joshi, et al., Cyber-attacks on smart farming infrastructure, in 2020 IEEE 6th International Conference on Collaboration and Internet Computing (CIC), 2020, pp. 135–143.

13 Cyberbullying
Understanding, Preventing, and Addressing Online Harassment

Syed Adnan Afaq, Mohammad Faisal, Saurabh Srivastava, Tasneem Ahmed, Gausiya Yasmeen, Soni Singh, and Saman Uzma

13.1 INTRODUCTION

Cyberbullying is a huge social concern in the digital age. Electronic communication is used to purposefully harm, harass, or scare others. This chapter aims to describe cyberbullying, how it manifests itself, and how it affects both victims and offenders. Cyberbullying affects people of all ages and backgrounds. It has become more common as a result of social reliance on Internet channels for communication and interaction. It is critical to prevent and treat it because it can harm one's mental health, well-being, and relationships. The chapter opens with defining cyberbullying and its prevalence online. We investigate the intricate digital relationships that enable cyberbullying and how it progresses. Because technology connects people and makes these things easier, we must understand how they work. In extreme cases, cyberbullying can result in emotional suffering, worse grades, social isolation, and self-harm or suicide. To address this problem, you must first understand its expressions. Hurting communications, cyberstalking, doxxing, impersonation, and exclusionary practices are examples. The next section delves into the reasons for cyberbullying, illuminating the complexities of digital human behavior. People become more aggressive online due to anonymity and the impact of disinhibition. Cyberbullying can easily traverse boundaries because of advances in technology and social media. Lack of empathy, peer pressure, and societal expectations all contribute to cyberbullying [1].

Cyberbullying is a pervasive and concerning issue that has permeated the fabric of our interconnected society. It encompasses a range of malicious behaviors where individuals exploit the anonymity and accessibility of the online world to intentionally harm, harass, or intimidate others [2]. The anonymity offered by the digital veil allows perpetrators to engage in hurtful actions without revealing their

DOI: 10.1201/9781003409571-13

identities, leading to a sense of detachment from the consequences of their actions. The screen becomes a mask behind which they can unleash vitriol and cruelty, perpetuating an environment of fear and insecurity for the victims [3]. Recent research reveals developing cyberbullying dynamics, stressing the role of social media and online gaming. Studies also stress the long-term psychological effects on victims, underlining the importance of focused therapy. Additionally, the study investigates novel preventive approaches, such as digital literacy initiatives. Some researchers study AI-based technologies and legal frameworks to address cyberbullying's technological challenges [4]. The intersectionality of cyberbullying, taking into account aspects such as gender and ethnicity, has also been explored. Expert perspectives provide guidance on future research directions. Investigations into cross-cultural variances in cyberbullying manifestations highlight the global nature of the issue. The literature emphasizes the importance of taking a holistic, interdisciplinary strategy to properly address the ever-changing digital context. Keeping up with the newest research in credible publications and databases is crucial for growing insights into cyberbullying [5].

The study's motivation is deeply rooted in the increasing significance of creating a secure and courteous online atmosphere. With the constant evolution of the digital realm, cyberbullying tactics and consequences also evolve. It is crucial to not view this phenomenon solely as an academic pursuit but as a responsibility towards society. By thoroughly understanding the complexities surrounding online harassment, we equip ourselves to implement proactive strategies that safeguard those who are vulnerable and foster an environment of empathy and acceptance in the virtual realm. The author presents a wealth of knowledge in this study. With experience in the field of cyber security, he devoted his attention to deciphering the complexity of cyberbullying. The author gives unique insights into the psychological, social, and technological elements of online harassment through thorough analysis and a commitment to social welfare. This chapter represents the author's motivation to make the digital world safer and contributes significantly to the continuing discussion about cyberbullying prevention and intervention.

13.2 FUNDAMENTAL OF CYBERBULLYING

Cyberbullying refers to the act of using digital communication tools, such as social media platforms, messaging apps, emails, or online forums, to harass, intimidate, or target individuals with hurtful, offensive, or threatening content [6, 7]. Unlike traditional bullying, cyberbullying occurs in the virtual realm, where perpetrators can remain anonymous or pseudonymous, making it difficult for victims to identify and protect themselves from the harassers [8]. The Cyberbullying Research Center, in operation since 2002, has conducted extensive research encompassing over 30,000 middle and high school students across the United States through 14 research projects. This report exclusively focuses on findings from the 12 most recent studies, shedding light on the prevalence of

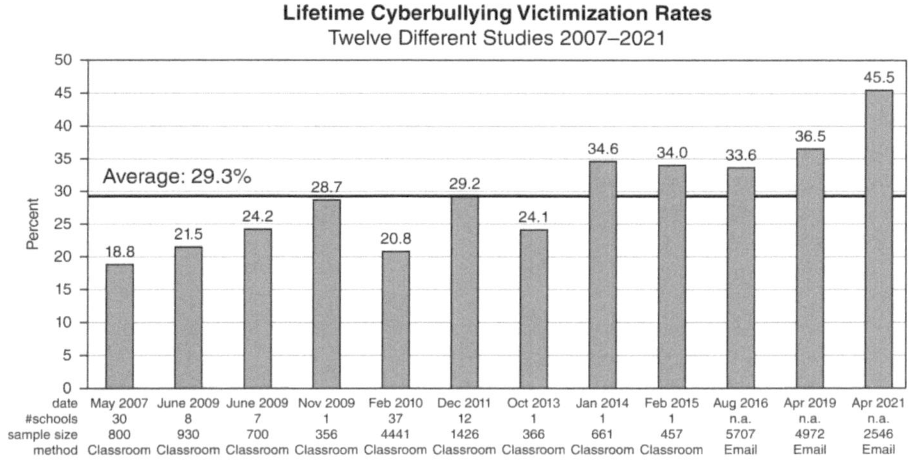

FIGURE 13.1 Cyberbullying victimization. (Source: Justin W. Patchin and Sameer Hinduja www. cyberbullying.org.)

cyberbullying among surveyed students. Figures 13.1 and 13.2 visually represent the data, showcasing the percentage of students who have experienced cyberbullying during their lifetime and trends observed across the 12 studies. These findings highlight the urgency of addressing cyberbullying to create safer digital environments for youth. The Cyberbullying Research Center remains dedicated to advancing understanding of this crucial issue and can be contacted for further information.

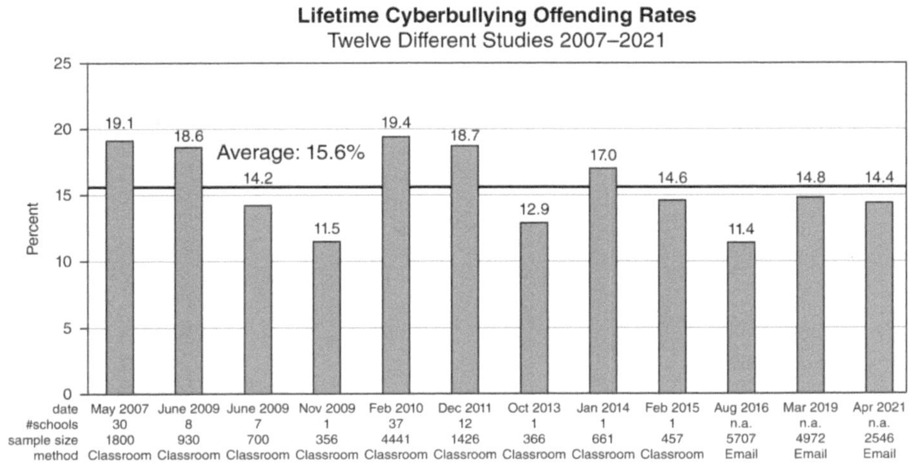

FIGURE 13.2 Cyberbullying offending. (Source: Justin W. Patchin and Sameer Hinduja www. cyberbullying.org.)

13.2.1 Types of Cyberbullying

There are many different manifestations of cyberbullying, such as sending abusive messages, spreading rumors or false information, sharing embarrassing images or films without the recipient's knowledge, engaging in online mobbing or collective harassment, and posting embarrassing photos or videos [9]. Anxiety, sadness, social isolation, and, tragically, even self-harm or suicide can be the result of cyberbullying, which can have a devastating impact on a person's emotional and psychological well-being [10]. Figure 13.3 refers the types of cyberbullying.

- *Harassment*: Sending offensive messages, threats, or spreading rumors to intentionally harm the target.
- *Flaming*: Engaging in online fights, arguments, or disputes with the purpose of provoking and upsetting others.
- *Impersonation*: Creating fake profiles or using someone else's identity to spread false information or damage their reputation.
- *Cyberstalking*: Repeatedly sending unwanted messages, monitoring someone's online activity, or making persistent threats.
- *Dissing*: It involves using disrespectful language and derogatory comments online to intentionally hurt or humiliate others, perpetuating a negative online environment.
- *Trickery*: It employs deceitful tactics such as impersonation or spreading false information to manipulate and harm individuals, amplifying the harmful impact of online harassment.

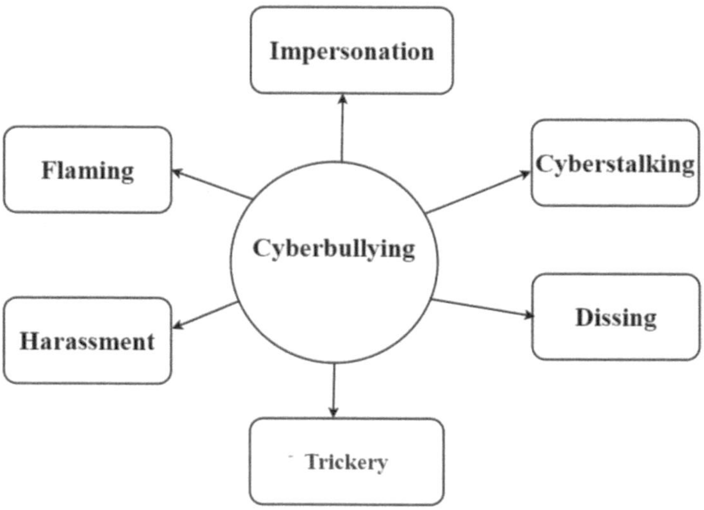

FIGURE 13.3 Types of cyberbullying.

The frequency of cyberbullying in online spaces is staggering, affecting individuals of all ages, genders, races, and socioeconomic backgrounds. This alarming trend has not spared the most vulnerable among us, as children, teenagers, and young adults increasingly find themselves entangled in the web of digital harassment. No longer confined to the schoolyard or workplace, cyberbullying has infiltrated the very spaces we once considered safe and private, encroaching upon our homes, social circles, and digital personas [11]. As cyberbullying becomes a growing concern for individuals, communities, policymakers, and technology companies, it is imperative to explore effective prevention and intervention strategies [12]. Empowering people with digital literacy and encouraging responsible online behavior can promote a culture of empathy and respect, lowering the likelihood of negative cyberbullying behavior. Creating a good and inclusive online environment is critical to building a digital landscape that promotes tolerance, generosity, and mutual support. As prominent administrators of online communication, social media companies carry a crucial duty in addressing cyberbullying. Implementing comprehensive reporting and moderation systems helps speed up the removal of bad information, sparing potential victims from additional harm. However, the effectiveness of these interventions is not without limitations, since the ever-changing nature of technology presents particular obstacles in effectively combating cyberbullying [13].

13.2.2 Prevalence and Impact of Cyberbullying

- *Prevalence*: The worldwide incidence of cyberbullying has escalated to frightening proportions and become a disturbing and widespread problem. The advent of the digital age has furnished cyberbullies with an extensive medium via which they can effortlessly target individuals across time zones and physical locations. Online contacts provide perpetrators with the ability to conceal their identities, so impeding victims' ability to recognize and confront their tormentors. The ability to remain anonymous provides cyberbullies with the confidence to partake in harmful actions without immediate consequences, hence increasing the prevalence and severity of cyberbullying occurrences [14].

  Cyberbullying impacts a considerable number of young people, according to research from many countries and organizations. According to the Cyberbullying Research Center, a poll of middle and high school students in the United States indicated that more than 37% of them had experienced cyberbullying at some time in their lives. Similarly, a survey conducted by EU Kids Online discovered that approximately 12% of European children aged 9 to 16 reported being cyberbullied [15–17]. The prevalence of cyberbullying is not limited to younger age groups; it also extends to adults. The workplace, social media platforms, and online forums have witnessed instances of cyberbullying among adults, coworkers, acquaintances, and even strangers engaging in harmful behaviors. The anonymity and distance provided by the digital environment often amplifies the severity of cyberbullying, as individuals may feel less restrained in expressing their hostile sentiments.

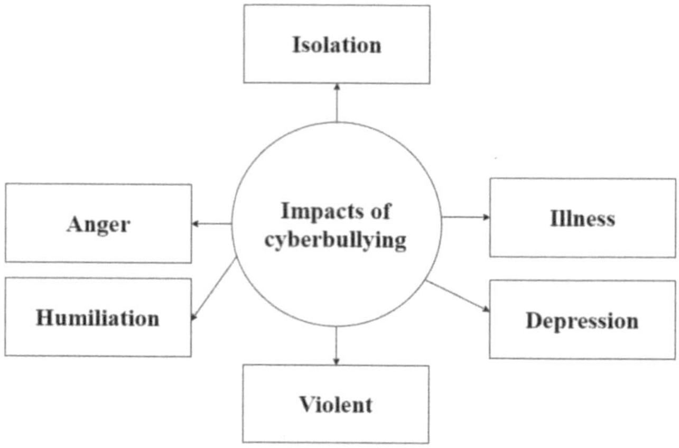

FIGURE 13.4 Impact of cyberbullying.

- *Impact*: The impact of cyberbullying extends beyond individual victims; it affects entire communities and societies. Schools and educational institutions witness a decline in the overall learning environment, as students fear being targeted and become hesitant to express themselves openly. Figure 13.4 shows the impact of cyberbullying.

Teachers and school staff also grapple with the aftermath of cyberbullying incidents, needing to address the emotional well-being of affected students while maintaining a safe learning environment for all. Furthermore, cyberbullying incidents can escalate into public spectacles, involving bystanders and witnesses who may participate or inadvertently contribute to the spread of harmful content [18]. The ripple effect of cyberbullying can lead to social discord and strained relationships within communities, eroding trust and cohesion. The impact of cyberbullying on victims' mental health and well-being cannot be understated. The persistent and widespread nature of online harassment leads to increased stress, anxiety, and feelings of helplessness. Victims often experience a loss of self-esteem and self-confidence, questioning their self-worth due to the relentless negative feedback they receive online. Moreover, cyberbullying can lead to social isolation, as victims may withdraw from online and offline interactions to avoid further harm. The lack of social support and the feeling of alienation can further exacerbate the negative emotional consequences of cyberbullying [19].

13.3 FORMS OF CYBERBULLYING

Cyberbullying encompasses a wide range of hurtful behaviors and tactics, all of which leverage digital technology and online platforms to harass, intimidate, or harm individuals. As technology continues to evolve, new forms of cyberbullying emerge, presenting unique challenges for both victims and those seeking to address and combat this pervasive issue. Below are some of the primary forms and manifestations of cyberbullying [20]:

Direct attacks involve sending hurtful or threatening messages directly to the victim through various digital channels, such as social media, email, instant messaging apps, or text messages. These messages may contain offensive language, derogatory remarks, or even threats of physical harm. Direct attacks can lead to emotional distress and fear for the victim, as they face constant online harassment.

Cyberstalking refers to the persistent and unwanted online attention directed at an individual. Cyberstalkers may monitor the victim's online activities, follow them across different platforms, and track their location or personal information. This type of cyberbullying can be very terrifying and intrusive because the victim feels continually monitored and harassed.

The person who *doxxes* someone else's private information, like their home address, phone number, or financial information, in order to hurt them is breaking the rules of cyberbullying. This information could be shared with the public or with certain groups, leaving the victim open to real-life threats like harm, identity theft, or harassment.

Cyberbullies may use photos or videos to humiliate, shame, or threaten the victim. This form of cyberbullying often involves the non-consensual distribution of explicit or compromising content, commonly known as "revenge porn." These images or videos can cause significant emotional and reputational damage to the victim [21]. In some cases, cyberbullies may engage in more sophisticated tactics, such as phishing or hacking, to gain access to the victim's personal information, accounts, or online activities. This invasion of privacy can lead to severe consequences for the victim's safety and security.

13.4 FACTORS CONTRIBUTING TO CYBERBULLYING

Cyberbullying is a complex social issue influenced by a combination of individual, interpersonal, and societal factors. Understanding these contributing factors is crucial for developing targeted prevention and intervention strategies. The following are some of the key factors that contribute to the prevalence of cyberbullying as shown in Figure 13.5.

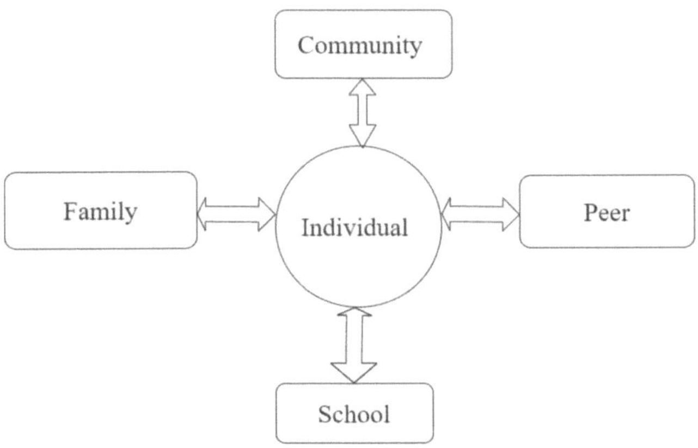

FIGURE 13.5 Risk factors for cyberbullying.

13.4.1 Anonymity and Disinhibition

The online environment provides a sense of anonymity, allowing individuals to conceal their identities and act without fear of immediate consequences. This anonymity can embolden cyberbullies to engage in hurtful behaviors they might not consider in face-to-face interactions. The lack of accountability and the perception of being invisible can foster a callous disregard for the impact of their actions on others.

13.4.2 Accessibility of Technology and Social Media

The widespread availability and accessibility of digital technology and social media platforms provide cyberbullies with the means to target victims across different contexts and geographic locations. The constant connectivity to the Internet allows cyberbullying incidents to occur at times, making it difficult for victims to escape the harassment.

13.4.3 Psychological and Sociocultural Factors

Individual psychological traits, such as low self-esteem, aggression, and a lack of empathy, can contribute to cyberbullying behavior. Additionally, sociocultural factors, including societal norms that tolerate or even glorify aggression, can shape the way individuals interact online and perpetrate cyberbullying [22].

13.4.4 Parental Supervision and Communication

The level of parental supervision and communication regarding online activities can play a significant role in preventing cyberbullying. Parents who are actively involved in their children's digital lives and promote open discussions about responsible online behavior are better equipped to address cyberbullying issues early on [23].

13.5 CYBERBULLYING PREVENTION AND INTERVENTION STRATEGIES

Cyberbullying prevention and intervention measures are critical in tackling this persistent problem and making the Internet a safer place for everyone. People, families, institutions, businesses, and industry as a whole should all be included in these measures. Here are some strategies for preventing and intervening with cyberbullying [5].

13.5.1 Education and Awareness

Create awareness of cyberbullying, its consequences, and the significance of appropriate online activity. Schools, companies, and organizations implement educational initiatives to encourage computer skills, empathy, and courteous digital communication [24].

13.5.2 Encourage Open Communication

Encourage open conversation regarding online experiences among parents, teachers, and children. Encourage victims to report instances without fear of being judged or retaliated against, and encourage bystander intervention to help individuals who are cyberbullied [25].

13.5.3 Empower Bystanders

Motivate bystanders to defend victims of cyberbullying and take action against it. Inform people on safe ways to step in and how to notify the appropriate authorities about instances of cyberbullying [26].

13.5.4 Provide Counseling and Mental Health Support

Provide victims of cyberbullying with mental health support services and counseling. Victims can develop resilience and manage the emotional toll with the support of professionals.

13.5.5 Training for Educators and Professionals

Establish training and professional development initiatives that equip educators, counselors, and other relevant personnel with the necessary skills to identify instances of cyberbullying and respond appropriately.

Through the integration of prevention, intervention, and support mechanisms, society can together confront the issue of cyberbullying and provide an online environment that is both secure and empathetic for all individuals [27].

13.6 CYBERBULLYING AND SOCIAL MEDIA PLATFORMS

Cyberbullying and social media platforms are closely intertwined, as these platforms often serve as the primary space where cyberbullying incidents occur. Social media platforms provide users with various tools and features for communication, interaction, and content sharing, but they also present opportunities for cyberbullies to target and harass others. Figure 13.6 shows the Percentage of Internet users in various countries who have ever experienced any cybercrime in 2022.

Here are some key points regarding the relationship between cyberbullying and social media platforms.

13.6.1 Amplification of Cyberbullying

Social media platforms have the potential to amplify cyberbullying incidents due to their widespread reach and instant communication capabilities. Hurtful messages, images, or videos can be shared and spread rapidly across multiple users, causing significant harm to the victim's reputation and emotional well-being.

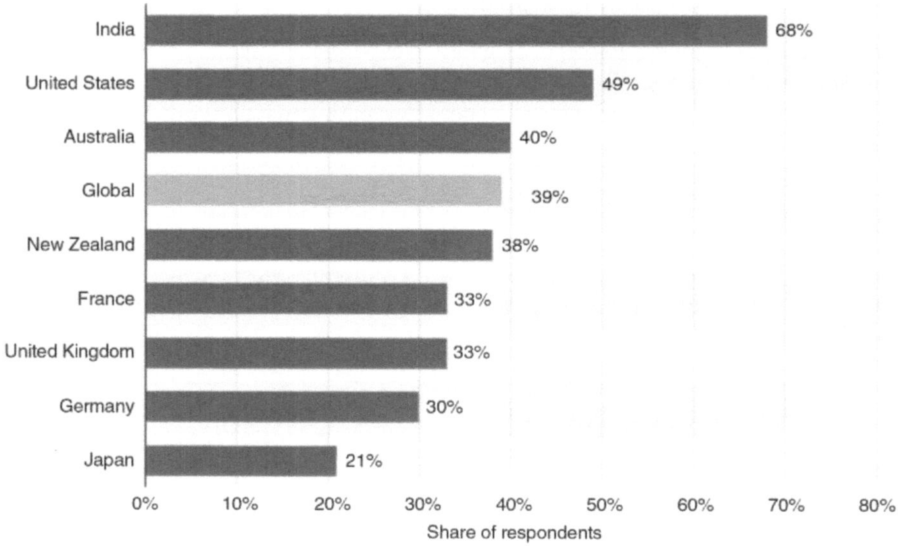

FIGURE 13.6 Percentage of Internet users in various countries who have ever experienced any cybercrime in 2022. (Source: https://www.statista.com/.)

13.6.2 Anonymity and Pseudonymity

Social media platforms often allow users to maintain a level of anonymity or pseudonymity, which can embolden cyberbullies to act without revealing their true identities. This anonymity can make it challenging to identify and hold perpetrators accountable for their actions.

13.6.3 Social Media Trolling

Trolling, a form of cyberbullying where individuals provoke or antagonize others for amusement, is prevalent on social media platforms. Trolls may target individuals based on their beliefs, appearance, or personal characteristics, leading to online harassment [28].

13.6.4 Cyberbullying in Private Chats and Groups

Social media platforms facilitate private group chats and communities, which can become breeding grounds for cyberbullying. Cyberbullies may use these secret spaces to harass and isolate people away from public attention.

Following are two case studies and real-life examples.

13.6.5 Case Study 1: Amanda Todd's Tragic Story

Amanda Todd, a Canadian teenager, became an iconic symbol of the devastating consequences of cyberbullying. In 2012, Amanda posted a YouTube video detailing her experiences with cyberbullying, which included online harassment, sextortion, and the circulation of explicit photos. She explained how her life had been plagued by relentless bullying, both online and offline since she was in seventh grade.

The video went viral, drawing attention to the harrowing effects of cyberbullying on mental health. Amanda had a difficult time coping with the emotional damage that was produced by the cyberbullying, despite her attempts to get care and support despite her efforts. In a tragic turn of events, Amanda committed suicide in October 2012, when she was only 15 years old. She left behind a legacy that seeks to raise public awareness about harassment and the impact it has on these young people.

13.6.6 Case Study 2: Tyler Clementi's Suicide

In 2010, Tyler Clementi, a college student in the United States who was 18 years old at the time, became a victim of cyberbullying he had experienced. Through the use of a camera, Tyler's roommate could covertly stream and publish a private experience between Tyler and another man on the Internet without his knowledge or consent. Additionally, the film was distributed to other kids, which resulted in Tyler being humiliated and embarrassed in front of his peers.

Tyler committed suicide by jumping off the George Washington Bridge, leaving behind a huge influence on his family as well as on society as a whole. He was devastated by the cyberbullying and invasion of privacy that he had just experienced. The untimely passing of this individual shed focus on the serious consequences of cyberbullying and the necessity of implementing more stringent measures to handle this type of harassment in educational institutions and online places.

13.7 DISCUSSION

The discussion part illuminates cyberbullying's pervasiveness and devastating effects on victims and perpetrators. Cyberbullying has become common in the digital age as people use Internet platforms to hurt, harass, and scare others. Defining cyberbullying and analyzing its prevalence online are the topics covered in this chapter. In order to gain a better understanding of the mechanisms behind cyberbullying, complex digital interactions that enable it are investigated. Since technology links people, cyberbullying is made easier. It is necessary to have an understanding of these digital dynamics to devise effective approaches for intervention and prevention. A person may experience emotional anguish, a decline in their academic performance, social isolation, and even self-harm or suicide as a result of cyberbullying. In order to establish focused methods for the prevention of cyberbullying, it is vital to have a solid understanding of various manifestations. A nuanced perspective on the multiple manifestations of cyberbullying is presented in this chapter. These manifestations include destructive communications, cyberstalking, doxxing, impersonation, and acts that encourage isolation.

13.8 CONCLUSION

To summarize, cyberbullying is a widespread and very worrying phenomenon that continues to affect people all around the world. As technology advances and online interactions become more intertwined in our lives, cyberbullying difficulties and trends grow, bringing additional complexities in tackling this detrimental behavior. To effectively combat cyberbullying, it is critical to know and comprehend the numerous forms and manifestations it can take, as well as the cyberbullies' worldview. Education, knowledge, and digital literacy are critical in equipping people to be responsible digital citizens and to develop empathy and respect online. In preventing and intervening in cyberbullying situations, the roles of social media platforms, technological corporations, educators, parents, and society as a whole are intertwined. To establish a safer digital environment in which individuals may engage without fear of harassment or violence, a collaborative strategy encompassing all stakeholders is required. While technological improvements provide new issues, they also provide the potential to develop innovative solutions to cyberbullying. Adopting AI-driven content moderation, enforcing stronger regulations against cyberbullying, and offering mental health care to victims are all potential solutions for effectively combating this issue. Combating cyberbullying is a never-ending journey that necessitates constant research, adaptability, and monitoring. We can develop a culture of empathy, respect, and responsible digital behavior by working together to make the Internet a safer and more compassionate place for everyone. Preventing cyberbullying is an ethical necessity that represents our commitment to the well-being and dignity of all individuals in the digital age.

REFERENCES

1. M. Kaur and M. Saini, "Indian Government initiatives on cyberbullying: A case study on cyberbullying in Indian higher education Institutions," *Educ. Inf. Technol.*, vol. 28, no. 1, pp. 581–615, 2023.
2. T. Holt and A. Bossler, *Cybercrime in progress: Theory and prevention of technology-enabled offenses.* Routledge, 2015.
3. N. Javed, T. Ahmed and M. Faisal, "A comprehensive study on prevalence of cyberbullying and its impact on youths and adults," *J. Stat. Manage. Syst.*, vol. 26, no. 7, pp. 1655–1672, 2023, doi: 10.47974/jsms-1132.
4. E. D. Ng, J. Y. X. Chua and S. Shorey, "The effectiveness of educational interventions on traditional bullying and cyberbullying among adolescents: A systematic review and meta-analysis," *Trauma, Violence, Abus*, vol. 23, no. 1, pp. 132–151, 2022.
5. N. Javed, T. Ahmed and M. Faisal, "Cyberbullying and The Laws : A Legal Barrier To Prevent Cyber Crimes Cyberbullying and The Laws : A Legal Barrier To Prevent Cyber Crimes," *Sustain. Environ. Manif. Augment.*, no. May, pp. 1–9, 2023, [Online]. Available: https://www.researchgate.net/profile/Dr-Tasneem-Ahmed/publication/370497145_Sustainable_Environment_Manifestation_and_Augmentation_Cyberbullying_and_The_Laws_A_Legal_Barrier_To_Prevent_Cyber_Crimes/links/645354c597449a0e1a762f95/Sustainable-Environment-Ma
6. N. Javed, T. Ahmed, M. Faisal, H. Sadia and E. Z. J. Sidaine-Daumiller, "Workplace cyberbullying in the remote-work era: A new dimension of cyberology," *Adv. Cyberology Advent Next-Gen Inf. Revolut.*, pp. 166–177, 2023, doi: 10.4018/978-1-6684-8133-2.ch009.
7. C. Langos, "Cyberbullying: The challenge to define," *Cyberpsychol. Behav. Soc. Netw*, vol. 15, no. 6, pp. 285–289, 2012.
8. G. W. Giumetti and R. M. Kowalski, "Cyberbullying via social media and well-being," *Curr. Opin. Psychol.*, vol. 45, p. 101314, 2022.
9. D. L. Espelage, J. S. Hong and A. Valido, "Cyberbullying," in *Handbook of educational psychology*, Routledge, 2024, pp. 654–673.

10. J. Chun, J. Lee, J. Kim and S. Lee, "An international systematic review of cyberbullying measurements," *Comput. Human Behav*, vol. 113, p. 106485, 2020.
11. S. A. Afaq, M. S. Husain, A. Bello and H. Sadia, "A critical analysis of cyber threats and their global impact," in *Computational intelligent security in wireless communications*, CRC Press, pp. 201–220, 2023.
12. D. L. Espelage, C. Torgal, A. Valido, L. E. Robinson and G. L. Clements, "Cyberbullying among youth," *Friend Or Foe Tackling Issue Soc. Media Sch.*, Vol. 14, p. 59, 2022.
13. C. Kaluarachchi, M. Warren and F. Jiang, "Responsible use of technology to combat cyberbullying among young people," 2020.
14. R. M. Kowalski, S. P. Limber and A. McCord, "A developmental approach to cyberbullying: Prevalence and protective factors," *Aggress. Violent Behav.*, vol. 45, pp. 20–32, 2019.
15. M. Vismara, N. Girone, D. Conti, G. Nicolini and B. Dell'Osso, "The current status of cyberbullying research: A short review of the literature," *Curr. Opin. Behav. Sci.*, vol. 46, p. 101152, 2022.
16. D. Kuzmanović, Z. Pavlović, D. Popadić and T. Milosevic, "Internet and digital technology use among children and youth in Serbia: EU Kids Online Survey Results, 2018," *Kuzmanović, D., Pavlov. Z., Popadić, D., Milosevic*, 2019.
17. D. Smahel *et al.*, "EU Kids Online 2020: Survey results from 19 countries," 2020.
18. P. Redmond, J. V. Lock and V. Smart, "Developing a cyberbullying conceptual framework for educators," *Technol. Soc.*, vol. 60, p. 101223, 2020.
19. M. R. Ohara, "The role of social media in educational communication management," *J. Contemp. Adm. Manage.*, vol. 1, no. 2, pp. 70–76, 2023.
20. J. W. Patchin and S. Hinduja, "Measuring cyberbullying: Implications for research," *Aggress. Violent Behav.*, vol. 23, pp. 69–74, 2015.
21. E. Reed, A. Wong and A. Raj, "Cyber sexual harassment: A summary of current measures and implications for future research," *Violence Against Women*, vol. 26, no. 12–13, pp. 1727–1740, 2020.
22. Y. Wang, "Understanding the role of social factors in cyberbullying at work," *Comput. Human Behav.*, vol. 134, p. 107325, 2022.
23. S. W. Azumah, V. Adewopo and Z. ElSayed, "A Secure Open-Source Intelligence Framework For Cyberbullying Investigation," *arXiv Prepr. arXiv2307.15225*, 2023.
24. M. Jitaru and O. Candel, "The role of school counsellors in cyberbullying prevention and intervention programmes," *Cyberbullying Educ. Context.*, vol. 3, pp. 137–150, 2022.
25. B. A. Spears, C. Taddeo and L.-A. Ey, "Using participatory design to inform cyber/bullying prevention and intervention practices: Evidence-informed insights and strategies," *J. Psychol. Couns. Sch.*, vol. 31, no. 2, pp. 159–171, 2021.
26. M. Heyeres, M.-A. Carter, S. M. Lui, Ä Low-Lim, S. Teo and K. Tsey, "Cyberbullying prevention and treatment interventions targeting young people: An umbrella review," *Pastor. Care Educ.*, vol. 39, no. 2, pp. 125–151, 2021.
27. C. B. Davison and C. H. Stein, "The dangers of cyberbullying," *N. Am. J. Psychol.*, vol. 16, no. 3, p. 595, 2014.
28. H. Jahankhani, A. Al-Nemrat and A. Hosseinian-Far, "Chapter 12 - Cybercrime classification and characteristics," B. Akhgar, A. Staniforth, and F. B. T.-C. C. and C. T. I. H. Bosco, Eds. Syngress, 2014, pp. 149–164. doi: 10.1016/B978-0-12-800743-3.00012-8.

14 The Perils of Cyberbullying on Social Media

Detection and Prevention with Artificial Intelligence

Neha Bhushan and Vipin Khattri

14.1 INTRODUCTION

Marshall McLuhan's concept of the global village was first introduced in the 1960s. He argued that the world was becoming increasingly interconnected due to the spread of electronic media, such as television and the Internet. McLuhan believed that this would lead to a more unified world, where people would be more aware of each other and their shared cultures [1]. In today's world, McLuhan's concept of the global village is more relevant than ever. The ubiquitous presence of smartphones and instant messaging apps has facilitated global connectivity, enabling individuals to remain in contact regardless of their geographical location. This has resulted in the formation of novel forms of social relationships, such as virtual friendships and online communities [2]. These emerging forms of social relationships encompass both advantageous and detrimental characteristics. Firstly, they can facilitate individuals in maintaining communication with distant friends and family members. Additionally, they can serve as a medium for individuals to establish connections with like-minded others who have similar interests. Conversely, these emerging types of social connections have the potential to cause isolation and dependence. In addition, they can be utilized to disseminate false information and engage in cyberbullying by promoting hate speech [2]. Dan Olweus was a pathfinder in defining bullying in schools as a harmful behavior and asserted that he emphasized that, contrary to common opinion, experiencing bullying in school is not a beneficial aspect of personal development [3]. Bullying is frequently distinguished from other forms of aggressiveness using the definitional criteria offered by Olweus which include the intention to cause harm, the act of repeating harmful actions, and an unequal distribution of power [4]. Bullying is not a new phenomenon, and as soon as digital technologies became the main forms of communication, cyberbullying began to appear. Both traditional bullying and cyberbullying are types of bullying, yet they differ significantly in some important ways. Cyberbullying, in contrast to traditional bullying, occurs online or using electronic devices. This indicates that, unlike traditional bullying, cyberbullying may occur around the clock and affect a larger demographic.

DOI: 10.1201/9781003409571-14

Cyberbullying can be classified into various forms, such as flaming, harassing, flooding, masquerade, impersonation, cyberstalking, denigration, outing, and stalking [5]. Each of these forms of cyberbullying possesses distinct characteristics [6].

- *Flooding*: Flooding is a form of cyberbullying characterized by the repetitive transmission of the same comments, making absurd remarks, and continually pushing the enter button to disrupt communication and undermine the role of the cyberbullying target [7].
- *Flaming*: Flaming is the act of sending angry, harsh, or vulgar remarks to a person or group of people either personally or to an online group [7].
- *Harassment*: Harassment is when someone receives unfavorable texts frequently [7].
- *Cyberstalking*: The act of repeatedly harassing an individual by electronic methods to cause anxiety or anguish [8].
- *Denigration*: Engaging in the act of disseminating or transmitting harmful and humiliating content, such as private text messages and photographs, about an individual to others, to degrade the person in the eyes of others [9].
- *Masquerade*: Using a false identity to send or publish anything that endangers someone or damages their reputation is known as masquerading [7].
- *Outing*: Disseminating or sharing personal information about someone without their consent, for causing embarrassment or injury to that individual [9].
- *Exclusion*: The deliberate and targeted act of eliminating an individual from an online group [9].
- *Impersonation*: The act of illicitly accessing someone's account, assuming their identity, and sending messages with the intention of tarnishing their image, causing harm or danger, or undermining their reputation and relationships [10].
- *Sexting*: Sexting is the act of sending obscene images of someone else without their consent [11].

Cyberbullying is a widespread social issue with a significant impact in the digital age, affecting people of all ages across the globe, including India. A renowned psychologist known for his work on social learning theory and self-efficacy, Albert Bandura said that deliberate actions are typically responsible for significant hazards to human well-being [12]. Out of several deliberate acts that can pose massive threats to human welfare, one can be cyberbullying. It is crucial to be alert to these dangers and to work to prevent them. In the mainstream press and literature, the word "cyberbullying" has been defined and described in a wide variety of contradicting ways. This should not be surprising given that experts cannot agree on how to spell cyberbullying correctly (cf. cyberbullying, cyberbullying). The interchangeability of terms like "cyberbullying," "online aggression," and "online harassment" further impedes understanding of this phenomenon [13]. According to UNICEF, cyberbullying is bullying that is carried out using technology online. It can take the form of recurrent acts meant to intimidate, irritate, or degrade the targets [14] Moreover, it could encompass the act of spreading false information about an individual or sharing humiliating visual content on the Internet. It also includes the act of sending harmful, menacing, or derogatory messages, visuals, or videos via messaging applications, assuming someone else's identity, and dispatching offensive messages to others either under their name or through fabricated accounts. Cyberbullying is defined as behavior on social media that is harmful to others as well as to the perpetrator and can include physical or verbal assault [15].

Cyberbullying can be distinguished from conventional bullying by several characteristics. These factors encompass the opportunity for the bully to remain unidentified, the existence of a wider potential viewership for the perpetuated abuse, and the challenge of exiting the online community [16]. The risk factors for traditional bullying are present in cyberbullying as well, but perhaps because of its unique characteristics, there are additional risks that should not be disregarded. One of these factors is the lack of control over personal data, which may be the result of a lack of knowledge about the dangers of sharing personal data online [17]. Cyberbullying and traditional bullying share some similarities, including the act of harshly insulting another person, a power imbalance where the perpetrator holds greater power, and a recurrent pattern of behavior [18]. Based on the review, the key distinctions between cyberbullying and traditional bullying are mentioned in Table 14.1.

A victim of traditional bullying can anticipate when they will be bullied (for example, at school or on the playground), but a victim of cyberbullying has no control over when, where, or how they will be bullied (for example, via smartphone or computer), which makes them feel even more helpless. Cyberbullying is a widespread societal problem that inflicts severe harm on its victims. The detection and prevention of this problem are difficult due to its covert character. Artificial intelligence (AI) has intriguing opportunities for addressing cyberbullying through its capacity to detect instances of bullying, harassment, and communications containing hate speech.

This chapter examines the capacity of AI to address the issue of cyberbullying. The published work in the area of supervised machine learning methods and natural language processing approaches, demonstrating the capacity to detect bullying, harassment, and texts related to hatred, has been explored. In addition, the

TABLE 14.1
Key Distinctions between Cyberbullying and Traditional Bullying

Characteristics	Traditional Bullying	Cyberbullying
What time does it occur?	In-person	Online, or with technology
Who is concerned?	Two or more people who are physically present	Two or more people who may or may not be physically present
What techniques are employed?	Verbal and physical abuse, as well as social exclusion	Spreading rumors, posting embarrassing photos or videos, sending mean text messages
What is the duration of it?	Can be on-going	It can be ongoing or can happen in a single incident
How easy is it to stop?	It might be challenging to stop, as the bully can easily find new ways to target the victim	This can be prevented with the help of technology, as the bully can be blocked or reported

Human-AI Collaborative Approach, emphasizing efficacy in establishing an online environment free from cyberbullying through the use of AI, has been evaluated. This chapter also stresses the importance of adopting a comprehensive approach that combines AI with other tactics, such as social media responsibility initiatives and digital literacy training.

14.2 REVIEW OF LITERATURE

14.2.1 Social media: A Platform for Cyberbullying

Today's generation resides in a society where advanced Information and Communication Technologies (ICTs) are integrated into daily activities like household chores, education, entertainment, and socializing [19]. According to Smith et al. (2008), cyberbullying happens when an individual is frequently and purposefully harmed through ICT and is unable to stop it [31]. Users' frequent and active use of digital technology makes them more susceptible to specific online interactions, despite the convenience, which can ultimately jeopardize their safety and emotional and mental well-being. Contemporary technology poses the risk of cyberbullying. It is recognized as one of the most prominent instances of technology abuse in the past ten years because of its negative and occasionally lethal impacts [20]. Social media refers to platforms and applications that let people produce and distribute content, such as text, pictures, and videos. Users of social media platforms can connect with friends and family and meet new individuals. They can also be used to follow companies, organizations, and public figures. A prevalent problem in online social media is cyberbullying, in which a bully mistreats a victim during a social media session [21]. The prevalence of suicidal ideation, intentions, and attempts is higher among elementary and high school kids who suffer cyberbullying, a prevalent and alarming feature of online activity and social media [22]. In a study conducted by Al-Rahmi et al. (2022), students responded well to the use of social media for group projects and classroom instructions, praising its apparent joy and convenience. Social media, on the other hand, has reportedly been linked to cyberbullying, cyberstalking, and addiction among students, all of which have a detrimental effect on their welfare [23]. The popularity of social media skyrocketed immediately after the creation and expansion of the Internet. Social networking sites are currently the most widely used for communication and information sharing. A typical social media session includes an initial post, image, or video, followed by a string of linked comments incorporating user interaction, user data, geographic location, and other social material [24]. In a research conducted by Abaido, most of the participants surveyed in his study believed that cyberbullying exists in the form of online harassment on social media platforms [20].

Some popular social media applications where cyberbullying is common are Snapchat, TikTok, Tumblr, Twitch, Twitter, WeChat, WhatsApp, Omegle, House Party, YouTube, Facebook, and Facebook Live [25]. Adolescents with advanced technological skills might effortlessly fabricate counterfeit social media profiles to torment others through posts, remarks, and messages while maintaining complete anonymity [18]. Stopbullying.gov suggests that the most vulnerable group who have experienced cyberbullying at some point are college students belonging to the

Lesbian, Gay, Bisexual, and Transgender (LGBTQ) community [25]. The main reasons for getting cyberbullied can be appearance, academic achievements, intelligence, race, sexuality, financial status, and religion. Many social media platforms, including Facebook, Instagram, and Twitter, have shown concern about cyberbullying and have taken preventive measures to protect their users [25]. Social media platforms, such as Twitter, have gained significant popularity as a tool for individuals of all age groups to establish connections, exchange information, and access content without being constrained by geographical or physical boundaries. Despite its significance as a near real-time communication tool, a study has found that Twitter is increasingly becoming a platform for cyberbullying [26].

14.2.2 Impact of Cyberbullying on Victims

Cyberbullying can affect mentally, emotionally, and physically. Cyberbullying is a pressing problem that needs a long-term solution, given the explosive rise of Internet usage and social media platforms. Numerous studies on cyberbullying have been conducted, but none of them have been able to offer a solid remedy. Research consistently verifies that bullying has a significant impact on the emotional well-being of children and young individuals. The social isolation and loneliness that victims experience as a result of being rejected by their peer groups. A negative sense of self-worth and sadness are likely to result from the young person's ensuing social retreat. Children who suffer cyberbullying feel angry, helpless, afraid, and depressed [27]. Bullying impacts every adolescent, encompassing both perpetrators, victims, and bystanders. Bullying might potentially have enduring consequences that last into adulthood. A young individual who bullies others does not fit into one particular profile. Adolescents who engage in bullying behavior may exhibit either strong social connections or experience social isolation. Additionally, they may also be victims of bullying themselves. Similarly, bullies can victimize others as well as affairs [25].

According to the UNESCO Institute of Statistics, in schools around the world, one-third of children endure bullying. Typically, males experience a higher degree of bullying in schools compared to females. However, in countries where bullying is more widespread, girls are more susceptible to it. Low socioeconomic position is the primary determinant of whether young adolescents in affluent countries may experience bullying in educational institutions [28]. Both instances of bullying and cyberbullying, regardless of whether they occur in person, indirectly, or through various technological means, have a detrimental effect on the majority of individuals who experience them. However, irrespective of the particular form of bullying encountered, not all victims are similarly impacted emotionally [29]. Cyberbullying has severe detrimental effects, including instances of suicidal behavior. It is imperative to implement intervention approaches and programs aimed at preventing cyberbullying [30].

14.2.3 The Challenges of Detecting and Preventing Cyberbullying

The detection of cyberbullying is a difficult and complex issue. Numerous techniques can be used to conceal or camouflage cyberbullying, making it challenging for computer programs or human reviewers to spot. Detecting cyberbullying is challenging and distinct from merely looking for offensive information [31]. It has certain unique issues, such as bullies may employ slang or coded language to hide their true intentions. For instance, they may utilize emojis to express negative thoughts or words like "ugly" or "loser" in a positive context. Furthermore, bullies may utilize apps that allow anonymous texting or phony profiles to conceal their identities. As a result, it is challenging to find them and hold them accountable. Cyberbullying frequently occurs in group conversations or private messaging, where it is difficult for others to see. This may make it more challenging to gather proof of the abuse and may make it more challenging for victims to receive assistance. Bullies may use comedy to mask their malicious aim. For instance, they can make comments about the looks or social standing of the victim. Additionally, cyberbullying occurs in groups, so it may be more challenging for victims to defend themselves. Bullies may group to target a victim, leaving them to feel alone and helpless. Many steps can be taken to prevent cyberbullying. These consist of teaching young people about staying safe online and preventing cyberbullying, educating kids and teenagers on how to spot and report online bullying, encouraging kids and teenagers to seek help from adults if they experience cyberbullying, and using technology to detect and prevent cyberbullying. There are also some technical challenges to detecting cyberbullying. For instance, there are many distinct platforms where cyberbullying can occur, each with its special characteristics and difficulties. Additionally, gathering and labeling sufficient data to train precise detection algorithms might be challenging. Despite these difficulties, there is an expanding amount of research on detecting cyberbullying. New techniques are being developed by researchers to spot cyberbullying in text, photos, and videos. Additionally, they are attempting to comprehend the social dynamics and environment of cyberbullying.

14.2.4 Initiative of SNS and Government to Prevent Cyberbullying

Facebook and Instagram have a Bullying Protection Hub, which provides users with a script to address the bullying situation. These platforms also provide reactive solutions to their users who experience cyberbullying, such as blocking and reporting bullies and their posts [32]. In order to shield users from widespread harassment and intimidation coming from several accounts, Meta (Facebook) introduced a policy on National Bullying Prevention and Awareness Day in the US in October 2021. These policies work in combination with the applications' built-in features to stop, prevent, and report online bullying and harassment. These technologies, according to Meta, enable people to control unpleasant or abusive interactions, such as barring or unfollowing someone on Facebook and Instagram and enforcing Instagram limits, hidden words, and restrictions. Twitter has additional proactive safeguards against cyberbullying in place [32]. Twitter strongly advises users to ban or report any cyberbullying they may encounter. Speech that supports terrorism, bigotry, and violence has been outlawed. Twitter has broadened its policy to encompass a prohibition on "coded language," also known as "dog whistles," that is employed to indirectly provoke violence. Furthermore, it implemented a regulation that forbids the conduct of endangering the integrity of residential homes and protective structures, or critical infrastructure necessary for everyday, communal, or commercial operations [33].

Stopbulling.gov suggests that the best practice for preventing cyberbullying is using social media privacy tools and settings [25]. The Indian Government has implemented several initiatives such as the Nirbhaya Scheme, CCPW Scheme, and I4C Scheme. They have also established online reporting portals like the National Cybercrime Reporting Portal, and helpline numbers for women and children, and made amendments to the necessary legal provisions of the IT Act and Indian Penal Code 1860 to address the issue of cyberbullying. Additionally, state governments have initiated numerous awareness initiatives. Following UGC regulations, educational institutions have also implemented their anti-bullying policies [34].

14.2.5 AI in Cyberbullying Prevention and Detection: The Future of Cyberbullying Detection and Prevention

Machine learning and AI may open up many possibilities to prevent cyberbullying. With the help of AI, the bully, the victim, and the bystanders can be determined in each situation, and the job of the human moderator of a website can become faster and more efficient.

Based on the most recent global survey, there is a consistent rise in instances of cyberbullying observed daily [21]. Several authors have proposed numerous natural language processing strategies to address this challenge (Table 14.2). However, these techniques are often time-consuming and lack automation. The progress of machine learning and AI has led to significant advancements. Models can be developed and automated detection mechanisms can be incorporated. In one of the studies conducted on the identification of cyberbullying on the social media website Twitter, the researchers conducted experiments utilizing both supervised and unsupervised machine learning approaches. The need to select appropriate keywords has been noted as a crucial aspect in achieving improved outcomes in sentiment analysis [21]. The findings suggest that the model developed has satisfactory performance and has the potential to be effectively utilized in the development of tangible monitoring apps aimed at addressing the significant societal issue of cyberbullying [21]. The experimental findings suggest that the approach based on Support Vector Machines (SVM) provides the highest level of accuracy. Furthermore, the performance of the system is enhanced when user-specific data is incorporated [31]. The superior performance of SVM in text classification may be attributed to many factors. Firstly, the high-dimensional input space of text datasets allows SVM to effectively capture the complex relationships between features. Additionally, SVM can handle datasets with few irrelevant characteristics, which is often the case in text classification tasks. Lastly, the linearly separable structure of text datasets further enhances the performance of SVM compared to other classification algorithms [31].

The study conducted by Dr. Sainju, A. Kuffour, L. Young, and N. Mishra focuses on analyzing the content of tweets connected to bullying. They demonstrate the use of computational techniques for gathering and analyzing data from social media platforms. This approach offers a potential solution to the inherent limitations of self-report biases commonly associated with conventional social scientific methodologies like surveys. A considerable quantity of tweets that exhibited indications of bullying were gathered, categorized based on their relevance to the victim, offender,

TABLE 14.2
Key Studies on AI-based Cyberbullying Prevention and Detection

Authors	Year	Findings
[21]	2023	The Social Media Session-based Cyberbullying Detection framework (SSCD) consists of four essential parts. The study emphasized the significance of taking repetition and power imbalance—two of the fundamental aspects of cyberbullying into account.
[31]	2021	Using supervised machine learning and natural language processing techniques, the author proposes a system that can properly identify cyberbullying as well as the themes and categories that are linked with it, such as racial, sexual, physical meanness, and swearing.
[35]	2022	Proposed an innovative piece of literature. The FSSDL-CBDC approach is a method for detecting and categorizing concept drift in social networks. It combines deep learning with feature subset selection. The technique involves several phases, such as preprocessing, feature selection, and classification.
[36]	2022	Emojis are very important in the process of identifying cyberbullying on social media. The most often utilized method for detecting cyberbullying is supervised learning (SL), in which all datasets are labeled either manually or automatically.
[37]	2023	This study suggests cyberbullying detection models (CDMs) that make use of emotional traits to improve the effectiveness of cyberbullying detection. The preprocessing of the data and the use of deep learning models were both taken into account in this work.
[38]	2023	The use of toxicity features to identify cyberbullying was pioneered in this study. This study is also the first to employ Empath's vocabulary and the most recent psycholinguistic features from the Linguistic Inquiry and Word (LIWC) 2022 tool to identify cyberbullying.
[39]	2022	The objective of this study was to create and improve a cyberbullying detection tool that may be utilized to identify and examine instances of online harassment perpetrated by social media users. Deep learning classifiers can be employed to develop cyberbullying detection systems on popular online social media platforms like Twitter and Facebook. These algorithms are capable of identifying abusive tweets and conversation topics before any cyberbullying occurs.
[40]	2022	The objective of this study was to ascertain the optimal model for forecasting image-based cyberbullying posts on social media sites through an analysis of deep learning and transfer learning frameworks.
[41]	2022	To identify cyberbullying in OSNs, this research has introduced a novel AI-enabled cyberbullying-free – Online Social Networks (AICBF-ONS) technique. Preprocessing, feature extraction, CSSO-based feature selection, SAE-based classification, and MFO-based parameter optimization are some of the various phases of operations that are included in the proposed AICBF-ONS technique.
[42]	2023	The researchers proposed a technique for identifying instances of bullying, harassment, and hate speech by utilizing supervised machine learning algorithms (SVM, Naive Bayes, Logistic regression, and Random Forest). Additionally, they employed unsupervised natural language processing, specifically latent Dirichlet allocation, to predict the topic associated with the text data.

or bystander, and afterward subjected to qualitative analysis to ascertain the distinctive attributes of the tweet content associated with each category. The results of this study offer valuable contributions to the field of social science by shedding light on the characteristics of cyberbullying incidents and the behaviors exhibited by each participant involved. Additionally, these findings can be utilized to enhance the development of classifiers that can more accurately detect instances of cyberbullying and identify bystanders who may be able to assist when implementing interventions to support victims [43]. The research findings of Dr. Kathleen Van Royen, Dr. Karolien Poels, Prof. Heidi Vandebosch, and Dr. Bieke Zaman delve into the concept of "reflective messages." This interface design incorporates AI to discourage users from engaging in cyber-harassment by encouraging them to pause and consider whether they genuinely intend to post such content. The authors offer valuable insights into the effectiveness and user experience of this approach. The article expands upon prior research about the efficacy of this method by particularly investigating how teenagers perceive and value these reflecting messages, while also assessing its pragmatic design and hedonic attributes. The study provides evidence that teenagers generally hold good views. Nevertheless, it is important to develop tactics that may effectively mitigate user tiredness over an extended period. The study offers a valuable analysis of the importance of ensuring user-friendliness and desirability of interventions for young users, contributing to the knowledge base for their effective implementation on online platforms [44].

Dr. Christopher E. Gomez, Dr. Marcelo O. Sztainberg, and Dr. Rachel E. Trana, who specialize in computational studies, addressed the limitations of human annotation, including inaccuracies resulting from cultural and language differences as well as subjective judgments. The researchers utilized AI methods, namely neural networks, which draw inspiration from the human brain. These algorithms are employed to offer valuable insights into the consistent labeling of data as either bullying or non-bullying. Furthermore, the researchers explored the potential of their approach to enhance the precision of annotations in comparable datasets [45]

Dr. Deborah Hall, Dr. Yasin Silva, Brittany Wheeler, Lu Cheng, and Kathleen Baumel elucidated the potential of machine learning models in detecting cyberbullying to enhance our comprehension of the psychological dimensions associated with the phenomenon. The study provides more emphasis on the practical consequences that machine learning techniques may have for families, doctors, and the broader domains of prevention and intervention [46]. Professor Eugenia Siapera offered a valuable and essential analysis of AI-driven content moderation, specifically examining its implications for addressing racist hate speech. Through a thorough examination of the readily accessible information regarding the design and execution of moderation practices, the study presents a persuasive assertion that how platforms address hate speech not only fails to consider the perspectives of racialized individuals but also appropriates their efforts without providing adequate recompense. Consequently, this approach inadvertently perpetuates racism rather than effectively eliminating it [47]. In the scholarly article authored by Dr. Semiu Salawu, the innovative cyberbullying detection application named Bull Stop was introduced. This application has been developed using a dataset obtained from Twitter, enabling it to effectively identify instances of cyberbullying. Many individuals may have come across applications in recent years that utilize AI to aid both children and adults in addressing cyberbullying. These apps employed various strategies, such as blocking abusive messages, implementing reflective messaging to discourage bullying behavior, and offering support and educational guidance to individuals involved in bullying incidents. The study elucidates the methodology employed in the development of the aforementioned application, highlighting its efforts to address the typical computational obstacles faced during this undertaking. Additionally, the study discussed the utilization of social scientific methodologies to include user feedback in the app's design [48].

The identification of bullying, harassment, and hate-related writings is achieved by the utilization of supervised machine learning algorithms and natural language processing techniques, AI-enabled cyberbullying – Online Social Networks, and Human AI-Collaborative Approaches are some of the promising methods for using AI to detect and prevent cyberbullying. These strategies could all be successful in identifying and preventing cyberbullying. At the same time, AI-based systems are susceptible to bias and can be tricked by cunning cyberbullies. AI should be used in conjunction with other strategies, such as social media responsibility campaigns and digital literacy education (Table 14.2).

14.3 THE POTENTIAL OF AI IN MANAGING CYBERBULLYING: DISCUSSION

This chapter explored the potential of AI to tackle the pervasive issue of cyberbullying on digital platforms. Novel approaches are required to tackle the complex task of identifying cyberbullying, which is frequently disguised or hidden. AI demonstrates promising approaches through the utilization of supervised machine learning algorithms and natural language processing techniques, effectively identifying instances of bullying, harassment, and hate-related content, despite the presence of certain challenges. The Human-AI Collaborative Approach showcases the effectiveness of combining human intelligence with AI to create online environments that are devoid of cyberbullying. However, the chapter acknowledges the potential for bias and manipulation in AI systems, emphasizing the need for a thorough plan. Employing AI alongside efforts to promote social media accountability and educate individuals on digital literacy is crucial for effectively tackling the problem of cyberbullying. Social media platforms may promote responsible online behavior through targeted advertising, while digital literacy education enables users to identify and report cases of cyberbullying.

This chapter's central message is that AI possesses significant potential as a tool to combat cyberbullying, provided that it is employed with ethical considerations and responsible practices. By integrating AI with other approaches, we can establish a more secure and constructive digital atmosphere for everyone. The study encompasses the recognition of AI's capacity to address cyberbullying by utilizing precise algorithms and procedures, the efficacy of the Human-AI Collaborative Approach, the necessity for responsible AI utilization, and a comprehensive strategy for preventing cyberbullying. Subsequent studies can explore the efficacy of various AI models in detecting cyberbullying and deploy AI-driven solutions to enable users to report and handle instances of cyberbullying.

14.4 CONCLUSION

AI has the potential to play a significant role in preventing cyberbullying. AI can identify cyberbullying at an early stage before it has an opportunity to become more serious. This is crucial because victims of cyberbullying may suffer from mental health issues, scholastic difficulties, or even commit suicide. AI can be used to monitor huge volumes of Internet content, which is something that humans are not capable of doing. This means that cyberbullying that might otherwise be missed can be found using AI. The ability to recognize cyberbullying can be taught to AI. This is extremely important because it lessens the likelihood of false positives, which may be harmful to both the victims and the people who are watching. AI is an intriguing technology that has the potential to offer analytics and intelligence to protect against cyberattacks that are continually evolving. This can be accomplished by rapidly analyzing millions of events and monitoring a wide range of cyber hazards to anticipate and manage any problems before they occur.

REFERENCES

1. Georgiadou, Elissavet. 2002. "McLuhan's Global Village and the Internet." *Proceedings of the 2nd International Conference on Typography and Visual Communication, June 2002, Thessaloniki, Greece.*, January. https://www.academia.edu/17449480/McLuhan_s_Global_Village_and_the_Internet.
2. Attwood, Maiken. 2015. "McLuhan's Global Village, Still Relevant Today." *Medium* (blog). August 5, 2015. https://medium.com/@maiken_louise/mcluhan-s-global-village-still-relevant-today-1bd4e3792b61.
3. Olweus, Dan. 1978. *Aggression in the Schools: Bullies and Whipping Boys*. Aggression in the Schools: Bullies and Whipping Boys. Oxford: Hemisphere.
4. Smith, Peter K. 2011. "Bullying in Schools: Thirty Years of Research." In *Bullying in Different Contexts*, edited by Claire P. Monks and Iain Coyne, 36–60. Cambridge: Cambridge University Press. https://doi.org/10.1017/CBO9780511921018.003.
5. Chakraborty, Arpita, Yue Zhang, and Arti Ramesh. 2018. "Understanding Types of Cyberbullying in an Anonymous Messaging Application." In *Companion of the Web Conference 2018 on The Web Conference 2018 - WWW '18*, 1001–5. Lyon, France: ACM Press.
6. Mahbub, Syed, Eric Pardede, and A. S. M. Kayes. 2021. "Detection of Harassment Type of Cyberbullying: A Dictionary of Approach Words and Its Impact." *Security and Communication Networks* 2021 (June): e5594175. https://doi.org/10.1155/2021/5594175.
7. Bilal, Alatas, and Can Umit. 2022. "Cyberbullying and Cyberstalking in Online Social Networks." in *Securing Social Networks in Cyberspace*, edited by Pathan, Al Sakib K, 141–145.CRC Press Taylor and Francis Group.
8. Worsley, Joanne D., Jacqueline M. Wheatcroft, Emma Short, and Rhiannon Corcoran. 2017. "Victims' Voices: Understanding the Emotional Impact of Cyberstalking and Individuals' Coping Responses." *SAGE Open* 7 (2): 2158244017710292. https://doi.org/10.1177/2158244017710292.
9. Chandrashekhar, A. M., Muktha, G, S., and Anjana, D, K. 2016. "Cyberstalking and Cyberbullying: Effects and prevention measures". *Imperial journal of interdisciplinary research*. (2): 95–102.
10. Mann, B. L. 2009. "Social Networking Websites – A Concatenation of Impersonation, Denigration, Sexual Aggressive Solicitation, Cyber-Bullying or Happy Slapping Videos." *International Journal of Law and Information Technology* 17 (3): 252–67. https://doi.org/10.1093/ijlit/ean008.

11. Siegle, Del. 2010. "Cyberbullying and Sexting: Technology Abuses of the 21st Century." *Gifted Child Today* 33 (2): 14–65. https://doi.org/10.1177/107621751003300206.
12. Bandura, Albert. 1978. "Social Learning Theory of Aggression." *Journal of Communication* 28 (3): 12–29. https://doi.org/10.1111/j.1460-2466.1978.tb01621.x.
13. Bauman, Sheri. 2012. *Principles of Cyberbullying Research: Definitions, Measures, and Methodology*. 1st ed. Routledge. https://doi.org/10.4324/9780203084601.
14. UNICEF. 2023. Cyberbullying: What is it and how to stop it. https://www.unicef.org/eca/cyberbullying-what-it-and-how-stop-it (accessed December 20, 2023).
15. Evangelio, Carlos, Pablo Rodríguez-González, Javier Fernández-Río, and Sixto Gonzalez-Villora. 2022. "Cyberbullying in Elementary and Middle School Students: A Systematic Review." *Computers & Education* 176 (January): 104356. https://doi.org/10.1016/j.compedu.2021.104356.
16. Juvonen, Jaana, and Elisheva F. Gross. 2008. "Extending the School Grounds?–Bullying Experiences in Cyberspace." *The Journal of School Health* 78 (9): 496–505. https://doi.org/10.1111/j.1746-1561.2008.00335.x.
17. Sengupta, Anirban, and Anoshua Chaudhuri. 2011. "Are Social Networking Sites a Source of Online Harassment for Teens? Evidence from Survey Data." *Children and Youth Services Review* 33 (2): 284–90. https://doi.org/10.1016/j.childyouth.2010.09.011.
18. Margolis, Jenna, and Dinara Amanbekova. 2023. "Social Media and Cyberbullying." In *Teens, Screens, and Social Connection: An Evidence-Based Guide to Key Problems and Solutions*, edited by Alma Spaniardi and Janki Modi Avari, 79–101. Cham: Springer International Publishing. https://doi.org/10.1007/978-3-031-24804-7_6.
19. Cross, Donna, Amy Barnes, Alana Papageorgiou, Kate Hadwen, Lydia Hearn, and Leanne Lester. 2015. "A Social–Ecological Framework for Understanding and Reducing Cyberbullying Behaviours." *Aggression and Violent Behavior*, Bullying, Cyberbullying, and Youth Violence: Facts, Prevention, and Intervention, 23 (July): 109–17. https://doi.org/10.1016/j.avb.2015.05.016.
20. Abaido, Ghada M. 2020. "Cyberbullying on Social Media Platforms Among University Students in the United Arab Emirates." *International Journal of Adolescence and Youth* 25 (1): 407–20. https://doi.org/10.1080/02673843.2019.1669059.
21. Yi, Peiling, and Arkaitz Zubiaga. 2023. "Session-Based Cyberbullying Detection in Social Media: A Survey." *Online Social Networks and Media* 36 (July): 100250. https://doi.org/10.1016/j.osnem.2023.100250.
22. Sampasa-Kanyinga, Hugues, Paul Roumeliotis, and Hao Xu. 2014. "Associations between Cyberbullying and School Bullying Victimization and Suicidal Ideation, Plans and Attempts Among Canadian Schoolchildren." *PLOS ONE* 9 (7): e102145. https://doi.org/10.1371/journal.pone.0102145.
23. Al-Rahmi, Waleed Mugahed, Noraffandy Yahaya, Uthman Alturki, Amen Alrobai, Ahmed A. Aldraiweesh, Alhuseen Omar Alsayed, and Yusri Bin Kamin. 2022. "Social Media – Based Collaborative Learning: The Effect on Learning Success With the Moderating Role of Cyberstalking and Cyberbullying." *Interactive Learning Environments* 30 (8): 1434–47. https://doi.org/10.1080/10494820.2020.1728342.
24. Cheng, Lu, Yasin N. Silva, Deborah Hall, and Huan Liu. 2021. "Session-Based Cyberbullying Detection: Problems and Challenges." *IEEE Internet Computing* 25 (2): 66–72. https://doi.org/10.1109/MIC.2020.3032930.
25. Affairs (ASPA), Assistant Secretary for Public. 2019. "Facts About Bullying." Text. September 24, 2019. https://www.stopbullying.gov/resources/facts.
26. Al-garadi, Mohammed Ali, KasturiDewi Varathan, and Sri Devi Ravana. 2016. "Cybercrime Detection in Online Communications: The Experimental Case of Cyberbullying Detection in the Twitter Network." *Computers in Human Behavior* 63 (October): 433–43. https://doi.org/10.1016/j.chb.2016.05.051.

27. Hoff, Dianne L., and Sidney N. Mitchell. 2009. "Cyberbullying: Causes, Effects, and Remedies." *Journal of Educational Administration* 47 (5): 652–65. https://doi.org/10.1108/09578230910981107.
28. UNESCO. 2023. "New Data Reveal That One out of Three Teens Is Bullied Worldwide | UNESCO. https://www.unesco.org/en/articles/new-data-reveal-one-out-three-teens-bullied-worldwide. (accessed January 26, 2024)
29. Ortega, Rosario, Paz Elipe, Joaquin A. Mora-Merchán, M. Luisa Genta, Antonella Brighi, Annalisa Guarini, Peter K. Smith, Fran Thompson, and Neil Tippett. 2012. "The Emotional Impact of Bullying and Cyberbullying on Victims: A European Cross-National Study." *Aggressive Behavior* 38 (5): 342–56. https://doi.org/10.1002/ab.21440.
30. Schenk, Allison M., and William J. Fremouw. 2012. "Prevalence, Psychological Impact, and Coping of Cyberbully Victims Among College Students." *Journal of School Violence* 11 (1): 21–37. https://doi.org/10.1080/15388220.2011.630310.
31. Smith, Peter K., Mahdavi, Jess., Carvalho, Manuel., Fisher, Sonja., Russell, Shanette., and Tippett, Neil. 2008. "Cyberbullying: its nature and impact in secondary school pupils". *Journal of Child Psychology and Psychiatry*, 49(4): 376–385. https://doi.org/10.1111/j.1469-7610.2007.01846.x
32. Andrea, Perera, and Pumudu Fernando. 2021. "Review of *Accurate Cyberbullying Detection and Prevention on Social Media*." www.elsevier.com/Locate/Procedia. Elsevier. 2021.
33. "Prevent Bullying & Harassment - Overview." n.d. About.meta.com. https://about.meta.com/actions/safety/topics/bullying-harassment. (accessed December 20, 2023)
34. "Twitter's New 'Violent Speech' Policy Similar to Past Rules - CBS San Francisco." 2023. Www.cbsnews.com. March 1, 2023. https://www.cbsnews.com/sanfrancisco/news/twitters-new-violent-speech-policy-similar-to-past-rules/.
35. Kaur, Manpreet, and Munish Saini. 2023. "Indian Government Initiatives on Cyberbullying: A Case Study on Cyberbullying in Indian Higher Education Institutions." *Education and Information Technologies* 28 (1): 581–615. https://doi.org/10.1007/s10639-022-11168-4.
36. Neelakandan, S, Sridevi M, Saravanan Chandrasekaran, Murugeswari K, Aditya Kumar Singh Pundir, Sridevi R, and T. BheemaLingaiah. 2022. "Deep Learning Approaches for Cyberbullying Detection and Classification on Social Media." Edited by Akshi Kumar. *Computational Intelligence and Neuroscience* 2022 (June): 1–13. https://doi.org/10.1155/2022/2163458.
37. Al-Harigy, Lulwah M., Hana A. Al-Nuaim, Naghmeh Moradpoor, and Zhiyuan Tan. 2022. "Building Towards Automated Cyberbullying Detection: A Comparative Analysis." Edited by Thippa Reddy G. *Computational Intelligence and Neuroscience* 2022 (June): 1–20. https://doi.org/10.1155/2022/4794227.
38. Al-Hashedi, Mohammed, Lay-Ki Soon, Hui-Ngo Goh, Amy Hui Lan Lim, and Eu-Gene Siew. 2023. "Cyberbullying Detection Based on Emotion." *IEEE Access* 11: 53907–18. https://doi.org/10.1109/ACCESS.2023.3280556.
39. Teng, Teoh Hwai, and KasturiDewi Varathan. 2023. "Cyberbullying Detection in Social Networks: A Comparison Between Machine Learning and Transfer Learning Approaches." *IEEE Access* 11: 55533–60. https://doi.org/10.1109/ACCESS.2023.3275130.
40. Aldhyani, Theyazn H. H., MoslehHmoud Al-Adhaileh, and Saleh NagiAlsubari. 2022. "Cyberbullying Identification System Based Deep Learning Algorithms." *Electronics* 11 (20): 3273. https://doi.org/10.3390/electronics11203273.
41. Roy, Pradeep Kumar, and FenishUmeshbhai Mali. 2022. "Cyberbullying Detection Using Deep Transfer Learning." *Complex & Intelligent Systems* 8 (6): 5449–67. https://doi.org/10.1007/s40747-022-00772-z.
42. Al-Marghilani, Abdulsamad. 2022. "Artificial Intelligence-Enabled Cyberbullying-Free Online Social Networks in Smart Cities." *International Journal of Computational Intelligence Systems* 15 (1): 9. https://doi.org/10.1007/s44196-022-00063-y.

43. Khalid M. O, Nahar., Mohammad Alauthman, Saud Yonbawi, and Ammar Almomani. 2023. "Cyberbullying Detection and Recognition With Type Determination Based on Machine Learning." *Computers, Materials & Continua* 75 (3): 5307–19. https://doi.org/10.32604/cmc.2023.031848.
44. Sainju, Karla Dhungana, Akosua Kuffour, Lisa Young, and Niti Mishra. 2022. "Bullying-Related Tweets: A Qualitative Examination of Perpetrators, Targets, and Helpers." *International Journal of Bullying Prevention: An Official Publication of the International Bullying Prevention Association* 4 (1): 6–22. https://doi.org/10.1007/s42380-021-00098-3.
45. Royen, Kathleen Van, Karolien Poels, Heidi Vandebosch, and Bieke Zaman. 2022. "Think Twice to Be Nice? A User Experience Study on a Reflective Interface to Reduce Cyber Harassment on Social Networking Sites." *International Journal of Bullying Prevention* 4 (1): 23–34. https://doi.org/10.1007/s42380-021-00101-x.
46. Gomez, Christopher E., Marcelo O. Sztainberg, and Rachel E. Trana. 2022. "Curating Cyberbullying Datasets: A Human-AI Collaborative Approach." *International Journal of Bullying Prevention: An Official Publication of the International Bullying Prevention Association* 4 (1): 35–46. https://doi.org/10.1007/s42380-021-00114-6.
47. Hall, Deborah L., Yasin N. Silva, Brittany Wheeler, Lu Cheng, and Katie Baumel. 2022. "Harnessing the Power of Interdisciplinary Research With Psychology-Informed Cyberbullying Detection Models." *International Journal of Bullying Prevention* 4 (1): 47–54. https://doi.org/10.1007/s42380-021-00107-5.
48. Siapera, Eugenia. 2022. "AI Content Moderation, Racism, and (De) Coloniality." *International Journal of Bullying Prevention* 4 (1): 55–65. https://doi.org/10.1007/s42380-021-00105-7.
49. Salawu, Semiu, Jo Lumsden, and Yulan He. 2022. "A Mobile-Based System for Preventing Online Abuse and Cyberbullying." *International Journal of Bullying Prevention* 4 (1): 66–88. https://doi.org/10.1007/s42380-021-00115-5

15 Cybersecurity Challenges and Countermeasures in Smart Cities for Sustainable Environment Development

A Comprehensive Analysis of Threats and Resilience Strategies

Saurabh Srivastava, Tasneem Ahmed, Nayyar Ali Usmani, Syed Adnan Afaq, and Mohammad Faisal

15.1 INTRODUCTION

A smart city guarantees high-quality maintenance across a range of industries, including healthcare, transportation, energy, and citizen safety and security. Furthermore, worries regarding privacy and data security in smart cities are intensifying [1]. Cities are considered engines of "economic growth," and without urbanization, no country has ever seen substantial growth in the industrial age. Rapid urbanization has surpassed all other global characteristics in recent years. A total of 9.2 billion people will inhabit the earth by 2050, with 6.4 billion of them residing in cities [2]. The world's urban population has grown dramatically during the past 50 years. Megacities have proliferated, mainly in less developed areas. Nothing separates India's urbanization trajectory from the rest of the world. India now holds the second-place position in the world for the size of its urban population, although experiencing a rise in this regard. By 2030, the world's biggest concentration of megacities will be found in India with seven megacities including Ahmedabad, Hyderabad, Mumbai, Delhi, Kolkata, Chennai, and Bangalore [3, 4]. Rapid urbanization in developing nations is causing significant challenges in urban planning, including mismatches between the supply and demand for infrastructure and utilities, traffic congestion, pollution, loss of natural green spaces, and the growth of peri-urban settlements that negatively impact both natural and

DOI: 10.1201/9781003409571-15

built environments [3]. Unreliable transportation, a lack of access to medical care, a lack of adequate water and electricity supplies, and other issues are all brought on by the rapid development in urbanization [2]. Urbanization is a significant urban development process that is linked to resource consumption, economic activity, and population concentration in metropolitan areas and is always growing in size and complexity [5]. The developments of sensor technologies and the Internet of Things (IoT) have accelerated, making it easier to construct IoT-based smart cities. However, due to their close connection with people's everyday lives, the security issues associated with smart cities have become the top concern of the people amid the rapid market expansion of the city [6]. Modern technology is utilized in the planning and implementation of thoughtful strategies for sustainable development, safe and secure communities, and equitable growth. Intelligent solutions that can accelerate economic growth by enhancing infrastructure and services, creating jobs, and raising the standard of living in smart cities are provided by digital technology and their interactions with physical systems [7]. It is anticipated that smart cities will improve urban process efficiency, promote sustainable development, and raise living standards. The widespread usage of smart devices has led to significant concerns about security and privacy, which require appropriate solutions. Nevertheless, the intelligent applications in smart cities are too diverse, too scalable, and too dynamic to directly employ classic cyber security protection strategies.

Cybercrime is constantly increasing in smart cities; cyber criminals are carrying out cybercrime using new technologies. Even today, there is a section of society that is not aware of modern technology, and cybercriminals target those people in carrying out their crimes. The main objective of this study is to highlight all those aspects, after knowing which a citizen can understand and avoid cybercrime. In this study, the author has highlighted all those aspects that can protect an aware citizen from cybercrime happening to them. This study explores several cyber security issues and solutions in smart cities for the development of a sustainable environment, which will aid in developing secure and reliable smart city design methods.

15.2 LITERATURE REVIEW

Smart cities, designed to enhance residents' quality of life, integrate intelligent transportation, automated traffic management, and electronic health services [8]. With a surge in global metropolitan population over the last 50 years, smart cities, predominantly in developing nations, have multiplied [9]. Urban resilience gains attention due to concerns like climate change and natural disasters [10]. While the concept's disciplinary implications are explored, research predominantly defines urban resilience rather than proposing implementation strategies [10]. Cities globally are developing smart plans to address challenges, promote well-being, and sustainably manage modern urban life [8]. Smart applications utilizing sensors and smartphone resources proliferate, yet security concerns arise [11]. This study delves into cybersecurity issues in smart cities, aiming to establish sustainable, secure solutions for urban planning [12].

15.3 SMART CITY AND ITS APPLICATION

15.3.1 Smart City

The term "smart city" became popular in the early 2010s to describe how modern technological advances and data can enable a more effectively managed metropolis. As a result, a smart city prioritizes information and communication technologies (ICTs) to encourage better communication with its citizens and ensure that its residents enjoy a better quality of life and environment despite the rapid development of the city [13]. Concern over major dangers like climate change, natural catastrophes, and other threats is growing, and this fascination with the idea of urban resilience is concomitant. The definition of urban resilience has attracted much attention from scholars as well as inquiries into its relevance and value [14]. Six pillars help to balance the smart city and are well defined by [13]. The detailed diagram of the six pillars of the smart city is shown in Figure 15.1.

- The smart economy is based on creativity and entrepreneurship, high productivity, labor market flexibility, openness to regional and international cooperation, and the ability to change [13].
- Smart citizens cover a wide range of topics, including social and ethnic diversity, open-mindedness, adaptability, creativity, and participation in public life. It also involves lifelong learning through the continuous enhancement of the capabilities and qualifications of citizens [13].
- Smart governance uses intelligent technology for intelligent public administration in which public input in decision-making and transparency of actions as well as the availability and quality of public services are given paramount importance.
- Smart mobility refers to the city's accessibility on a national and worldwide scale, as well as the use of ICTs that can improve traffic flow and transportation systems both inside and outside the city by fostering innovation, sustainability, and safety [13].

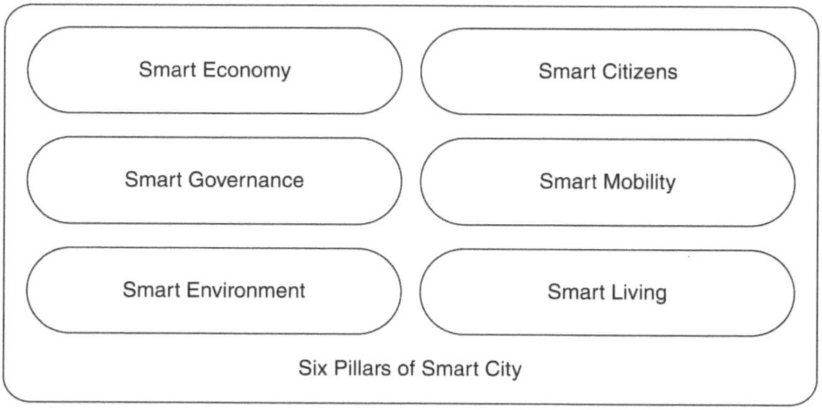

FIGURE 15.1 Six pillars of a smart city.

- The smart environment attempts to optimize the smart city. It uses several technologies to raise public awareness of environmental issues and services, including energy, water, and gas. A smart environment aims to change people's behavior, cut down on waste, and increase resource efficiency.
- Smart living involves creating an effective network of top urban public spaces where educational facilities can be established by building top-tier colleges and institutes. To support a healthy lifestyle for every citizen, it should offer world-class hospitals and tourist attractions as well as all the latest technology-enabled technologies and equipment. Additionally, citizens should enjoy social cohesion and access to high-quality housing [13].

15.3.2 Applications of Smart Cities

Various smart city applications are available that provide different kinds of services that make human life easy. Some important applications that are frequently used by smart citizens are shown in Figure 15.2.

- *Air quality management*: Smart cities are putting in place technologies to record pollution data in real-time and forecast emissions. Poor air quality can lead to ill health, early mortality, and damage to crops, structures, and ecosystems, among other things. The effects of bad air are more severe in urban areas because they are home to the bulk of people. The vast majority of cities therefore have faith in air quality monitoring systems. Currently, big, expensive sensing stations with sensors deployed in strategic areas monitor the air quality of smart cities. These stations collect data on variables like particulate matter (PM), nitrogen dioxide (NO_2), carbon dioxide (CO_2), and ozone (O_3), allowing for precise monitoring but only in select regions [15].

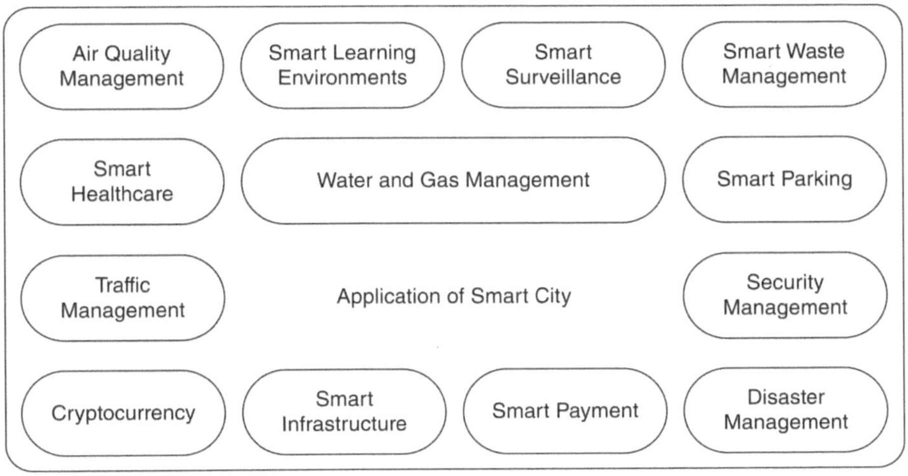

FIGURE 15.2 Various applications of smart city.

- *Smart payment*: A smart payment system enhances and streamlines payment processes. A digital wallet can be used by one party to exchange units of digital currency for products and services with another party in an electronic transaction. This can involve shopping online or offline (in physical stores) using NFC (Near Field Communication), QR codes, and other technologies that link cell phones to the real world. Digital wallets are useful for a variety of transactions, including paying bills (electricity, water, fees, etc.), purchasing online, reserving seats on trains and airplanes, and more [13].
- *Smart infrastructure*: Modern technologies are emerging as a significant source for cities due to their capacity to foster ongoing development. Buildings and urban infrastructures should be constructed efficiently and sustainably as a result. Cities must also invest in electric and self-propelled vehicles to cut CO_2 emissions. Intelligent technology may enable the construction of a sustainable infrastructure and the achievement of efficient energy consumption [15].
- *Cryptocurrency*: Companies like Tesla, Microsoft, and Pavilion Hotels & Resorts are just a few that are accepting cryptocurrency as a legitimate form of payment for their products and services. So, one of the blockchain's innovations that is frequently employed as decentralized digital money is cryptocurrency. Cryptocurrency refers to a type of digital money with no physical counterpart; examples include Bitcoin, Ethereum, Litecoin, Monero, and many more [13].
- *Traffic management*: Cities are looking for the best traffic-improvement solutions to solve one of their main issues with traffic. Intelligent transportation technologies have been used in numerous smart cities across the globe to manage traffic. Real-time updates on traffic flow are sent via the embedded pavement sensors to the central traffic management platform. This platform analyzes the data and promptly modifies the signal lights to correspond with the traffic flow. Parallel to this, projections of traffic flow are formed using the collected data; none of these processes call for human involvement [15].
- *Smart healthcare*: IoT and mobile Internet are used by "smart healthcare" systems to dynamically access information and link individuals and resources in the healthcare industry. Healthcare is intelligently managed and responsive to the needs of the medical ecology. IoT and related technologies can be used for patient monitoring and alert generation in real time, telemedicine, which is the practice of providing healthcare to patients remotely, and home healthcare for the elderly that can be set up in homes [13].
- *Smart parking*: The goal of smart parking-management systems is to control traffic flows effectively so that they can alert citizens when there are free parking spaces nearby. They work in conjunction with traffic monitoring systems. Intelligent parking solutions are being utilized by numerous cities. The ground has sensors installed to help detect the vehicle that has left the parking space. Through a mobile app, drivers are informed where available parking spaces are by sensors [15].

- *Smart learning environments*: Technology-supported smart learning environments enable students to interact with learning systems from any location at any time, access digital resources, and receive needed instructional support, supplemental materials, and learning recommendations at the appropriate time, location, and format. Numerous hardware and software innovations have been used to advance and enhance learning. Hardware includes physical objects such as e-bags, smart tables, and interactive whiteboards. The software offers a variety of educational tools and systems, as well as Internet resources and virtual reality [13].
- *Smart surveillance*: Smart cities must have strong urban security as citizens are very concerned about their safety. Nowadays, many cities use sophisticated surveillance systems that rely on cameras located almost everywhere in the city. Cameras installed by the police are capable of real-time monitoring of the most important locations, while some private cameras installed by private organizations are also available to combat crime [15].
- *Smart waste management*: The majority of the environmental issues related to inefficient garbage collection are more efficiently handled by waste (garbage) management solutions, which also increase the effectiveness of rubbish collection and reduce operating costs. In these approaches, a level sensor is installed in the trash can, and when a certain level is reached, the truck driver receives a notification on their smartphone via the control platform [15]. IoT-enabled municipal infrastructure can help with smart waste management to make cities cleaner and healthier. The IoT platform provides a management system that uses sensors to gather data from garbage cans and sends information about filled trash cans to waste management vehicles or trucks. It helps waste management services, offices, and connected businesses manage their time more effectively and gives information about the cleanliness, health, and rubbish state of the neighborhood [16].
- *Water and gas management*: Smart cities managing gas and water supplies are considered important as managing electricity and energy supplies. Gas and water suppliers and customers can communicate with each other in both directions using IoT network-enabled water and gas supply systems. Additionally, IoT management and control devices like the Raspberry Pi 3 can analyze real-time information about the needs of any area, home, building, or city. Water wastage can be reduced by monitoring the fluid level in the water tank inside the house building. Gas and water smart meters are linked to supply providers for scheduling, invoicing, and data collecting. In an intelligent dwelling, water and gas waste is avoided through automated valve control and closure, as well as motor, stove, and heater switching [16].
- *Security management*: Real-time monitoring is an advantage of the proposed system, and users can be remotely notified about events by email or SMS. IoT technology improves the ability of security systems to detect incidents such as accidents, terrorism, and strange occurrences. A decentralized architecture that manages drone cameras using IoT protocols to identify warning events in places like train stations, bus terminals, and airports has been designed with high scalability and fast detection [16].

- *Disaster management/emergency services*: Although disasters cannot be predicted, preparedness can be taken. Relief efforts after a disaster are much different and more challenging than aid distributions made by government organizations under normal circumstances [17]. The purpose of disaster management is to lessen the effect of catastrophes by the organization and administration of resources and responsibilities for managing all humanitarian elements of emergencies, especially state, response, and recovery [17]. In an IoT-based catastrophe monitoring and management system for dams, a mechatronics system with an IoT app is introduced to monitor and control the real-time status of water level, gate opening and closing, and water discharge [16].

15.4 CYBER CONCERNS WITH SMART CITY APPLICATIONS

For almost two decades, the Internet has been an indispensable tool for global communication and has grown increasingly integrated into people's daily lives. There are currently about 3 billion Internet users worldwide as a result of advancements and low prices in this industry that have significantly enhanced the Internet's availability, use, and functionality [18]. IoT-based applications of smart cities have quickly infiltrated human life to give more convenience, comfort, and efficiency due to the fast growth of the IoT and sensor technologies [6]. The urban environment of the future is extremely concerning since rising inequality and a lack of basic facilities are expected to lead to issues with environmental degradation as well as a rise in both individual and collective violence [2]. A crucial frontier factor in public safety resilience offers a methodical framework for tackling issues related to urban safety and strengthening cities' ability to withstand disasters as urbanization picks up speed. As a result, improving urban resilience has emerged as an important component of achieving sustainable development in the face of increasing urban stress, and resilience theory is increasingly integrated into urban planning [5]. The idea of resilience has caught the attention of the general public and is being used more frequently concerning cities and urban regions. Growing awareness and worry about risks including terrorism, disease outbreaks, natural calamities, and climatic change appears to have coincided with recent interest [14]. Significant scholarly effort has been made in analyzing the idea of the various disciplinary meanings of urban resilience, criticizing the ambiguity of the term, and discussing its applicability. Despite the increasing emphasis on improving the resilience of cities, most research has focused on defining urban resilience, rather than figuring out how to create policies that support urban resilience [14]. Smart city administrators, designers, integrators, and organizations face major political, technological, and economic hurdles as a result of the interconnected and complex nature of these new entities [19]. There is a growing body of research focusing on security, privacy, and vulnerabilities within smart city infrastructure in relation to information security risks and issues [19]. There are lots of cyber concerns that

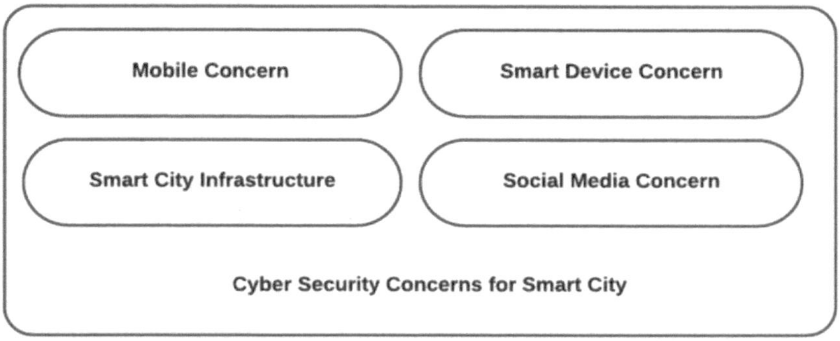

FIGURE 15.3 Cyber security concerns for smart city.

may harm smart city applications but the most dangerous concerns are shown in Figure 15.3.

- *Mobile concern*: Although mobile devices are essential for communicating with smart city network infrastructure, they also put users' security and privacy at serious risk by disclosing private information to outside attackers. One potential cryptographic method to provide authentication and privacy protection features for smart cities is to use privacy-preserving authentication (PPA) protocols for mobile services [19].
- *Smart device concern*: Most of the application data in IoT applications comes from human concerns, businesses, or industries. Data in these programs must be secured and protected to prevent obfuscation, theft, and tampering. The communication connection's security is crucial. Even though they are the main focus of IoT, there are numerous problems with application availability, reaction time, and scalability [20].
- *Smart city infrastructure*: The IoT is essential to the infrastructure of smart cities because it offers the network architecture needed to collect and analyze data from dispersed sensors and smart devices. According to studies, there are two types of assaults on IoT devices: external and internal. The vulnerability of IoT-based applications is directly related to the network paradigm, which uses physical objects like sensor-based devices to communicate over wired or wireless connections and gather information on significant network interactions [19]. Information technology component security is one of the important security elements of IoT infrastructure, and confidentiality, honesty, and availability are three important elements that are referred to as the CIA triumvirate (i.e. confidentiality, integrity, and availability) [20].
- *Social media concern*: Government, politicians, authorities, and commercial businesses can exploit the social, economic, and cultural data gathered via online social networks (such as Facebook, Instagram, and LinkedIn).

The use of open data sources may aid in their better understanding of market trends and behavioral patterns that affect personal dynamics. However, there are privacy risks associated with using online social networks. According to the survey, the primary dangers associated with using online social networks include those that pose a threat to an individual's identity, anonymity, personal space, privacy, and communication, as well as security hazards brought on by outside parties [19].

15.5 CHALLENGES IN SMART CITIES DEVELOPMENT

Globally, the emergence of smart cities has enhanced people's quality of life and economic success by enabling them to communicate with several tiers of government, obtain services, and optimize system interaction efficiency and efficacy. However, from a variety of stakeholder viewpoints, there are still big challenges with privacy, security, and risk [19]. A lot of challenges may arise during the implementation of smart city application development which may be related to trust, technology, operations sustainability, etc. as shown in Figure 15.4.

- *Trust challenge*: Human beings must learn to work with others because they are fundamentally social creatures. In a city, people live and work together with people who have different needs, wants, beliefs, and hobbies. It is far from clear. For example, not all interests coincide with those on whom they depend for support. Cities depend heavily on the ability of residents to live closely with strangers, who may have vastly different views but can still agree to maintain basic political standards [21]. Trust is essential in the context of smart cities since the integrated design and underlying technical architecture depend on the safe and effective flow of massive amounts of data. These data are now being used by brand-new programs and services that let people communicate with different levels of government. There are serious security and privacy issues when GPS-based surveillance data is joined with precise personal information about a person's location, interests, and purchase preferences and communicated through IoT-based infrastructure [19].

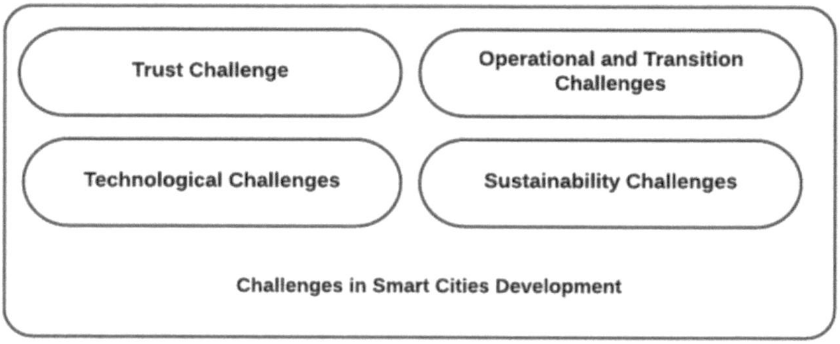

FIGURE 15.4 Challenges in smart cities development.

- *Operational and transition challenges*: There is no such city in the world that is smart from the beginning. The truth is that a major issue facing many cities worldwide is creating a deliberate and well-thought-out transition to smart capabilities while limiting threats to data security and system integrity and causing the least amount of disturbance to present stakeholders [19]. Smart city operations require a very different engagement model, where components collect usage-based data such as user feedback and use that data to make predictions, enhance current services, uncover needs, and develop new initiatives. Must adopt and use a wide range of technology to learn about and leverage various services, including interoperable digital platforms, design/agile thinking, smartphones, and IoT technologies to address implementation difficulties [22].
- *Technological challenges*: In order to become smart, cities must implement new smart technology. Cybercriminals have fresh opportunities with each new technology implemented in urban systems [23]. Wireless and sensor-based technologies have made it possible for IoT-based solutions to be widely deployed in smart city environments. IoT, big data, sensors, machine learning, and GPS-based apps are some of the key technologies that must be integrated for a smart city to function properly. All these technologies pose serious risks to the security and integrity of data relating to citizens. The system should be technically hardened and have adequate security measures to protect against data breaches and disclosed vulnerabilities [19].
- *Sustainability challenges*: More than half the world's population lives in cities, where increasing amounts of energy, land, and other resources are being used. Due to the continued concentration of people in metropolitan areas around the world, these areas are considered more and more important when it comes to addressing challenges related to sustainable development [24]. Nonetheless, low-carbon transportation infrastructure may be included in smart cities to accommodate citizens' flexible mobility needs. IoT-enabled waste management infrastructure enables efficient material collection, transportation, and recycling because many areas of smart city government have gone digital. Smart settings can better address the challenges posed by growing urbanization if they have more of an emphasis on sustainability, energy distribution, mobility, health, and safety [19].

15.6 DISCUSSION

This chapter highlights the growing importance of smart cities in addressing urban challenges and improving the quality of life for residents. The integration of technology, such as intelligent transportation and health tracking systems, plays a central role in the development of smart cities. The chapter also points out a crucial aspect concerning cybersecurity concerns in the context of smart cities. The use of applications that rely on smart sensors and technologies raises potential risks related to the security and privacy of user information. The study aims to explore these cybersecurity concerns and proposes countermeasures to ensure the creation of sustainable environments and secure solutions for smart city planning.

15.7 CONCLUSION

Urban planning issues like an imbalance between supply and demand for infrastructure services and utilities, traffic jams, pollution, a loss of natural green cover, and the typology of peri-urban settlements are a result of the rapid urbanization that is taking place in urban areas of developing countries. These issues have an adverse effect on both the natural and built environments. Over the past 50 years, urban populations around the world have increased significantly. Megacities have significantly increased in number, mostly in less developed regions. There is nothing that sets India's urbanization trajectory apart from the rest of the world. India currently holds the second-largest urban population in the world. Over the past 20 years, the Internet has become increasingly integrated into people's daily lives and has become a key component of international communication. Currently, there are around 3 billion Internet users worldwide. This is because the accessibility, usability, and performance of the Internet have significantly increased thanks to innovations and low costs in this industry. Since the Internet has created such a vast global network, the global economy now generates billions of dollars annually. Cities employ technology to develop and carry out smart plans for inclusive growth, efficient neighborhood infrastructure, and healthy, sustainable ecosystems.

REFERENCES

1. A. Al Omar *et al.*, "A Transparent and Privacy-Preserving Healthcare Platform With Novel Smart Contract for Smart Cities," *IEEE Access*, vol. 9, pp. 90738–90749, 2021, doi: 10.1109/ACCESS.2021.3089601.
2. T. R. Bhattacharya, A. Bhattacharya, B. Mclellan, and T. Tezuka, "Sustainable Smart City Development Framework for Developing Countries," *Urban Research & Practice*, vol. 13, no. 2, pp. 180–212, Mar. 2020. doi: 10.1080/17535069.2018.1537003.
3. A. Randhawa, and A. Kumar, "Exploring Sustainability of Smart Development Initiatives in India," *International Journal of Sustainable Built Environment*, vol. 6, no. 2. Elsevier B.V., pp. 701–710, Dec. 01, 2017. doi: 10.1016/j.ijsbe.2017.08.002.
4. S. Srivastava, and T. Ahmed, "Correlation Between Built-Up and Land Surface Temperature Using Sentinel-2 and Landsat-8 Images through Semi-Supervised Deep Learning Model for Efficient Land Surface Monitoring.", 2022. [Online]. Available :https://www.researchgate.net/publication/363715336
5. L. Kong, X. Mu, G. Hu, and Z. Zhang, "The Application of Resilience Theory in Urban Development: a Literature Review," *Environmental Science and Pollution Research*, vol. 29, no. 33. Springer Science and Business Media Deutschland GmbH, pp. 49651–49671, Jul. 01, 2022. doi: 10.1007/s11356-022-20891-x.
6. J. Yang, and L. Sun, "A Comprehensive Survey of Security Issues of Smart Home System: 'Spear' and 'Shields,' Theory and Practice," *IEEE Access*, vol. 10, pp. 124167–124192, 2022. doi: 10.1109/ACCESS.2022.3224806.
7. A. Pundir, S. Singh, M. Kumar, A. Bafila, and G. J. Saxena, "Cyber-Physical Systems Enabled Transport Networks in Smart Cities: Challenges and Enabling Technologies of the New Mobility Era," *IEEE Access*, vol. 10, pp. 16350–16364, 2022. doi: 10.1109/ACCESS.2022.3147323.
8. J. A. Smith, "Smart Cities: Enhancing Quality of Life Through Integrated Technologies," *Journal of Urban Planning and Development*, vol. 146, no. 3, pp. 245–262, 2020.
9. K. S. J. Garcia, "Smart City Initiatives for Sustainable Urban Development: A Comparative Study of Global Cases," *Sustainability*, vol. 10, no. 2, p. 391, 2018.

10. J. R. M. Brown, "Urban Resilience: A Comprehensive Review of Interdisciplinary Perspectives," *Sustainable Cities and Society*, vol. 47, p. 101482, 2019.
11. W. Y. Chen, "Towards Secure Smart Cities: A Comprehensive Analysis of Cybersecurity Challenges and Solutions," *Computers, Materials & Continua*, vol. 68, no. 3, pp. 2745–2760, 2021.
12. Z. Y. Liu, "Cybersecurity Challenges in Smart Cities: A Review," *Procedia Computer Science*, vol. 109, pp. 877–882, 2017.
13. O. Saber, and T. Mazri, "Smart City Security Issues: The Main Attacks and Countermeasures," in *International Archives of the Photogrammetry, Remote Sensing and Spatial Information Sciences - ISPRS Archives*, International Society for Photogrammetry and Remote Sensing, Dec. 2021, pp. 465–472. doi: 10.5194/isprs-Archives-XLVI-4-W5-2021-465-2021.
14. S. Shamsuddin, "Resilience Resistance: The Challenges and Implications of Urban Resilience Implementation," *Cities*, vol. 103, Aug. 2020, doi: 10.1016/j.cities.2020.102763.
15. S. Srivastava, T. Ahmed, and A. Saxena, "An Approach to Secure IoT Applications of Smart City using Blockchain Technology," 2023.
16. I. Rafiq, A. Mahmood, S. Razzaq, S. H. M. Jafri, and I. Aziz, "IoT Applications and Challenges in Smart Cities and Services," *The Journal of Engineering*, vol. 2023, no. 4, Apr. 2023, doi: 10.1049/tje2.12262.
17. H. N. Saha *et al.*, "Disaster management using Internet of Things," in *2017 8th Industrial Automation and Electromechanical Engineering Conference, IEMECON 2017*, Institute of Electrical and Electronics Engineers Inc., Oct. 2017, pp. 81–85. doi: 10.1109/IEMECON.2017.8079566.
18. Y. Li, and Q. Liu, "A Comprehensive Review Study of Cyber-Attacks and Cyber Security; Emerging Trends and Recent Developments," *Energy Reports*, vol. 7, pp. 8176–8186, Nov. 2021. doi: 10.1016/j.egyr.2021.08.126.
19. E. Ismagilova, L. Hughes, N. P. Rana, and Y. K. Dwivedi, "Security, Privacy and Risks Within Smart Cities: Literature Review and Development of a Smart City Interaction Framework," *Information Systems Frontiers*, vol. 24, no. 2, pp. 393–414, Apr. 2022. doi: 10.1007/s10796-020-10044-1.
20. D. Singh, B. Pati, C. R. Panigrahi, and S. Swagatika, "Security issues in IoT and their countermeasures in smart City applications," in *Advances in Intelligent Systems and Computing*, Springer, 2020, pp. 301–313. doi: 10.1007/978-981-15-1483-8_26.
21. E. Keymolen, and A. Voorwinden, "Can We Negotiate? Trust and the Rule of Law in the Smart City Paradigm," *International Review of Law, Computers and Technology*, vol. 34, no. 3, pp. 233–253, Sep. 2020. doi: 10.1080/13600869.2019.1588844.
22. S. Hasija, Z. J. M. Shen, and C. P. Teo, "Smart City Operations: Modeling Challenges and Opportunities," *Manufacturing and Service Operations Management*, vol. 22, no. 1, pp. 203–213, 2020. doi: 10.1287/msom.2019.0823.
23. C. Ma, "Smart City and Cyber-Security; Technologies Used, Leading Challenges and Future Recommendations," *Energy Reports*, vol. 7, pp. 7999–8012, Nov. 2021. doi: 10.1016/j.egyr.2021.08.124.
24. M. Höjer, and J. Wangel, "Smart Sustainable Cities: Definition and Challenges," *Advances in Intelligent Systems and Computing*, vol. 310, pp. 333–349, 2014. doi: 10.1007/978-3-319-09228-7_20.

16 Cyber Risk Management in IoT Security Technologies

Amrit Suman, Preetam Suman, Sasmita Padhy, and Shahana Gajala Qureshi

16.1 INTRODUCTION

The Internet of Things has made it possible for a network of machines and gadgets to communicate and operate together [1]. Businesses are being compelled to innovate new processes due to this new paradigm. IoT cybersecurity is being targeted by attacks that are becoming more common and widespread. Compliance, reputation, finances, and business operations have been affected by these issues for individuals and organizations. The sharp increase in attacks is caused by the astonishing amount of IoT devices in sectors like smart grids, ecological surveillance, healthcare surveillance mechanisms, intelligent production, and transportation. IoT safety maintenance is challenging due to the constantly changing and ephemeral architecture of the connectivity among devices [2], the variety of entities that might communicate with IoT networks [3], and limited resources [4]. The Internet has seen a rise in IoT attacks from 2018 to 2023 due to the development of IoT safety laws and growing security concerns. An average annual growth rate of 33.7% [5] is predicted for the global IoT protection market. A previous survey indicated that IoT-based attacks will increase in both quantity and intensity, and upper management will need to pay more attention to these hazards [6]. The percentage of respondents who have implemented their IoT security plan is only 28% compared to 35% who claim to have one in place [7]. Existing risk assessment methodologies are inappropriate for fluid-based networks like the IoT, notwithstanding the lack of proper security mechanisms [8]. Due to the complexity of healthcare IoT systems, which exposes various assault points to hackers, current risk evaluation approaches, for example, are not well suited for risks related to cybersecurity evaluation [9]. Building IoT devices on a common platform can make it easier for companies to adopt IoT security measures without unintentionally increasing safety hazards [10]. The aim of cybersecurity in IoT is to reduce cybersecurity risks for both consumers and companies by securing IoT infrastructure and user privacy. Despite this, 26% of the enterprises were not equipped with security defenses. Businesses need to make proactive investments in IoT cybersecurity, as demonstrated by these two studies many IoT devices have security flaws.

Connecting devices and instruments and cooperating is now possible with a network of devices and instruments. Businesses are being pushed to create new processes by the paradigm. The cyclical shifting and transient nature of devices make

DOI: 10.1201/9781003409571-16

IoT security management difficult. IoT privacy regulations will lead to a 33.7% CAGR increase in the global IoT health market from 2018 to 2023, security-related issues, and IoT attacks through the internet. Despite this, only 28% stated that they had implemented it. According to another survey, 80% of firms stated that their IoT devices were attacked by hackers in the previous year. It has been shown that 26% of firms lack security defense measures. Businesses should actively engage in IoT cybersecurity due to the widespread security flaws revealed in the two polls mentioned above. The effectiveness of current risk assessment methodologies is lacking in IoT systems that are constantly evolving despite the absence of precautionary measures. The methods currently used for risk assessment are not effective in evaluating the risks associated with cyber security. Healthcare IoT devices are still complex and susceptible to various attack vectors. Businesses can unknowingly increase cyber vulnerabilities by building IoT systems on a standard platform while simplifying the deployment of IoT security measures.

This chapter is dedicated to IoT privacy technology and strategies for managing security risks. The discussion includes the shortcomings of IoT risk management for cybersecurity, as well as a four-layer structure for IoT risk management for cybersecurity. The topic of linear programming (LP) in this chapter is the best way to allocate funds among multiple IoT vulnerability projects. The demonstration of Internet of Things cyber threat management for a smart hotel accommodation is explained to support risk assessment through proof-of-concept validation.

16.2 LITERATURE REVIEW

Considering the specific form underlying cyber hazards and the weaknesses of IoT networks, understanding the safeguards of IoT devices and presenting cyber hazard approaches to management is essential for establishing a new IoT cyber risk prevention paradigm.

16.2.1 IoT Architecture Cyber-Security

Every stage of the planning process must take privacy considerations into account since every element of IoT technology communicates through various levels and presents specific security issues [11]. We could potentially possible to gain systematic and in-depth knowledge of IoT security measures by conducting a review of the research on security measures through the IoT infrastructure. Copying gadget circuits for intrusions is the main safety concern at the behavioral layer. Distributed denial-of-service (DDoS) attacks can be initiated with the use of counterfeit RFID tags. For the production, identification, and authentication of chip-based cryptographic keys, physical unclonable functionalities (PUFs) have been used [12]. By identifying devices, authenticating users, avoiding device cloning, and repelling manipulation, PUF chips enhance security [13]. Because IoT device components are typically deployed on platforms with restricted resources, lightweight PUF designs are required [14–17].

To boost IoT cyber-security, manual processes must be carried out, staff experience and access to tools must be increased, and concerns must be raised with manufacturers and other outside organizations [18]. IoT devices need to have a certified security level to be accepted. However, creating a cybersecurity accreditation framework is difficult from a technical and legal perspective due to the frequently changing and heterogeneous form of IoT gadgets [19].

16.2.2 Network-Layer Cyber-Security

The efficient operation of devices and compute nodes, which are part of the complete IoT system, requires safe information transit through networks. The network layer plays a crucial role in the overall effectiveness of IoT security. Attacks are recognized, defenses are put in place, and packets are tracked using intrusion detection systems, or IDSs [20]. Numerous methods for methodologies detection are used by the IDS, including statistical evaluation to detect irregularities [21, 22], and process identification for determining unusual activity [23]. The random forest method and deep learning can aid in the identification of network breach patterns.

Deep learning algorithms for identifying the existence of DDoS assaults display good results, with a maximum accuracy of 97.16% [24]. Using hybrid techniques that combine dimension reduction and machine learning algorithms for classification to find malicious behavior on IoT infrastructure also produces encouraging outcomes [25].

16.2.3 Processing-Layer Security

Cloud computing is a commonplace method for keeping and analyzing massive data streams generated by several IoT devices simultaneously at the processing layer. Processing data obtained via network connections while accounting for latency is known as fog computing [26].

By releasing and maintaining data as an open record for every user or node in the framework, the layer that processes information can take advantage of the blockchain system [27]. Distribution networks, intelligent power, and medical are just a few of the blockchain's intriguing IoT applications [28]. In order to provide IoT security certificates, blockchain technology may also be used [29]. This would enable the safe and automated deployment of IoT gadgets.

16.2.4 Application-Layer Cybersecurity

Enterprise Internet of Things applications are frequently used for supervision and oversight, large-scale research, sharing and collaborating knowledge, and large gathering of data [30]. Among the application fields that need a range of security management solutions are smart grids, home automation, intelligent transportation, and intelligent healthcare [31]. When working with highly individualized data in smart health, for example, the highest level of safety and confidentiality is necessary [32, 33]. Numerous methods have been developed to address the aforementioned security challenges, including Methods for secrecy protection, key supervisors, privilege over access, unsuitable connection authorization, and information safety.

16.2.5 Service Management Layer Cybersecurity

Perceptions of value and risk are influenced by trust perceptions, which in turn affect perceptions of IoT adoption [34, 35], as reported in a study on agricultural technological advances. The privacy of customers must be protected by safeguarding their devices and data when processing personally identifiable information (PII) Confidentiality safety mechanisms must be implemented from the start of IoT development to build confidence and support the widespread usage of IoT systems [36]. But because the majority of IoT devices are often small, light, and low power, it can be difficult to maintain security and privacy [37]. Luo et al. [38] offer the Privacy Protector architecture to guard against complex attacks that involve conspiracy attacks and data leaks in healthcare networks based on the Internet of Things.

16.2.6 Review of the Literature on Cybersecurity Risk Management

There isn't much academic or commercial research on IoT security risk management. As a result, this literature evaluation covers the security issues.

16.2.6.1 Analytical Approaches for Handling Cyberspace Risk

Different risk management concepts coexist and vie for attention. Organizations were considered when creating this high-level, readable information [39]. The specific cybersecurity skills displayed by a company's cybersecurity management practices are identified by implementation levels. To improve their cybersecurity status, corporations can use the framework profile to identify opportunities. The cybersecurity framework created by NIST only briefly addresses risk management concerns, acknowledging that every business handles cybersecurity risk differently. Collaboration between the International Organisation for Standardisation (ISO) and the Worldwide Electrotechnical Commission (IEC) led to the creation of the ISO/IEC 27005 series of specifications. Rules and techniques are used by management to develop and manage risks related to information security [40].

There are seven steps or chains in the Cyber Killing Chain (CKC) structure [41]. The model examines the steps a cyber-attacker takes to achieve their goals. The defense is advised to employ strategies to disrupt the chain both immediately and at a later date. The Security Assessment Methodology 'Functioning Critical Threat, Asset, and Vulnerability Evaluation' (OCTAVE) was developed at the CERT(R) Coordination Centre in Pittsburgh, Pennsylvania, in the United States.

Identification and ranking of essential IT assets, as well as evaluating threats, are assisted by OCTAVE. Furthermore, it evaluates vulnerabilities and their repercussions and establishes security priorities [42]. An organization's evaluation process is conducted by a small, multidisciplinary team overseen by internal departments. OCTAVE Allegro provides a thorough security evaluation without requiring a comprehensive understanding of risk assessment. The Cyberactivity Platform has a plan of a personalized list of tasks ranked according to the risks that are most important to the enterprise [43]. A security gap analysis is performed by the Cyberactivity Platforms to compare the current maturity to the intended maturity and prioritize security actions. Organizations should implement the 20 important safety measures based on the Consensus Audit Guidelines (CAG).To prevent or reduce intrusions, the Centre for Internet Security (CIS) has released this [44]. Implementation Groups (IGs) are part of CAG and can be compared to the implementation layers of the NIST Cybersecurity Framework.IG 1's understanding of finance and cybersecurity is the lowest, while IG 3's understanding is the highest. The goal of the IGs is to help companies determine their level of counterintelligence maturity, prioritize control applications, and create effective cybersecurity programs.

16.2.6.2 Quantitative Methods for Handling Cyberspace Risk

The study that employs statistical methods to manage risks related to cybersecurity is reviewed in the following paragraphs. Applying quantitative techniques to a cyber-risk assessment usually limits the study's scope. To raise the security level, a set of crucial processes must be correctly executed in the framework; risk assessment, mitigation, validation, and monitoring are all aspects of risk management. BDN (Bayesian decision network) [45] models the data necessary to control security risks. The cost-benefit of risk mitigation strategies can be evaluated using modern Bayesian inference methods. Their analysis indicates that their approach significantly enhances network security by analyzing risks accurately and minimizing risks efficiently. AVARCIBER [46] is a new risk evaluation method that expands on the specifics of ISO 27005. A variety of procedures are used to put the method into reality, such as starting the risk evaluation, determining and evaluating assets, determining privacy and security dangers, quantifying the risk, and taking precautions. When compared to ISO 27005, the whole procedure is conventional. Nonetheless, there are additional tasks and useful manuals offered. Through an instance study, the effectiveness of applying the framework was evaluated.

The NIST Cybersecurity Strategy now includes an assessment of costs and benefits because it has been determined that there had been no workable economic technique for justifying cybersecurity projects [47]. The research establishes a link between the NIST Implementation Tier levels and the proper amount of cash that a corporation should set aside for cybersecurity expenditures and shows how a cost-benefit evaluation may assist companies in choosing the best NIST Implementing Tier level. Nevertheless, neither the research nor the explanation of how financial resources might be distributed between different cybersecurity programs offers any useful information for budgeting choices.

16.2.7 Literature Evaluation

High-level concepts are the main emphasis of subjective methods of cyber risk mitigation, according to a survey of the literature on both qualitative and quantitative methodology. These subjective standards adhere to a common cybersecurity risk management procedure while respecting the validity of earlier standards [47]. According to the literature review, IoT cybersecurity risk management has not received much attention. The complex web of connections between the numerous actors and equipment cannot be handled by the current methodologies for risk assessment [3]. Our motivation is to develop an IoT security risk management system that improves the current frameworks and quantitative methods for risk evaluation by analyzing previous studies. Efficient risk administration requires the four-layer architecture for IoT cyber risk assessment that this chapter proposes, which combines both quantitative and qualitative methods.

16.2.8 The Four-layer Framework for Handling Cyberspace Risk in IoT

To assist IoT security executives in creating an approachable risk management and cybersecurity strategy, a framework is made available (Figure 16.1). In order to plan and examine risk mitigation actions layer by layer while keeping the bigger cybersecurity issues in mind, the suggested structure analyses the key factors determining risks associated with cybersecurity and groups these features into four distinct layers. The 4-components that together make up the four-layer framework for IoT cyber-hazard control represented in Figure 16.1 include the IoT cyberspace ecological system, IoT cyberspace structures, IoT cyber threat evaluation, and IoT cyber performances.

The IoT cybersecurity ecosystems layer serves as the foundation for the risk monitoring architecture. An enterprise can recognize and comprehend the dynamics and functions of its constituents by evaluating the ecosystem's constituent parts. The layer of the IoT cyber ecosystem that monitors and assesses the outside world regularly and/or constantly transmits results to other relevant levels. The company evaluates the current state of the security infrastructure at the IoT layer by examining employee roles and duties, organization rules, and deployment of IoT security technology.

16.2.8.1 IoT Cyber Ecosystem Layer

Connected devices have interacted with each other by participants in the IoT cyber ecosystem in both collaborative and competitive ways. Developers of IoT cybersecurity systems are the dominant force in the ecosystem. Outside consumers and clients, enemies, authorities, and standardization groups are part of this. In order to

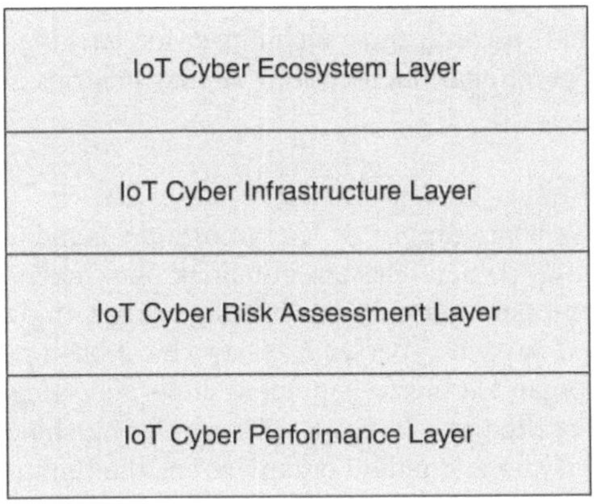

FIGURE 16.1 A framework for managing cybersecurity risks in IoT.

protect IoT systems, cybersecurity directors must respond quickly to changes in the IoT cyber ecosystem.

16.2.8.1.1 Creators of IoT Cybersecurity Technology

The expertise required to create IoT security solutions is not typically available to administrators of IoT systems. As a result, IoT administrators must be kept up-to-date with the advancements in the IoT communities of developers. The IoT and IoT cybersecurity technology markets are growing in importance. The most recent technologies are being integrated into IoT devices by cybersecurity innovators to protect data flow, stop hacking, and respect privacy requirements [48]. Servers calculating, edge/fog IoT devices, and 5G ought to be able to offer greater protection from a variety of safety risks. Real-time prevention, detection, and recovery methods that are more accurate than current approaches can be offered to IoT devices through machine learning and artificial intelligence. In order to develop secure Internet of Things systems integrating, IoT security platforms must offer complete solutions.

16.2.8.1.2 Customers and Outside Users

IoT innovations cannot completely realize their advantages without the support of external users and clients. For IoT applications to be successful, users' and clients' approval of IoT services is essential. A study [49] found that while reported enjoyment and usability of IoT apps have a beneficial impact on usage behavior, perceived privacy risk has a negative influence on IoT adoption. IoT managers must first identify external IoT users and consumers, analyze their behavior and choices, and address their confidentiality and safety issues in order to develop an IoT use model that emphasizes these concerns.

16.2.8.1.3 Opponents

The digital environment accounts for adversaries like invasions and hacks who pose dangers online for monetary profit or for unlawful motives. The risk presented by assailants who are better informed in cybersecurity hasn't been completely addressed by the adoption of improved wireless security protocols and cryptography [50]. The company must constantly monitor for new attackers, understand why they are breaking into networks, and research their techniques for hacking, data theft, malware installation, and/or operational interception. The adversaries are always coming up with new tactics.

16.2.8.1.4 Authorities

To comply with laws that impose severe minimum standards, businesses must invest in cybersecurity [51, 52]. These regulations may potentially have an impact on mobile healthcare operations and have a significant impact on how hospitals exchange patient data. Several cybersecurity laws have been passed in the US as of July 2020 to protect organizations' computer systems and safeguard data from hackers. The Cyberspace of Medical Things Resilience Partnership Act of 2017 created a consortium of both private and public organizations that aim to establish voluntary norms and procedures [53-56]. Security technologies are encouraged by authorities through the creation of laws and various taxes to address security issues associated with the use of online services. Companies engaged in Innovations for safeguarding IoT goods, facilities, or procedures should keep an eye on how government rules change, and determine whether they will affect the expansion of their IoT networks, before making any required investments.

16.2.8.1.5 Organizations for Standardization

There are no guidelines to help cybersecurity management select the best cybersecurity solutions regardless of increased expenditure in information assurance. The need for fast deployment of safe IoT goods and services has motivated recent initiatives to standardize IoT security. The discussion afterward examines the initiatives taken by three standardizing organizations to standardize IoT cybersecurity. The ITU-T Recommendations were created by the Study Groups of the ITU's Telecommunication Standardisation Sector and detail the components of the global ICT infrastructure. The IoT-GSI sponsored a coordinated strategy for the creation of technological standards to support the IoT's cross-border operations. The first international standard for consumer IoT security was published by the European Telecommunications Standards Institute (ETSI), and it is designed to meet high-security principles. According to TS 103 645, developers must have an attack reporting procedure and avoid using global passwords that are defaults.

16.2.8.2 IoT Cyber Infrastructure Layer

Assessing themselves is necessary for companies to create a plan to mitigate cyber risks. Many studies have focused on the technology elements involved in managing risks to ensure the security of IT assets and services.

16.2.8.2.1 Employees and Internal Users

The focus of this section is on the variables that influence users' attitudes, behaviors, and satisfaction with cybersecurity technologies, as well as their compliance with security and privacy laws. The efficacy of managing cybersecurity depends on promoting best practices and increasing public knowledge of cybersecurity issues. Over 85% of managers rank employee irresponsibility as one of their top security concerns, according to Shred-it research.

16.2.8.2.2 Structure

Executive assistance, governance of security, expenditures in protection, security education, and other administrative initiatives are the main themes of the business aspect. Because they are recognized to have a beneficial impact on the mindsets and actions of stakeholders, these activities are essential for improving IoT cybersecurity effectiveness. Enterprise risk management now includes the ability of an organization to respond to cyberattacks. Successful security plan development and implementation require strong senior management backing.

Organizations must update their cybersecurity plans. Connecting IoT cyber risk management with IT risk management and corporate risk administration requires the creation of cybersecurity oversight and functional strategies. An organization's cybersecurity abilities determine the four implementation tiers recommended by the Cybersecurity Framework developed by NIST [39]. Safety protocols must be designed to match the intended degree of Deployment Tier based on the organization's current cybersecurity capacity.

16.2.8.2.3 Cybersecurity Technologies

In order to find and deploy reliable IoT security measures, the organization could profit from guidance from the IoT cyber innovators in the IoT cyberspace ecosystems. Domestic technology for cybersecurity should serve both the business's overall privacy objectives and the IoT safety goals. Technology that ensures information confidentiality, technology that detects and counters online dangers and vulnerabilities, and technology that detects and counters criminality are Hildebrandt's three key categories of cybersecurity technologies. Cyber-security technology comprises verification, which also includes certification and the management of credentials. Firewalls that are DDoS mitigation, analytics for security, response to incidents systems, device identification and leadership, and safe communication.

16.2.8.3 A Layer for Assessing the Risk of IoT Cyber

After determining their current IoT cybersecurity infrastructure, the organization must assess their current and potential future cyber risks. The IoT cyber risk assessment layer description can be found below:

16.2.8.3.1 Risk Assessment

All risk identification steps [46] involve the identification of IoT assets, vulnerabilities, and cyber threats. Each IoT asset has a list of vulnerabilities and threat categories (e.g., IoT device 1 has vulnerabilities for one type of threat, IoT device 1 has vulnerabilities for two types of threats, and so on. In order to identify IoT risks, one must comprehend how hackers carry out cyberattacks. Exploratory and exploitative mindsets are the two main ways intruders think. Intruders frequently employ intentional and intuitive reasoning throughout the exploratory stages and heavily rely on experimenting. The CKC architecture is widely used to identify and thwart attacks. Defending against the adversary's hostile actions can be more effective with early link breakage. Identifying specific threats can be done by using the seven steps in the CKC framework.

16.2.8.3.2 Quantification of Risk

For risk quantification, the frequency of attacks refers to the projected number of cybersecurity attacks over a specific time period. Quantifying the projected impact of a cyber-attack during a particular period of time can be done by using multiple metrics, including the number of exposed information files and/or economic impairment. The likelihood that an attack will be stopped is also computed; threat type A, for instance, has a defense probability of 0.9. A Gaussian likelihood density operation, which is frequently employed for a range of approach projects, including hackers, can be utilized to represent the occurrence of attacks as an arbitrary method of arriving at the initial stage in risk assessment. For researchers to comprehend the potential danger status of all IoT materials, weaknesses, and hazards, an assessment matrix is also built. Two components make up the risk matrices. One aspect is the expected monetary loss per infiltrated attack, while

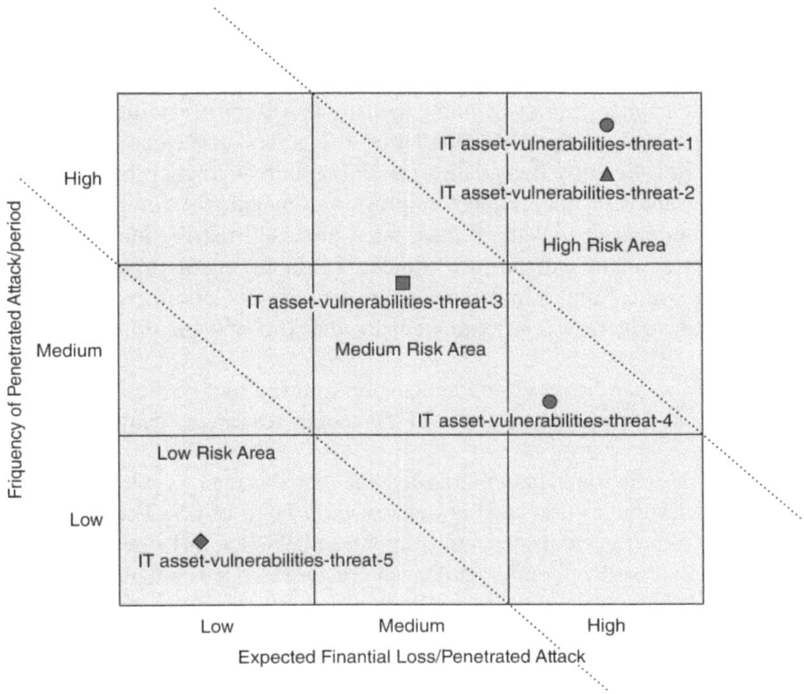

FIGURE 16.2 Risk matrix.

another is the average number of infiltrated attacks for each IoT asset, fragility, and threat. A failed resistance is a weak one. IT asset vulnerabilities and threats are regarded as particularly important because they create a likelihood of risks. An illustration of a risk matrix is shown in Figure 16.2. A high, medium, and low-risk area is arbitrarily separated using dotted lines for demonstration purposes. Risks 1 and 2 related to IoT device weaknesses are depicted in the graphic as being in great danger, while threats 3 and 4 are depicted as moderate to low risk. IoT device weaknesses provide a low-risk danger. Management can use the risk hierarchy to prioritize wealth, weaknesses, and risks while allocating resources for effective risk control. To protect a company from evolving attack patterns, the risk matrix should be revised regularly, even though it's not an easy task.

16.2.8.3.3 Resource Allocation

The resource allocation model relies on the frequency of penetrating attacks each period. The Poisson probability distribution is a specific probability distribution that cyberattacks follow, as previously mentioned. The defender can achieve their goal of increasing the defense likelihood by investing in a way that is comparable. Defense chances are binomial, meaning they can either be successful or unsuccessful. To compute the percentage of infiltrated attacks, n * (1 p) is used if n represents the number of attacks throughout the time and p represents the chance of a successful defense. The total economic loss is calculated by multiplying n* (1 p) * m. Assume the case when the predicted present defense likelihood for the IoT equipment weaknesses threat-1 is 0.9. presuming if the IoT equipment weaknesses threat #1 has 5000 occurrences per cycle and expenses $1000 for every intrusion attempt. Organizational assets are formed when labor, IoT assets, and capital are converted into currency. To efficiently distribute financial resources to each hazard posed by IoT asset vulnerabilities, the chapter employs a conventional computing strategy. It's crucial to emphasize that there have been no asset distribution methods used in cybersecurity to make expenditure choices. LP is a computational model methodology that is logical and simple to use for a variety of commercial challenges. Operational and budgetary restrictions ensure that the best possible outcomes are guaranteed.

The LP method can be an effective decision-making tool because of the numerous cybersecurity projects and limited financial resources available for their administration.

Figure 16.3 demonstrates how increasing the distribution of capital will improve the defense likelihood with a starting defense likelihood of 0.9. The defense probability ranges on the lateral dimension from 0.9 to 0.99. The IoT device weaknesses risk will result in lesser economic damage as greater resources are assigned because of the enhanced probability of defense. Combining the monetary damage brought on by successful attacks with the allotted price for the intended defense probability's cyber expenditure yields the total expenditure.

Regarding the sake of this instance, the quadratic curve around 0.9 and 0.99% are taken into account for both the economic damage and the expenses of the cyber expenditure for defensive possibilities of 0.99 and higher, a rising trend in the expense of making an expenditure in online is projected due to the quick rise technology cost. The monetary damage is determined as FL = $5L x$50L* (x 0.9), where x is the

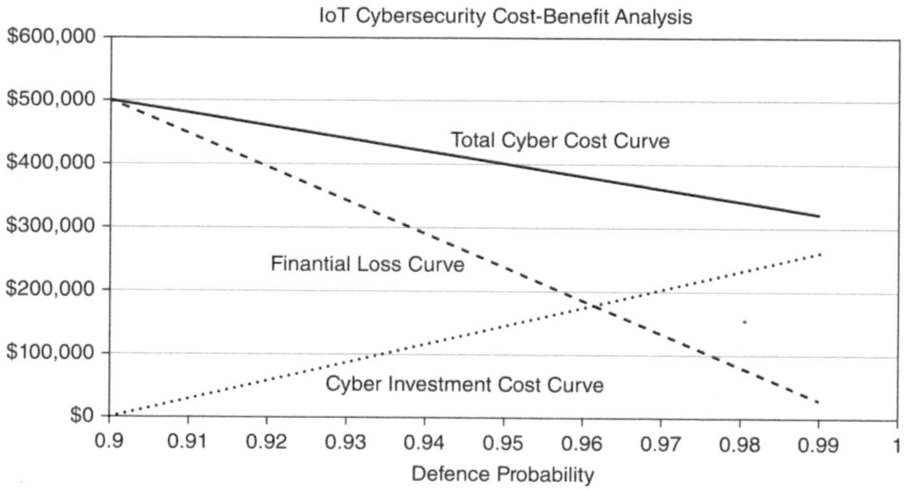

FIGURE 16.3 A cost-benefit analysis of IoT cybersecurity.

likelihood of security. The total price of investing is determined as IC = $30L* (x 0.9). Integrating FL and IC results in the overall cost, which is computed as follows: TC = FL + IC = $5Lx $20Lx (x 0.9).

16.2.8.3.4 The Constraints of IoT Asset Vulnerabilities and Threats

The entire cost decreases up to 0.99, as shown in Figure 16.3. It can be challenging to decide how to allocate the firm's assets of a business among a variety of IoT security initiatives because the expenditure is subject to a number of operating and monetary limitations and has a limited amount of administrative assets. The NIST's four Implementation Tiers are broken down into multiple stages for each organization, and every IoT asset vulnerability threat is likely governed by a separate set of operational and financial restrictions, thus limiting the available possibilities.

In accordance with these limitations, we aim to reduce the total cost of cybersecurity for all IoT assets, vulnerabilities, and threats. The current research aims to develop a LP model that takes into account economic and operational constraints. The goal function is a mathematical representation of the variables used to make a decision to reduce overall cyber expenses. A set of functions of equality or inequalities that indicates economic and technological issues are included in the restraints, and functional restrictions on what amount of money can be allocated to the option that is chosen, and the defense likelihoods of IT asset vulnerabilities and threats. All calculations involving the goal functional and restrictions have been done in linear forms. The mathematical description of the LP framework falls beyond the purview of this investigation; readers seeking the details should consult Chakraborty, Chandru, and Rao.

Constraints guarantee that each IoT asset, vulnerability, and threat has a maximum defense probability. For instance, they might hold the belief that the company lacks the necessary personnel or expertise to create a security plan that is highly likely to succeed. A lack of manpower or experience may prevent the company from implementing a complex technology.

The canonical version is known as the following:

$$\begin{aligned} &\text{Minimize}\, d^R y \\ &\text{Subject to}\, Ay \leq s \\ &\text{And}\, 1 \geq y \geq 0 \end{aligned}$$

where y is a variable of probability, A, d, and s are coefficients, and $(.)^R$ denotes the transposition of the matrix.

The differences In order to minimize the goal variable for the total expense of online communities, the limitations $Ay \leq s$ and $1 \geq y \geq 0$ construct a convex polytope.

16.2.8.4 The IoT Cybersecurity Performance Layer

After allocating resources and detecting a cyber-response, the IoT cyber efficiency actions are initiated. Despite this level being essential for all aspects of managing risks, there will not be a lot of debate since decisions established at the risk evaluation layer determine the majority of duties. Execution, tracking and oversight, and ongoing enhancement are the primary activities at the IoT cyber security level.

16.2.8.4.1 Implementation

To install the new IoT cyber architecture, it is necessary to create, test, install, develop new regulations, teach, and educate users of IoT cyber security technologies. To compare and select cybersecurity technologies in the IoT cyber ecosystems layer, organizations need to establish selection standards. The security of both the IoT network and individual devices can be guaranteed through a gateway-based architecture.

16.2.8.4.2 Monitoring and Control

To address varying threats, prevention, detection, and recovery are the main tasks of the monitoring and control stage, which are carried out simultaneously. Through prevention and detection operations, data on the equipment's operation, unusual user behavior, and unauthorized access to data and applications are collected. Real-time resolution is possible through recovery actions.

16.2.8.4.3 Ongoing Development

Key performance indicators that the organization must track for ongoing improvement include cyberattack frequency, source, and impact of penetration. The time of activities and information that they are interested in is the primary difference between tracking, oversight, and constant enhancement. Constant growth makes use of past information to identify trends and assess permanence over time. Real-time data is collected by surveillance and control to track cyberattacks and respond quickly, on the flip side. Creating frequent performance reports and setting quantitative performance goals are necessary for sustainable development activities.

16.2.8.4.4 A Smart Room Scenario Is Used to Illustrate the Risk Assessment

The focus of this section is on a genuine smart room scenario as a demonstration of IoT cyber risk assessment.

Intelligent rooms powered by the IoT are becoming more popular in hotels, clinics, private residences, and other institutions. In a connected smart room that is connected to the Internet, the IoT risk assessment can be presented. In intelligent spaces, complex IoT assets, vulnerabilities, and threats are present where multiple diverse gadgets and technology frameworks communicate with each other. Due to the rising competition in the hotel industry and the declining cost of IoT equipment, smart hotel rooms have become the norm. Smart device users are still concerned about privacy infringement, though. The proposed LP model is used to simulate a hypothetical smart room as a proof of concept. Although this scenario is probable to happen in any upscale hotel room and is realistic, this example does not suggest that any real hotel chain was engaged in the creation of this scenario. Types v and w may be independent and capable of choosing their own resource allocation strategies. The network types x, y, and z require optimized resource distribution.

The LP model's hypothesizes

1. The factors that influence decision-making for types x, y, and z in a consistent manner.
2. The chance that a defense will achieve success is x = 0.92, y = 0.92, and z = 0.92.
3. All x, y, and z must have a maximum defense probability of 0.99.
4. The financial constraints that hinder the achievement of the 0.94 defense probability are:

$$\$40Lx + \$6Ly \leq \$4,290K,$$
$$\$40Lx + \$6Ly \geq \$4,150K,$$
$$\$6Ly + \$30Lz \leq \$3,350K,$$
$$\$40Lx + \$30Lz \leq \$5,700K,$$
$$\$40Lx + \$30Lz \geq \$5,500K,$$

5. Defense likelihood to y and z: y + z <= 1.92 operational feasibility limitation.

Using the presumptions as a foundation, the following LP model is developed:

GTC minimization equals 7,165K x 1,600K y 4,000K z
Depending on

$$\$40Lx + \$6Ly \leq \$4,290K,$$
$$\$40Lx + \$6Ly \geq \$4,150K,$$
$$\$6Ly + \$30Lz \leq \$3,350K,$$
$$\$40Lx + \$30Lz \leq \$5,700K,$$
$$\$40Lx + \$30Lz \geq \$5,500K,$$
$$y + z \leq 1.92$$

The outcome of employing a free online LP to implement this LP model is discussed below.

Having kind y provided the defensive likelihood of 0.98, kind x provided the protection possibility of 0.933, and kind z allotted the protection possibility of 0.93, figure 3 demonstrates that Option 3 is the most expensive option. $304K is the ideal total investment cost. Despite having a lower beginning cost, Option 3 incurs fewer expenses overall than Option 2 when using the LP framework. This example demonstrates where the LP model may be helpful when choosing how to distribute difficult IoT cyber capabilities. By altering the premises, a number of sensitivity tests can be carried out, giving supervisors more assurance when deciding how to split resources while employing the LP framework.

16.3 DISCUSSION

For example, driverless cars, drones, smart grids, smart manufacturing, and smart health all have their roots in the IoT. IoT networks experience a significant increase in potential security risks when more connected devices of all types and sizes are added. Security in IoT systems makes it easier to access sensitive data and vital infrastructure. Investing in and managing IoT cybersecurity risks effectively is a challenge for organizations without a framework for managing IoT cybersecurity risks. To identify, measure, and prioritize IoT cyber threats, the IoT cyber risk assessment layer is responsible. The creation of an LP model was necessary to distribute resources among competing IoT security initiatives. An LP model was used as a demonstration of the principle to evaluate cybersecurity risks for IoT. This study aims to fill a gap in cybersecurity risk management for IoT and attract more people to the field. The existing frameworks did not provide managers with any resource allocation techniques, for instance. There were no techniques for allocating resources, so cyber investments were based solely on intuition and weak justification. This flaw has been addressed with the design of a four-layer IoT cyber risk prevention architecture. It is possible to expand the LP framework to deal with larger Internet of Things (IoT) networks that have dozens of decision-making variables, including those used in intelligent production and intelligent mobility.

Companies need to continuously evaluate technological advancements to respond quickly to hacking attacks and assaults. Various 5G-enabled IoT applications, such as autonomous cars, augmented reality, and intelligent patient tracking, will see a rise in hacker activity. In addition, to the advancements of 5G and the related 5G-enabled IoT innovations. If executives want to be more prepared to defend against evolving adversary attacks, they must understand risk mitigation practices and the rapid adoption of IoT apps. Offering a strategy for managing IoT risks that can be utilized to design and deploy security IoT devices. This study adds to the existing body of information that is already available.

REFERENCES

1. Lee, I. The Internet of things for enterprises: An ecosystem, architecture, and IoT service business model. *Internet Things Eng. Cyber Phys. Hum. Syst.* 2019, *7*, 100078.
2. Atzori, L.; Iera, A.; Morabito, G. The Internet of Things: A survey. *Comput. Netw.* 2010, *54*, 2787–2805.
3. Nurse, J.R.C.; Creese, S.; de Roure, D. Security risk assessment in Internet of Things systems. *IT Prof.* 2017, *19*, 20–26.
4. Malik, V.; Singh, S. Security risk management in IoT environment. *J. Discret. Math. Sci. Cryptogr.* 2019, *22*, 697–709.
5. MarketsandMarkets. IoT Security Market Worth $35.2 Billion by 2023. 2019. Available online: https://www.marketsandmarkets.com/PressReleases/iot-security.asp.
6. PwC. Managing Emerging Risks from the Internet of Things. 2016. Available online: https://www.pwc.com/us/en/services/consulting/cybersecurity/library/broader-perspectives/managing-iot-risks.html.
7. Irdeto. New 2019 Global Survey: IoT-Focused Cyberattacks Are the New Normal. 2019. Available online: https://resources.irdeto.com/global-connected-industries-cybersecurity-survey/new-2019-global-survey-iot-focused-cyberattacks-are-the-new-normal.

8. Aldmour, R.; Burnap, P.; Lakoju, M. Risk assessment methods for converged IoT and SCADA systems: Review and recommendations. In Proceedings of the Living in the Internet of Things (IoT 2019), London, UK, 1–2 May 2019; p. 6.
9. Rao, A.; Carreón, N.; Lysecky, R.; Rozenblit, J. Probabilistic threat detection for risk management in cyber-physical medical systems. *IEEE Softw.* 2018, *35*, 38–43.
10. Deloitte. Secure IoT by Design. 2018. Available online: https://www2.deloitte.com/us/en/pages/operations/articles/iot-platform-security.html (accessed on 17 September 2020).
11. Bendavid, Y.; Bagheri, N.; Safkhani, M.; Rostampour, S. IoT device security: Challenging "A lightweight RFID mutual authentication protocol based on physical unclonable function. *Sensors.* 2018, *18*, 4444.
12. Gao, Y.; Ranasinghe, D.C.; Al-Sarawi, S.F.; Kavehei, O.; Abbott, D. Emerging physical unclonable functions with nanotechnology. *IEEE Access.* 2016, *4*, 61–80.
13. O'Neill, M. Insecurity by design: Today's IoT device security problem. *Engineering.* 2016, *2*, 48–49.
14. Mukhopadhyay, D. PUFs as promising tools for security in Internet of Things. *IEEE Des. Test.* 2016, *33*, 103–115.
15. Kulseng, L.; Yu, Z.; Wei, Y.; Guan, Y. Lightweight Mutual Authentication and Ownership Transfer for RFID Systems. In Proceedings of the 2010 IEEE INFOCOM, San Diego, CA, USA, 15–19 March 2010; pp. 1–5.
16. Xu, H.; Ding, J.; Li, P.; Zhu, F.; Wang, R. A lightweight RFID mutual authentication protocol based on physical unclonable function. *Sensors.* 2018, *18*, 760.
17. Zhu, F.; Li, P.; Xu, H.; Wang, R. A lightweight RFID mutual authentication protocol with PUF. *Sensors.* 2019, *19*, 2957.
18. Boeckl, K.R.; Fagan, M.J.; Fisher, W.J.; Lefkovitz, N.B.; Megas, K.N.; Nadeau, E.M.; Piccarreta, B.M.; O'Rourke, D.G.; Scarfone, K.A. Considerations for managing Internet of Things (IoT) cybersecurity and privacy risks. *NISTIR.* 2019, 8228. Available online: https://nvlpubs.nist.gov/nistpubs/ir/2019/NIST.IR.8228.pdf.
19. Matheu, S.N.; Hernandez-Ramos, J.L.; Skarmeta, A.F. Toward a cybersecurity certification framework for the Internet of Things. *IEEE Secur. Priv.* 2019, *17*, 66–76.
20. Hodo, E.; Xavier Bellekens, X.; Hamilton, A.; Dubouilh, P.-L.; Iorkyase, E.; Tachtatzis, C.; Atkinson, R. Threat analysis of IoT Networks Using Artificial Neural Network Intrusion Detection System. In Proceedings of the 2016 International Symposium on Networks, Computers and Communications (ISNCC), YasmineHammamet, Tunisia, 11–13 May 2016; pp. 1–6.
21. Pacheco, J.; Benitez, V.; Félix, L.. Anomaly Behavior Analysis for IoT Network Nodes. In Proceedings of the 3rd International Conference on Future Networks and Distributed Systems, Paris, France, 1–2 July 2019; pp. 1–6.
22. Li, J.; Zhao, Z.; Li, R.; Zhang, H. AI-based two-stage intrusion detection for software defined IoT networks. *IEEE Internet Things J.* 2019, *6*, 2093–2102.
23. Subasi, A.; Al-Marwani, K.; Alghamdi, R.; Kwairanga, A.; Qaisar, S.M.; Al-Nory, M.; Rambo, K.A. Intrusion Detection in Smart Grid Using Data Mining Techniques. In Proceedings of the 2018 21st Saudi Computer Society National Computer Conference (NCC), Riyadh, Saudi Arabia, 25–26 April 2018; pp. 1–6.
24. Roopak, M.; Tian, G.Y.; Chambers, J. Deep Learning Models for Cyber Security in IoT Networks. In Proceedings of the 2019 IEEE 9th Annual Computing and Communication Workshop and Conference (CCWC), Las Vegas, NV, USA, 7–9 January 2019; pp. 452–457.
25. Pajouh, H.H.; Javidan, R.; Khayami, R.; Dehghantanha, A.; Choo, K.R. A two-layer dimension reduction and two-tier classification model for anomaly-based intrusion detection in IoT backbone networks. *IEEE Trans. Emerg. Top. Comput.* 2019, *7*, 314–323.

26. Sohal, A.S.; Sandhu, R.; Sood, S.K.; Chang, V. A cybersecurity framework to identify malicious edge device in fog computing and cloud-of-things environments. *Comput. Secur.* 2018, *74*, 340–354.
27. Minoli, D.; Occhiogrosso, B. Blockchain mechanisms for IoT security. *Internet Things.* 2018, *1*, 1–13.
28. Rao, A.R.; Clarke, D. Perspectives on emerging directions in using IoT devices in blockchain applications. *Internet Things.* 2020, *10*, 100079.
29. Neisse, R.; Hernández-Ramos, J.L.; Matheu, S.N.; Baldini, G.; Skarmeta, A. Toward a Blockchain-Based Platform to Manage Cybersecurity Certification of IoT devices. In Proceedings of the 2019 IEEE Conference on Standards for Communications and Networking (CSCN), Granada, Spain, 28–30 October 2019; pp. 1–6.
30. Lee, I.; Lee, K. The Internet of things (IoT): Applications, investments and challenges for enterprises. *Bus. Horiz.* 2015, *58*, 431–440.
31. Puthal, D.; Nepal, S.; Ranjan, R.; Chen, J. Threats to networking cloud and edge datacenters in the Internet of Things. *IEEE Cloud Comput.* 2016, *3*, 64–71.
32. Almulhim, M.; Zaman, N. Proposing Secure and Lightweight Authentication Scheme for IoT Based E-Health Applications. In Proceedings of the 2018 20th International Conference on Advanced Communication Technology (ICACT), Chuncheon, Korea, 11–14 February 2018; pp. 481–487.
33. Tweneboah-Koduah, S.; Skouby, K.E.; Tadayoni, R. Cyber security threats to IoT applications and service domains. *Wirel. Pers. Commun. Int. J.* 2017, *95*, 169–185.
34. Jayashankar, P.; Nilakanta, S.; Johnston, W.J.; Gill, P.; Burres, R. IoT adoption in agriculture: The role of trust, perceived value and risk. *J. Bus. Ind. Mark.* 2018, *33*, 804–821.
35. Zhou, J.; Cao, Z.; Dong, X.; Vasilakos, A.V. Security and privacy for cloud-based IoT: Challenges. *IEEE Commun. Mag.* 2017, *55*, 26–33.
36. Sicari, S.; Rizzardi, A.; Grieco, L.A.; Coen-Porisini, A. Security, privacy and trust in Internet of Things: The road ahead. *Comput. Netw.* 2015, *76*, 146–164.
37. Dorri, A.; Kanhere, S.S.; Jurdak, R.; Gauravaram, P. Blockchain for IoT Security and Privacy: The Case Study of a Smart Home. In Proceedings of the 2017 IEEE International Conference on Pervasive Computing and Communications Workshops, Kona, HI, USA, 13–17 March 2017; pp. 618–623.
38. Luo, E.; Bhuiyan, M.Z.A.; Wang, G.; Rahman, M.A.; Wu, J.; Atiquzzaman, M. PrivacyProtector: Privacy-protected patient data collection in IoT-based healthcare systems. *IEEE Commun. Mag.* 2018, *56*, 163–168.
39. NIST. Cybersecurity Framework. 2018. Available online: https://www.nist.gov/cyberframework.
40. ISO/IEC. ISO/IEC 27005:2018(en) Information Technology—Security Techniques—Information Security Risk Management. 2018. Available online: https://www.iso.org/obp/ui/#iso:std:iso-iec:27005:ed-3:v1:en.
41. Martin, L.. Cyber Kill Chain®. 2009. Available online: https://www.lockheedmartin.com/en-us/capabilities/cyber/cyber-kill-chain.html.
42. Caralli, R.A.; Stevens, J.F.; Young, L.R.; William R Wilson, W.R. TECHNICAL REPORT CMU/SEI-2007-TR-012 ESC-TR-2007-012 CERT Program. 2007. Available online: https://resources.sei.cmu.edu/asset_files/TechnicalReport/2007_005_001_14885.pdf.
43. CMMI Institute LLC. Over 1/2 of Cyber Professionals Expect a Cyber Attack within 12 Months. 2020. Available online: https://cmmiinstitute.com/products/cybermaturity.
44. CIS. CIS Controls® V7.1. 2019. Available online: https://www.cisecurity.org/controls/.
45. Khosravi-Farmad, M.; Ghaemi-Bafghi, A. Bayesian decision network-based security risk management framework. *J. Netw. Syst. Manage.* 2020, 28, 1794–1819.
46. Rea-Guaman, A.M.; Mejía, J.; San Feliu, T.; Calvo-Manzano, J.A. AVARCIBER: A framework for assessing cybersecurity risks. *Clust. Comput.* 2020, 23, 1827–1843.

47. Gordon, L.A.; Loeb, M.P.; Zhou, L. Integrating cost–benefit analysis into the NIST cybersecurity framework via the Gordon-Loeb model. *J. Cybersecur.* 2020, *6*, tyaa005.
48. Thomas, M. 13 IOT security companies you should know. 2019. Available online: https://builtin.com/internet-things/iot-security-companies-startups.
49. Hsu, C.-L.; Lin, J.C.-C. Exploring factors affecting the adoption of Internet of Things services. *J. Comput. Inf. Syst.* 2018, *58*, 49–57.
50. Das, R.; Gadre, A.; Zhang, S.; Kumar, S.; Moura, J.M.F. A Deep Learning Approach to IoT Authentication. In Proceedings of the 2018 IEEE International Conference on Communications (ICC), Kansas City, MO, USA, 20–24 May 2018; pp. 1–6.
51. U.S. Department of Health and Human Services. Health Insurance Portability and Accountability Act of 1996. 1996. Available online: https://aspe.hhs.gov/report/health-insurance-portability-and-accountability-act-1996.
52. U.S. Department of Health & Human Services. HITECH Act Enforcement Interim Final Rule. 2009. Available online: https://www.hhs.gov/hipaa/for-professionals/special-topics/hitech-act-enforcement-interim-final-rule/index.html.
53. US Congress. H.R.3985—Internet of Medical Things Resilience Partnership Act of 2017. 2017. Available online: https://www.congress.gov/bill/115th-congress/house-bill/3985/text?format=txt.
54. US Congress. S.734—Internet of Things Cybersecurity Improvement Act of 2019. 2019. Available online: https://www.congress.gov/bill/116th-congress.
55. European Union. General Data Protection Regulation GDPR. 2016. Available online: https://gdpr-info.eu/.
56. European Union. The EU Cybersecurity Act. 2019. Available online: https://ec.europa.eu/digital-single-market/en/eu-cybersecurity-act.

17 Landscape of Cyber Terrorism on Internet and AI-Powered Countermeasures

Vipin Khattri, Neha Bhushan, Deepak Kumar Singh, and Mohammad Shiblee

17.1 INTRODUCTION

Terrorism is such a name that after hearing it, a feeling of fear comes to mind. Numerous incidents of terrorism exist such as the 9/11 attack and the Kandahar plane hijacking incident. The combination of the Internet and cyberspace is involved with terrorism and creates cyber terrorism [1]. The Internet and cyberspace are the foundation of cyber terrorism. Therefore, it becomes more dangerous [2] because the accessibility of the Internet is to all people, and no one can differentiate between a common person and a cyber-terrorist. There was a problem with carrying out terrorist activities when there was no or less internet availability as the terrorist was involved in real terms. Cyber terrorism is not a new form of crime. Rather, traditional terrorism is presented with the Internet and cyberspace [1]. In cyber terrorism, cyber terrorists first select their target, and all the facts and statistics about the target are taken from cyberspace through the Internet. After that cyber terrorists design cyber-attacks from any place and complete the task without being caught by the police. The terrorist organization needs money to survive and execute their goal. For this reason, the terrorist organization takes the help of cyber terrorism so that funding can be raised [3]. The goal of terrorists is to make a presence around the world. Therefore, cyber terrorists use social media networks to promote their propaganda with the help of videos, photos, stories, and messages. Planning of cyber-attacks and execution of plans can be communicated among terrorists using social media networks [4].

Cyber terrorists carry out lots of illegal business with the help of the Internet and cyberspace [3]. These illegal businesses include smuggling, drug trafficking, human trafficking, and buying and selling weapons. Cyber terrorists use the dark web platforms for these illegal businesses [5]. The dark web is the most appropriate place for terrorists to purchase or sell weapons and money transfers. Terrorists consider the dark web as heaven [6] for buying and selling illegal items anonymously. Cyberspace provides an open environment for cyber terrorism because of anonymity [7], time efficiency, and secrecy of donor and recipient [8]. Recruitment of terrorists is also done through the Internet and cyberspace. In cyber terrorism, terrorists use cyber hackers for cyber-attacks, and fundraising and terror can be augmented among civilians [3].

 DOI: 10.1201/9781003409571-17

In today's information and communication technology era, every business organization is being automated, for example, oil supply, food supply, electricity control, traffic control, air space control, and many more. This whole system is linked with the economy; therefore, the threat of cyber terrorism remains [9]. Terrorist organizations use cyberspace and the Internet for drug trafficking, human trafficking, money laundering, extortion, arms trafficking, and online training to involve people in terrorist activities [10]. Cyberspace and the Internet are open to all and the option of cyber terrorism increases for the country. Therefore, the responsibility of all countries is increased to obstruct cyber terrorism. The main goal of this study is to prepare everyone for the danger of cyber terrorism. The goal will be completed using the following objectives which are as follows and make everyone aware of the following:

- Cyber terrorism
- Working of cyber terrorism
- Threat of cyber terrorism on social media
- Illegal business on the dark web
- Artificial intelligence (AI) techniques to prevent cyber terrorism

To achieve this goal, this study is divided into seven sections, each focusing on the working and different areas of cyber terrorism. The second section explores the details of cyber terrorism. The third section provides the details of fundraising for carrying out activities of cyber terrorism. The fourth section completes the details of cyber terrorism using social media. The fifth section details the dark web used in cyber terrorism. In the sixth section, the use of various artificial techniques to prevent cyber terrorism is provided. The seventh section gives the concluding remarks on cyber terrorism.

17.2 CYBER TERRORISM

Cyberspace provides a platform that removes the location barrier to perform criminal activities. Terrorist organizations use this cyber platform for various illegal purposes, including drug trafficking, human trafficking, money laundering, extortion, arms trafficking, and recruiting individuals into terrorist activities [11]. Cyber terrorism is not a different form of crime; instead, it emerges over time cyber terrorism from traditional terrorism. The cyber platform has provided opportunities to achieve the objectives of cyber terrorism. The first opportunity involves fundraising, while the second is planning and executing attacks on a country's economy through the use of the Internet and cyberspace [12]. The challenges against countries are becoming more prominent due to the rise of cyber terrorism [13]. Terrorist organizations control cyber terrorism and use the Internet to propagate their propaganda and hate speech [14]. Cyber terrorism uses the Internet and social media as tools to influence individuals, promoting violence against civilians [14]. Cyber terrorism creates political instability and promotes violence through social media [13, 15].

There are numerous central investigating agencies, and the task of investigation agencies is to monitor any terrorist activities. Europol found that the Internet is the most important part of terrorist activities for planning and execution of terror attack plans [16]. This can be easily understood that terrorist websites are available on the web. Before coming to the Internet, terrorists were dependent on traditional media, and that was a very cumbersome process. After the emergence of digital technology, terrorists built websites for their communication and publicity worldwide. Now, complete control is taken by the terrorist group. Somalia, Nigeria, Syria, Afghanistan, Iraq, Germany, Spain, the UK, and France are the most targeted countries on the tip of terrorists for attacks in the last few decades [16]. Cyber terrorism performs digital or cyber-attacks that can come from various sources, whether internal or external. Cyber terrorism is motivated by a variety of factors [17]. Terrorist groups increasingly use the Internet, cyberspace, and social media to propagate their ideologies, achieve their objectives, and enhance operational efficiency [17]. Cyber terrorism creates five distinct forms [17], generating various cyber threats.

- *Publicity*: Terrorist organizations are using virtual platforms like Twitter, WhatsApp, and Instagram to promote their ideologies and acquire sympathy on a global scale.
- *Funding*: Cyber terrorist groups initiate fundraising campaigns using charity, acquire funds to support their organization, and carry out their attack plans.
- *Training*: Terrorist organizations find recruits by accessing social media platforms such as Facebook and Instagram and identifying suitable people for recruitment. Social media platforms are used by terrorists for training purposes, including activities like weapon deployments for attacks.
- *Planning*: Terrorist group members are dispersed around the world. Cyber terrorist groups use social media as a platform for sharing their plans for terrorist attacks due to the end-to-end encryption provided.
- *Execution*: Cyber terrorist activities are also executed with the help of social media platforms.

17.3 FUNDRAISING OF CYBER TERRORISM

Digital technology creates a path for the smooth execution of tasks in human life to progress. However, terrorists exploit the positive aspects of digital technology for negative purposes [12]. Cyber terrorism is influenced by cyber criminals [18]. Therefore, there is a rise of cyber-attacks to acquire funds for terrorists or terrorist organizations. Terrorist organizations of any size require financial resources to survive and execute cyber terrorism functions [3]. The fundraising efforts are done for two different purposes. The first purpose is for terrorist attacks and the second purpose is for the operation of terrorist organizations. The fundraising campaigns of a terrorist group create financial support [19]. The Islamic State of Iraq and Levant required a monthly budget of $5 million for its operational needs [20]. The financing of the Irish Loyalist Paramilitaries involved numerous legal or illegal business activities such as legal businesses, extortion, smuggling, robbery, and various other illegal tasks [21]. Terrorist organizations are involved in legal businesses and acquire donations with numerous advantages such as decreased identification, reduced suspicion, and difficulty to recognize. Charities or donations perform outside developed financial regulatory frameworks, which challenge counter-terrorism agencies to track whether the funds are being used for legal or illegal purposes [22]. Funding can be acquired from numerous sources, which are influenced by

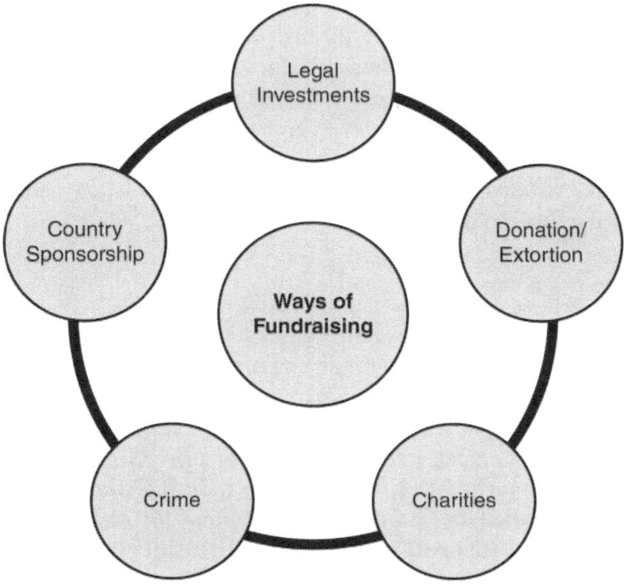

FIGURE 17.1 Ways of fundraising of terrorist group.

ideological, political, and religious influences [22]. The fundraising sources for terrorist organizations depend on numerous factors, including politics, law enforcement, the size of the organization, the skills of its members, and various other factors. Few terrorist organizations are involved in illegal activities such as drug smuggling and fuel laundering concerning their fundraising. Small terrorist groups or individuals are involved in fundraising using one or two sources. Terrorist groups are established by recruiting or combining groups of members, and as a result, organizations can vary from small to large. Cyber-attacks initiated by individual groups require a small budget for execution, whereas cyber-attacks of larger organizations need a higher budget because of more complex targets. This observation can be seen with the help of the following cyber-attacks and related funding requirements [20].

- Boston Bombing (few hundred US$)
- Bali Bombing (US$50,000)
- 9/11 Attack (US$500,000)

Terrorist organizations use five different ways (Figure 17.1) for fundraising.

17.3.1 Donation/Extortion (Contribution/Threats)

Donation or contribution and extortion or threats represent two different purposes of the same concept. Contribution involves voluntary participation, and extortion involves illegal demands [20]. Funding can be sourced from legal and illegal business activities, and the Internet plays a significant role in fundraising [23]. Terrorist organizations apply extortion fees on legal and illegal businesses to ensure their smooth execution. The terrorist organization raised funds through extortion from various businesses such as nightclubs or pubs [20]. When a business does not obey the instructions of a terrorist organization, the organization applies various threats, including violence, property destruction, and kidnapping. The terrorist groups use the funds acquired from extortion to purchase weapons for acts of terrorism. In Iraq, a terrorist organization issued threats against Christians, demanding their conversion to Islam under the threat of death or requiring them to pay a tax.

17.3.2 Charities

Charities are an easy way for terrorist organizations to collect funds. Charities are helping to assist individuals on humanitarian grounds. However, if information about charities is not transparent, there is a risk of transferring funds toward illegal terrorist activities. Terrorist organizations use charities for procurement including weapons, expert skill development, and recruitment of terrorists for attacks or threats. Charities are used by terrorist organizations to hide the source and movement of funds [24]. A group called the "Safa" group, including 100 charity organizations but many of these charities exist only on paper [20]. The goal of the terrorist group is to hide funds and transfer processes to other terrorist organizations [23]. The terrorist group involves charities to operate hospitals, schools, and orphanages to build strong political relationships within the country [25].

17.3.3 Crime

The size of an organization in criminal activities can vary from less organized to highly organized crime. Smaller terrorist groups are involved in less organized criminal activities, such as street crime, because of limited resources [23]. Larger terrorist organizations require highly organized criminal activities such as robbery, drug trafficking, human trafficking, arms smuggling, and money laundering [26]. The risk associated with engaging in organized crime is high because law enforcement agencies (LEA) at all levels closely monitor such activities. Organized crime activities impact the political environment and individuals' assets and properties [27]. Crime activities can be categorized into two types: operational and symbolic. The country's economic strength can be weakened when there is smuggling of taxable goods. Similarly, a country's security can be weakened when incidents of kidnapping of civilians and tourists occur.

17.3.4 Legal Business

Larger terrorist organizations develop legal, legitimate businesses intending to generate funds for their activities and survival. Terrorist groups are highly organized when it comes to fundraising through legal channels [28]. There are various reasons behind using the legal businesses by terrorist groups. The main reasons for legal businesses are high safety, security, credibility, and reduced risk compared to illegal activities [29]. Another reason for legal businesses is covering criminal activities, allowing illegal money for money laundering [23]. Legal businesses can also provide function support by facilitating activities like smuggling and procuring weapons [26].

The decision of legal businesses depends on the availability of capital and the skill within the terrorist group [30].

17.3.5 Country Contribution

Several countries provide different types of help such as training, logistical aid, arms, and financial support to terrorist organizations [24]. The connection [31] between terrorist organizations and countries is examined such as the connection between Iraq and Hezbollah, Pakistan and Lashkar-e-Taiba, as well as Syria and the PKK.

17.4 SOCIAL MEDIA

Social media serves as an efficient, widely accepted mode of sharing information on the Internet. It offers a flexible and user-friendly platform for communication [17]. The reach of social media covers across the world [14]. As of July 2023, social media platforms have a user base of 4.9 billion [32]. Social media can be seen as a catalyst that facilitates and increases cyber terrorism activities. Terrorist groups examined various social media platforms and found that the social media platforms are very secure for communicating on the Internet. Terrorist organizations develop videos to promote their agendas using social media [14]. The excessive use of social media by cyber terrorists has impacted national security [17]. Europol reported that terrorists use about 150 platforms of social media for executing their activities all over the world. Intending to hide their identity, location, and talks, terrorists use end-to-end encryption social media platforms such as WhatsApp, Twitter, Instagram, and telegrams. Terrorists are very relaxed and do not care about tracing while communicating with each other using social media platforms [33]. Social media is considered a favorite tool for cyber terrorists due to its worldwide acceptance [17]. Terrorists know the workings of social media platforms and are used for various illegal activities such as cyber-attacks, online fraud, and money laundering [17]. Cyber terrorists use social media for numerous purposes such as planning, training, financing, propagating their ideologies, and executing cyber-attacks [17].

Terrorist groups develop videos of their ideology in different languages and distribute them across the world using social media [14]. Cyber terrorists have the technical skills to perform the activities on social media platforms [14]. Terrorist organizations utilize social media platforms for campaigning like the "One Billion campaign," to motivate young Muslims to join their organization [34]. Al-Qaeda made its own media control named As-Sahab to publicize, communicate, and motivate others based on their ideology. The Al Hayat Media Centre serves as the social media platform for ISIS, where they communicate their propaganda messages and videos [14]. ISIS created its mobile application, which was used to keep updated with the latest news and campaign information [14]. Members of ISIS are involved in creating videos and messages with the intention of propagating propaganda [35] that shows themselves as generous personalities among all other terrorist groups. The social media platform i.e., Twitter, is mostly used by terrorist organizations [4] for communicating their propaganda. In propaganda videos, terrorists are showing themselves as celebrities or stars. Social media is utilized by cyber terrorism as a means for conducting psychological warfare [14]. In the propaganda videos, terrorists are showing their positivity in the form of hospital visits, distributing sweets to children, and protecting civilians [14]. Terrorist organizations are using chat rooms on social media to recruit new members in the terrorist group [14].

17.4.1 End-to-End Encryption Tool for Cyber Terrorism

Numerous messaging tools are available on social media platforms that provide the facility of end-to-end encryption. Terrorist organizations are using encrypted tools (Table 17.1) at a very high rate for communication among terrorist organizations and

TABLE 17.1
End-to-End Encryption Tool for Cyber Terrorism

Encryption Tool	Description

Telegram	• Terrorist organizations use the social media application known as Telegram for their communication purpose.
	• The Terrorist groups use the Telegram application for various purposes such as:
	  • End-to-End Encryption
	  • Exclusion from Search Engine Indexing
	  • Highly Secure Communication
	  • Member Non-Suspension
	  • Unlimited Messaging (text, image, and video)

Facebook	• Terrorist groups use Facebook for various purposes such as recruitment, training, fundraising campaigning, propaganda dissemination, and performing cybercrime activities. • Terrorists create fake profiles and send friend invitations to various users who have similar ideologies on Facebook. • Once users accept the friend request, terrorists proceed to share illegal materials of their organization and are motivated to join as supporters of the terrorist group.
YouTube	• This is a social media application designed for accessing and uploading media content. • Terrorist organizations make frequent use of this platform. • Terrorist organizations use the YouTube application to create fear among civilians and nations, train the members for both cyber and physical attacks, and acquire sympathy from their home countries. This is done by showing propaganda videos including the distribution of items like sweets, clothing, and medicine to civilians. • Terrorist organizations use various media platforms to acquire new techniques and raise a propaganda war.
Instagram	• Instagram is a worldwide highly popular social media application and is used by numerous terrorist organizations. • Terrorist organizations use Instagram for recruitment, finding profiles with similar ideologies, communicating propaganda campaigns, and raising funds.
Tumblr	• Tumblr is a micro-blogging platform that is an integral part of social networking. • Terrorist organizations use Tumblr for accessing blogs because it does not require sign-in or sign-up procedures to access content.

terrorists [36]. The high use of end-to-end encryption tools by terrorist organizations is due to global acceptance. Terrorist organizations utilize social media for various purposes [37] such as recruitment, internal communication among their members, and spreading propaganda to motivate jihadists.

17.4.2 Personal Media of Terrorist Organization

Numerous communication options are accessible through various social media platforms, websites, and mobile applications, but LEA closely monitor and track all these communication platforms. As a result, terrorist organizations may find it challenging to maintain secure communication due to the risk of exposing their identities. While numerous tools and applications offer end-to-end encryption and terrorist organizations are using them. Considering the tracking of law enforcement, terrorist organizations have created their secure communication media [37, 38]. Below are a few examples (Table 17.2).

TABLE 17.2
Secure Personal Media Tools Developed by Terrorist Organizations [38]

Personal Media Name	Description
Alemarah	• It was a mobile application and was created by a terrorist organization. • It is used to show the news of terrorist action. • The purpose of the mobile application was to message terrorist groups. • Initially, it was available on Google Play Store but after suspicion, it was removed from Google Play Store.
Amn al-Mujahid	• It is an encryption program to hide the digital footprint while performing online activity. • It is the most secure and highly encrypted program and is used by terrorist organizations. • It removes the online tracking identity.
Tashfeer al-Jawwal	• Terrorist organizations developed their encryption application for secure communication and propagating propaganda on the web because they knew that with encryption communication, information could not be traced by law enforcement agencies.
Asrar al-Dardashah	• Terrorist organizations were using messenger applications for communication. However, messenger applications did not support encryption, and secure communication was impossible. • Terrorist organizations created their encryption plugin and easily worked with messenger applications. • It was used to encrypt the chat messages for secure communication.

17.5 DARK WEB

The dark web is a hidden part of the Internet. Terrorists and terrorist organizations use it to hide their identity on the web, and all the work is completed anonymously [6]. The basic purpose of using the dark web is to hide the identity, which anyone cannot trace [6]. The other purpose of using the dark web is to communicate with other terrorist groups, transfer money, and sell and purchase illegal items [39]. The dark web is a treasured place for cyber terrorists, and on the dark web, terrorists can buy and sell illegal items anonymously. In view of completing the attack goal, the dark web is the most appropriate place for purchasing weapons [40]. Terrorists use the dark web to sell (drugs), purchase (weapons) and transfer funds to other parties. In fundraising activity, cryptocurrency such as Bitcoin is very important for buying and selling illegal items on the Dark web [22].

17.5.1 Secure Communication with Dark Web

LEA always keep track the terrorist activities on the web. Terrorist organizations use different tools for secure communication and a digital footprint is left after their work. Based on digital footprint, law enforcement can capture terrorist activities. Terrorist organizations handle this situation by using numerous tools that are available on the dark web. The encrypted tools of the dark web support anonymity without any identification and non-discoverable messages. The email services (Table 17.3) [41] which are used by terrorist organizations [37] are the following:

TABLE 17.3
Email Services Available on the Dark Web

Email Service	Description
Mail2Tor	• This is an email service that can be accessed through the Tor network using Tor software. • It becomes functional once you have downloaded it using Tor software. • Terrorist organizations use it for anonymity and to hide network surveillance. • In addition, it is used in conjunction with Gmail or Yahoo Mail. • It does not store any data on its servers, and encryption is applied to all communications.
SecTor.City	• This is an anonymous email service that can be accessed through the Tor Network. • It retains privacy, does not store any log information, and deletes all data from the server after processing.
DNMX	• This is an email service provider with the anonymity that operates on the darknet. • Individuals can create an account and use this email service on the darknet without the need to provide identity information. • The data of email service is encrypted, and it can be accessed on the Tor network to ensure anonymity and security.
Proton Mail	• This is an email service that offers end-to-end encryption. • It converts email data on the client side before transmitting it to the email server. • It can be accessed through the Tor network or a webmail interface, and it maintains a high level of security on both platforms such as hardware and software. • It has its own data centers.

17.6 COMBATING CYBER TERRORISM USING ARTIFICIAL INTELLIGENCE

The Internet is a fundamental part of global communication and provides connections among people worldwide. The importance of communication on the Internet has increased due to social media. Social media is used for connecting and sharing information. In the meantime, terrorists are using the Internet as a platform for performing cyber terrorism [42]. In the earlier stage, there was an assumption that terrorists were using the Internet for their functioning. However, following the 9/11 terrorist attack, it shows that terrorists were using the Internet to achieve the objectives of their organizations. Terrorists are using the Internet for numerous activities, including planning, recruitment, training, propaganda communication, jihadi activities, fundraising, and secure communication. This situation creates alarm for LEA and National Security Agencies (NSA).

A gap exists between the LEA and NSA in combating terrorist activities. Presently, the NSA and LEA are facing challenges due to the increase in cyber terrorism and the absence of physical evidence. The NSA and LEA analyzed and reviewed various digital techniques and technology to combat cyber terrorism [11]. They find that AI plays a crucial role in monitoring and combating cyber terrorism. AI is implemented in both public and private sectors in different fields because of intelligence, visual recognition, translation, speech recognition, problem-solving, and decision-making capabilities that are similar to humans [7]. In addition, AI shows the ability to manage vast amounts of data, expertise in pattern recognition, and quickly discover patterns with high accuracy that cannot be handled by humans [43]. Numerous applications of AI exist in every field such as manufacturing, cyber security, fashion design, space exploration, healthcare, medicine, agriculture, electronics, electrical engineering, mechanical engineering, media, law, and numerous others. AI-driven sentiment analysis is used to monitor social media platforms to find terrorist activities [4]. AI-based image analysis is used to detect hidden messages and patterns within images concerning finding and monitoring terrorist activity plans and communications [44]. They are developing automated tools for the surveillance and monitoring of terrorist activities on the Internet. A survey conducted by UNICR and Interpol found that LEA and NSA worldwide are utilizing AI knowledge and techniques (Figure 17.2) in their operations [43].

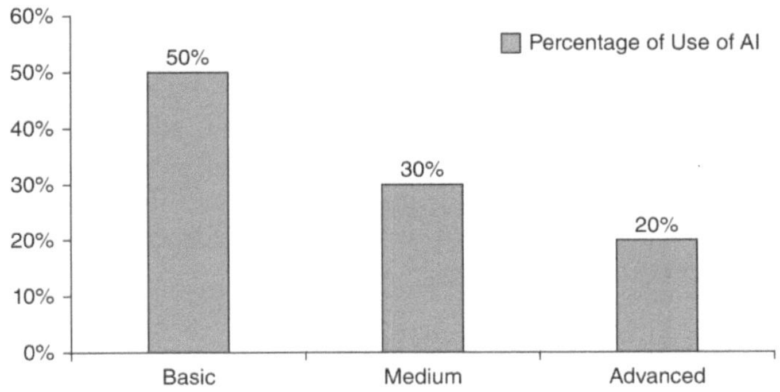

FIGURE 17.2 Different levels of AI used by counter terrorism agencies [43].

Counter-terrorism agencies employ various components, techniques, and methodologies to combat cyber terrorism. These may include [45] the following:

- AI and Machine Learning (ML)
- Past Historical Data
- ML
- Deep Learning (DL)
- Natural Language Processing (NLP)
- Object Recognition
- Predictive Analytics
- Social Networking Analytics
- Content Matching Techniques
- Open Source Intelligence
- Social Media Intelligence

The following AI-implemented applications are used to combat cyber terrorism.

17.6.1 Prediction of Future Activities of Terrorists

LEA and NSA are monitoring the behavioral activities of numerous terrorist groups to identify the threats. This monitoring includes the analysis of online and offline behavior and communication patterns of terrorist activities. The objective of LEA and NSA is to prevent cyber terrorism and protect national security. Manually predicting the future activities of terrorists is a very challenging job. LEA and NSA have recognized the power of AI and are using it to enhance their capabilities in monitoring, analyzing, and predicting terrorist activities [43]. Counter-terrorism agencies are collecting historical records of the behaviors of terrorists and terrorist groups, and this information can be acquired from recorded incidents, social media platforms, and governmental historical archives. Using the collected data, an AI model is created with exhaustive training and testing. This model is designed to analyze and recognize patterns based on the historical behavior of terrorists and terrorist groups. Based on the high accuracy performance, the AI model is implemented to reveal the future activities of terrorists and terrorist groups. This prediction is made by analyzing the patterns of previous activities. Predictive analysis [39] is used to issue warnings, formulate strategic decisions, and develop new policies to combat cyber terrorism.

17.6.2 Identification of Radicalization

NLP is an ML technique within the field of AI that LEA, NSA, and other counter-terrorism organizations use. It identifies keywords that can promote radicalization on social media using NLP and also aids in identifying individuals who show an interest in searching for content related to terrorist activities, particularly content associated with radicalization [43].

17.6.3 Automated Content Moderation

Social media is a platform that enables people to communicate with each other. However, terrorists misuse the freedom offered by social media platforms for negative purposes. They use social media as a tool of communication in order to perform their activities and propagate their propaganda. Social media platforms handle this situation using AI innovations. AI techniques are implemented in analyzing the content posted on social media platforms. If any content is found to deviate the standards then content will be removed or modified to restrict the propagation of terrorist ideologies. In more serious cases, social media accounts may be marked as suspicious, suspended, or permanently deleted [43].

17.7 CONCLUSION

Cyber terrorism is a severe activity of terrorist groups in a hidden pattern and it creates a threat against everyone as well as every nation. It creates challenges for each country. The goal of the study is achieved by exploring the landscape of cyber terrorism and its countermeasures. In this regard, the authors reviewed the concept of cyber terrorism as well as its elaboration and penetration into the world. Fundraising activity was also reviewed including five different categories. The authors also reviewed the use of social media platforms in terrorism for performing terrorist activities, promoting propaganda, and jihads. The authors also reviewed the hidden platform, i.e., the dark web of terrorist activities and various email services used on the dark web. The authors reviewed the power of AI, which was used to combat cyber terrorism and terrorist activities.

REFERENCES

1. JangadaCorreia, V. 2021. An Explorative Study into the Importance of Defining and Classifying Cyber Terrorism in the United Kingdom. *SN Computer Science* 3, no. 84: 1–31. https://doi.org/10.1007/s42979-021-00962-5
2. Golose, P. R. 2022. Cyber Terrorism – A Perspective of Policy Analysis. *International Journal of Cyber Criminology* 16, no, 2: 149–61. https://cybercrimejournal.com/menu-script/index.php/cybercrimejournal/article/view/118/38
3. Margiati, T., and U. Qodarsasi. 2022. Cyber-Terrorism of ISIS and the Challenge Towards Global Security. *Politea: JurnalPemikiranPolitik Islam* 5, no. 2: 224–40. https://doi.org/10.21043/politea.v5i2.17369
4. Saidi, F., Z. Trabelsi, and E. Thangaraj. 2022. A Novel Framework for Semantic Classification of Cyber Terrorist Communities on Twitter. *Engineering Applications of Artificial Intelligence* 115: 105271. https://doi.org/10.1016/j.engappai.2022.105271
5. Rawat, R., V. Mahor, M. Chouhan, K. Pachlasiya, S. Telang, and B. Garg. 2022. Systematic Literature Review (SLR) on Social Media and the Digital Transformation of Drug Trafficking on Darkweb. In Proceedings of International Conference on Network Security and Blockchain Technology, ed. D. Giri, J. K. Mandal, K. Sakurai, and D. De, 181–205. Singapore: Springer Nature. https://doi.org/10.1007/978-981-19-3182-6_15
6. Shaffer, R. 2023. Drone Activity and Cyber Terrorism. *Terrorism and Political Violence*: 1–6. https://doi.org/10.1080/09546553.2023.2204033
7. Jaladanki, R., S. I. Patel, I. Khan, K. I. A. Mohammed, M. Thenmozhi, and A. K. Tripathi. 2022. Machine Learning Models in Detecting Cyber Crimes and Cyber Terrorism in India. In Dark Web Pattern Recognition and Crime Analysis Using Machine Intelligence, ed. R. Rawat, S. Telang, P. William, U. Kaur, and O.K.C. U., 47–61. IGI Global. https://doi.org/10.4018/978-1-6684-3942-5.ch004

8. Weimann, G. 2016. Going Dark: Terrorism on the Dark Web. *Studies in Conflict & Terrorism* 39, no. 3: 195–206. https://doi.org/10.1080/1057610X.2015.1119546
9. Onat, I., M. F. Bastug, A. Guler, and S. Kula. 2022. Fears of Cyberterrorism, Terrorism, and Terrorist Attacks: An Empirical Comparison. *Behavioral Sciences of Terrorism and Political Aggression*: 1–17. https://doi.org/10.1080/19434472.2022.2046625
10. Lee, H., and K. S. Choi. 2022. Interrelationship between Bitcoin, Ransomware, and Terrorist Activities: Criminal Opportunity Assessment via Cyber-Routine Activities Theoretical Framework. In The New Technology of Financial Crime, ed. D. Rebovich, and J. M. Byrne, 82–103. London: Routledge.
11. Prasad, S., and A. Kumar. 2022. Cyber Terrorism: A Growing Threat to India's Cyber Security. In Nontraditional Security Concerns in India: Issues and Challenges, ed. S. K. Singh, and S.P. Singh, 53–73. Singapore: Springer. https://doi.org/10.1007/978-981-16-3735-3_4
12. Smith, K. T., L. M. Smith, M. Burger, and E. S. Boyle. 2023. Cyber Terrorism Cases and Stock Market Valuation Effects. *Information & Computer Security* 31, no. 4: 385–403. https://doi.org/10.1108/ICS-09-2022-0147
13. Asongu, S. A., S. M. I. Orim, and R. T. Nting. 2019. Terrorism and Social Media: Global Evidence. *Journal of Global Information Technology Management* 22, no. 3: 208–28. https://doi.org/10.1080/1097198X.2019.1642023
14. Awan, I. 2017. Cyber-Extremism: Isis and the Power of Social Media. *Society* 54, no. 2: 138–49. https://doi.org/10.1007/s12115-017-0114-0
15. Bejan, V., M. Hickman, W. S. Parkin, and V. F. Pozo. 2018. Primed for Death: Law Enforcement-Citizen Homicides, Social Media, and Retaliatory Violence. *PLOS ONE* 13, no. 1: e0190571. https://doi.org/10.1371/journal.pone.0190571
16. Lakomy, M. 2023. Dark Web Jihad: Exploring the Militant Islamist Information Ecosystem on the Onion Router. *Behavioral Sciences of Terrorism and Political Aggression*: 1–20. https://doi.org/10.1080/19434472.2022.2164326
17. Parlakkiliç, A. 2018. Cyber Terrorism through Social Media: A Categorical Based Preventive Approach. *International Journal of Information Security Science* 7, no. 4: 172–78. https://dergipark.org.tr/en/download/article-file/2160190
18. Qabooli, R. A., and J. Setiyono. 2022. Law Enforcement Policy for Mayantara Crime (Cyber Crime) in Indonesia. *International Journal of Social Science Research and Review* 5, no. 11: 426–34.
19. Carroll, P., and J. Windle. 2018. Cyber as an Enabler of Terrorism Financing, Now and in the Future. *Journal of Policing, Intelligence and Counter Terrorism* 13, no. 3: 285–300. https://doi.org/10.1080/18335330.2018.1506149
20. Windle, J. 2018. Fundraising, Organised Crime and Terrorist Financing. In The Routledge Handbook of Terrorism and Counter-Terrorism, ed. A. Silke, 195–206. New York: Routledge.
21. Venkatachary, S. K., J. Prasad, and R. Samikannu. 2018. Cybersecurity and Cyber Terrorism – in Energy Sector – A Review. *Journal of Cyber Security Technology* 2, no. 3–4: 111–30. https://doi.org/10.1080/23742917.2018.1518057
22. Ginting, J., and P. Talbot. 2023. Fundraising Aspect of International Terrorism Organization in ASEAN: Legal and Political Aspects. *LexScientia Law Review* 7, no. 1: 1–30. https://doi.org/10.15294/lesrev.v7i1.60074
23. Comras, V. 2007. Al Qaeda Finances and Funding to Affiliated Groups. In Terrorism Financing and State Responses: A Comparative Perspective, ed. H. A. Trinkunas, and J. K. Giraldo, 115–33. Stanford: Stanford University Press.
24. Levitt, M. 2007. Hezbollah Finances: Funding the Party of God. In Terrorism Financing and State Responses: A Comparative Perspective, ed. J. K. Giraldo, and H. A. Trinkunas, 134–51. Stanford: Stanford University Press.

25. Financial Action Task Force. 2015. Financing of the Terrorist Organisation Islamic State in Iraq and the Levant (ISIL). https://www.fatf-gafi.org/content/dam/fatf-gafi/reports/Financing-of-the-terrorist-organisation-ISIL.pdf (accessed October 10, 2023).
26. Roth, M. P., and M. Sever. 2007. The Kurdish Workers Party (PKK) as Criminal Syndicate: Funding Terrorism through Organized Crime, A Case Study. *Studies in Conflict & Terrorism* 30, no. 10: 901–20. https://doi.org/10.1080/10576100701558620
27. Felbab-Brown, V. 2009. Shooting Up: Counterinsurgency and the War on Drugs. Washington: Booking Institution Press.
28. Passas, N. 2007. Terrorism Financing Mechanisms and Policy Dilemmas. In Terrorism Financing and State Responses: A Comparative Perspective, ed. J. K. Giraldo, and H. A. Trinkunas, 21–38. Stanford: Stanford University Press.
29. Napoleoni, L., 2007. Terrorism Financing in Europe. Terrorism Financing and State Responses: A Comparative Perspective, ed. J. K. Giraldo, and H. A. Trinkunas, 171–84. Stanford: Stanford University Press.
30. Ryder, C. 2008. Organised Crime and Racketeering in Northern Ireland. In Combating Terrorism in Northern Ireland, ed. J. Dingely, 128–36. London and New York: Routledge.
31. Freeman, M. 2012. Sources of Terrorist Financing: Theory and Typology. In Financing Terrorism, ed. M. Freeman, 7–26. London: Routledge.
32. Wong, B. 2023. Top Social Media Statistics and Trends of 2023. *Forbes Advisor*. https://www.forbes.com/advisor/business/social-media-statistics/#:~:text=Key%20Social%20Media%20Statistics%201%20In%202023%2C%20an,social%20media%20site%20in%20America%20…%20More%20items (accessed 20 October 2023).
33. Albahar, M. 2019. Cyber Attacks and Terrorism: A Twenty-First Century Conundrum. *Science and Engineering Ethics* 25, no. 4: 993–1006. https://doi.org/10.1007/s11948-016-9864-0
34. Irshaid, F. 2014. How Isis Is Spreading Its Message Online. *BBC News*. http://www.bbc.co.uk/news/world-middle-east-27912569 (accessed October 12, 2023).
35. Klausen, J. 2015. Tweeting the Jihad: Social Media Networks of Western Foreign Fighters in Syria and Iraq. *Studies in Conflict & Terrorism* 38, no. 1: 1–22. https://doi.org/10.1080/1057610X.2014.974948
36. Al-Masaeed, S., and H. Yaseen. 2023. Social Media Effect on National Security. *Journal of Southwest Jiaotong University* 58, no. 1: 63–75. http://jsju.org/index.php/journal/article/view/1490/1480
37. Lissaris, E., G. Giataganas, D. Kavallieros, D. Myttas, and E. Kermitsis. 2021. Terrorist Activities in the Dark and the Surface Web. In Dark Web Investigation, ed. B. Akhgar, H. Gibson, M. Gercke, and S. Vrochidis, 49–84. Switzerland: Springer, Cham. https://doi.org/10.1007/978-3-030-55343-2_3
38. Paganini, P. 2014. Al-Qaeda Is Developing New Encryption Tools in Response to NSA Surveillance. *Cyber Defense Magazine (blog)*. https://www.cyberdefensemagazine.com/al-qaeda-is-developing-new-encryption-tools-in-response-to-nsa-surveillance/ (accessed October 15, 2023).
39. Rawat, R. 2023. Logical Concept Mapping and Social Media Analytics Relating to Cyber Criminal Activities for Ontology Creation. *International Journal of Information Technology* 15, no. 2: 893–903. https://doi.org/10.1007/s41870-022-00934-9
40. Denić, N. V., and S. Devetak. 2023. Dark Web – As Challenge of the Contemporary Information Age. *TRAMES* 27, no. 2: 115–26. https://doi.org/10.3176/tr.2023.2.02
41. Stosh, B. 2023. List of Secure Dark Web Email Providers in 2023. *Freedom Hacker*. https://freedomhacker.net/list-of-secure-dark-web-email-providers-in-2016-4946/ (accessed October 20, 2023).

42. Bianchi, S., M. Mancuso, and C. Paternoster, et al. 2023. Artificial Intelligence to Counter Cyber-Terrorism. In Proceedings of the International Conference on Cybersecurity and Cybercrime – 2023, ed. I.C. Mihai, C. CIUCHi, and G. Petrică, 12–20. AsociatiaRomanapentruAsigurareaSecuritatiiInformatiei. https://www.ceeol.com/search/chapter-detail?id=1123458
43. Voronkov, V., and A. Marie De Meo. 2021. Countering Terrorism Online with Artificial Intelligence. United Nations Interregional Crime and Justice Research Institute (UNICRI). https://www.un.org/counterterrorism/sites/www.un.org.counterterrorism/files/countering-terrorism-online-with-ai-uncct-unicri-report-web.pdf (accessed 20 October 2023).
44. Lv, Z., D. Chen, R. Lou, and A. Alazab. 2021. Artificial Intelligence for Securing Industrial-Based Cyber–Physical Systems. *Future Generation Computer Systems* 117: 291–98. https://doi.org/10.1016/j.future.2020.12.001
45. Yamin, M. M., M. Ullah, H. Ullah, and B. Katt. 2021. Weaponized AI for Cyber Attacks. *Journal of Information Security and Applications* 57: 102722. https://doi.org/10.1016/j.jisa.2020.102722

18 Hybrid Secure Clustered Nature-Inspired Decision Support System for Secure Data Communication in Cyberspace

Shahana Gajala Qureshi, Sasmita Padhy, Preetam Suman, and Amrit Suman

18.1 INTRODUCTION

To increase the lifespan of the wireless sensor network (WSN), several clustering algorithms have been implemented so far [1]. Along with this in WSNs, there are so many challenges exist. To deal with such challenges as security, less energy consumption, prolong the lifetime of the cyber networks many proactive methods are implemented by the researchers [2, 3]. Organizations, industries, and researchers are continuing to implement and enhance the safety measures for Internet of Things (IoT) tools and IoT networks such as WSNs. The researchers have discussed many technical challenges and issues related to algorithms, hardware, protocols, and architecture in WSNs [4]. The WSNs are implemented to gather, transmit, and sense data from the ground to the remote stations to monitor activities. For IoT security, Securing the WSNs is a major concern [5]. The study says the technology of WSN is resource-constrained in terms of energy. The longevity and durability of the WSN are directly influenced by the connection between each sensor node's protocol stack. As a solution, the researchers implemented an energy-efficient surveillance system through simulation [6]. The sensor devices use energy control to apply practical energy in WSNs. The task management feature regulates the programming of the sensing jobs to a specific sensing device inside the site of appearance. The mobility management task was used to track the sensor nodes' movements in order to keep the information about their neighbors' current for a variety of applications [7]. Because of these management features unified into the WSN protocol stack, networking within the many sensor devices has become sought. It makes it possible for sensor nodes to collaborate, route data effectively, use resources effectively, etc. Software of the application can be created and used at the application layer in WSN for applications like e-healthcare. The transport

DOI: 10.1201/9781003409571-18

layer has provided consistent data communication between a source node and the sink node. Routing for data from the transport layer has been impacted by the network layer. The data link layer covers methods for establishing a solid connection between two communicating sensor nodes. The data link layer controls energy awareness, propagation delay, error control, and media access. The physical layer eventually resolves issues like frequency selection, modulation, encryption, and signal detection [8].

The aforementioned layers of the protocol stack in cyber networks should be constructed vigilantly because they are highly prone to several cyber attacks [9]. As a result of cyber threats, security is a major issue for WSNs together with energy management [10]. It has become more popular to protect cyber networks from several cyber threats using decision support systems (DSSs) that meet WSN security and energy efficiency standards. Clustering and data transmission are still two aspects of WSNs that require high energy efficiency and cybersecurity protection. A number of clustering solutions [11–15] have been presented to increase network lifetime. Using nature-inspired algorithms for Quality of Service (QoS), but still ensuring optimally secured clustering is a challenging task in cyberspace. Recent research indicates that in cyber networks like WSN, providing better security, and maintaining the minimum energy consumption are serious issues for investigation relating to the many sorts of attacks. WSNs have a simple and lightweight trust model, but a trust-based DSS is necessary to categorize sensor nodes as authentic or malicious accurately and reliably. When the multiple factors that affect a genuine node's behavior are not taken into account when calculating trust scores, it can result in erroneous DSS for WSNs. The aforementioned techniques illustrated various trust metrics that were employed to assess the credibility of sensor nodes. A research issue for cyber networks is still how to implement an adequate trust model to estimate trust value and utilize those values in DSS for the segregation of legitimate and attacker nodes and secure optimal cluster head (CH) selection, which enables safe and energy-efficient data transfer with the least amount of communication overhead. The main purpose of this research is to develop a DSS. In the proposed secured clustered nature-inspired DSS (SCNIDSS) method, the following contributions are included:

- Determining numerous trust-related elements to create a strong hybrid trust model to provide safe communication for cyber threats. During the process of choosing optimal CH, the trust values are considered for each sensor node to identify malicious nodes.
- The hybrid trust model incorporates a fitness function in an algorithm that is inspired by nature and has an optimized DSS.
- Malicious nodes are blocked in a multi-hop manner using the shortest path tree (SPT) technique.
- A thorough analysis of the SCNIDSS protocol's performance takes into account the variations in attack and node densities. The SCNIDSS protocol's performance is compared to that of more contemporary and traditional methods.

The rest of the chapter is organized as follows: Section 8.2 presents a concise survey of existing. The proposed methodology is presented in Section 8.3. Network scenarios, simulation findings, and performance assessments are covered in Section 8.4 while Section 8.5 describes the summary.

18.2 LITERATURE REVIEW

The majority of security strategies were developed utilizing algorithms for optimization and fuzzy logic that were inspired by natural phenomena. The trust evaluation on the basis of sensor node metrics in terms of the number of sensor nodes, residual energy, and distance to the base station (BS); is computed in existing techniques for secured energy-efficient clustered WSNs. The studies say the trust management framework and the trust model were first developed in [16] for secured data transmission in WSN. They found various trust factors that affect the performance of WSNs and used those trust factors to manage trust in the face of security threats. A secure clustering algorithm was implemented by the authors [17] for secure CH selection using trust assessments. For WSN, a unique clustering protocol was presented in [18] to improve network stability and energy efficiency while maintaining optimum network connectivity. Every clustering round involved the adaptive and proportionate CH choosing. They simulated the trust-based energy-efficient and secure clustering protocol against various security risks and recommended it for WSNs [19]. In Ref. [20], the Honey Bee Mating tool (BMA) was developed as an optimization tool for trust-based clustering in WSNs. For precise network life span computation, they built the pragmatist energy consumption framework. The best and most secure CH section for WSN is carried out using the genetic algorithm (GA) in [21]. Commonly utilized inputs for GA were distance, residual energy, and CH count, and many researchers used five different parameters for the trust score computation. The trust value of every sensor node was taken into consideration for choosing the best CH method [22]. Through the application of the trust computation methodology, the ant bee colony (ABC) method was applied for the best CH selection in multi-hop WSN [23]. They assessed security measures against online dangers like DoS, black holes, selfish nodes, and malevolent nodes. Another older method for safe and effective clustering in WSNs was published in [24]. To provide security during clustering and data transmission, an energy-efficient clustering solution, and network operations, they built a secure hardware system. The first clustering protocol employing a differential evaluation (DE) algorithm with natural inspiration was put forth in [25]. Introduced is a technique for estimating the best clustering and data transmission for extending network lifetime. The combination of the DE nature-inspired approach and the diverse vicinity approach resulted in minimizing energy consumption and data transmission delay. Additionally, the best CH choice was made in Ref. [26] utilizing brand-new swarm intelligence approaches. They evaluated the outcomes of the suggested formulation using a variety of metrics and compared it to Low Energy Adaptive Clustering Hierarchy (LEACH) protocol.

The reliability has not been attained since malicious operations can still be carried out at many different stages of the protocol stack, despite the fact that the aforementioned approaches used various parameters to identify/determine whether the nodes were legitimate or attackers for the selection of secure CH or secure data transmissions in WSN. The approaches also don't deal with evaluations of clustering complexity or communication overhead; they are merely introductions. The proposed SCNIDSS method incorporates a hybrid lightweight trust estimation model to enhance performance in terms of QoS, energy efficiency, and dependability.

In order to achieve optimal CH selection or secure data transfer, it is necessary to determine if a node is a non-attacker or an attacker, one of the most crucial factors was reliability, as the methods listed above used similar factors. Because malicious operations can be carried out over multiple layers of the protocol stack, reliability cannot be guaranteed. Additionally, these approaches do not take transmission overhead and clustering complexity into account. To enhance reliability, energy efficiency, and QoS performance in WSN, the hybrid approach, incorporates the DE algorithm. Recently, a new CH selection technique that was both secure and optimal [27] was presented, taking into account the weight of each sensor node, to address the challenges of high energy consumption and decreased security in cyber networks. Trust metrics were taken into account to calculate the trust score of every sensor node to segregate legitimate nodes and attacker nodes for secure CH. In the work, moth flame optimization (MFO) [28] was implemented for data communication between CHs and sink nodes to minimize energy consumption and find the optimal path by using an objective function. Besides this for energy saving various clustered-based approaches incorporated with nature-inspired algorithms were mentioned in [29]. Along with this a nature-inspired energy-efficient clustering approach was developed for WSN. For clustering c-means approach was taken and for optimal result DE algorithm was incorporated. Along with this to build an accurate decision-making model fuzzy technique was implemented [30]. In Ref. [31], the researchers presented the adaptive PSO (particle swarm optimization) approach in the SCBRP (secure cluster based routing protocol) protocol for WSN. In order to save energy, an optimized firefly population-based algorithm was integrated with clustering in WSN. A method for aggregating and transmitting data in clustered WSNs in [32] was proposed based on trust. Utilizing compressed sensing, data aggregation is carried out with the least amount of transmission overhead. They utilized a natural method to achieve performance trade-offs. Another new lightweight DSS technique based on trust for WSN clustering was propounded [33]. Another trust method was implemented to compute trust score and the MFO along with GA was used in conjunction with the energy and security-aware clustering methodology. A fitness function was used to evaluate the fitness score for each sensor node. The optimal CH was picked by using the MFO approach. The addition of GA and MFO could lead to an increase in computational overhead [34]. Another approach based on cluster and routing was presented for smart farming applications employing a nature-inspired algorithm with cross-layer WSN parameters. The detailed review, hybridization of nature-inspired approaches with clustering, to design accurate DSS fuzzy technique was incorporated, and for segregation legitimate and attacker nodes trust model was presented in [35–37]. Recently researchers have implemented various security techniques [38–41] for the cyber physical system (CPS). Blockchain techniques are implemented for data security. Vectorized elliptic curve cryptography is also implemented to provide security for data communication in CPS along with this the comprehensive review of CPS is also presented by them.

These most recent researches succeeded remarkably well in achieving security while controlling WSN-specific problems like more energy consumption and greater communication overhead. However, by using nature-inspired methodologies and cutting-edge optimization, these works also open up new avenues for more complex research in this field. This inspired the current study to experiment with the recognized nature-inspired methodologies while putting a strong emphasis on data security.

18.3 SCNIDSS METHODOLOGY

18.3.1 Assumptions and the System Model

The suggested system model and functionalities are presented in this section and can be seen in Figure 18.1. It demonstrates how an optimization-based DSS system is created to choose the best and most secure CH for each clustering after the network is deployed and the initial clustering.

The proposed system's primary component is a hybrid-trust model which is implemented for safe and effective CH selection. The DE algorithm uses this model as a fitness function (Figure 18.1). The proposed DSS includes the following:

1. Classifying the nodes as either legitimate or attacker.
2. Choosing the right CH.

The DSS was categorized as an optimization problem in clustering for optimal CH selection in this study. The clusters formed around the chosen CHs once the ideal CH was chosen. To preserve network dependability and lengthen the lifetime of the WSN, the processes of re-clustering and cluster maintenance will occur on a periodic basis in accordance with TDMA (time division multiple access) channel scheduling.

The SCNIDSS model is described as follows: Let's assume that a network of X-Y dimensions contains N sensor nodes, where S = s1, s2,..., sN. After deployment, the clustering k-means algorithm, i.e. C = c1, c2,..., cM, produced the initial M number groups. The suggested secure and ideal CH selection procedure is used with the aim to enhance the QoS of the cyber network and prolong the lifetime of the network in the existence of T number of malicious attackers A = a1, a2,..., aT in the network for every jth, j M cluster cj. Other fundamental tenets on which the proposed SCNIDSS design is predicated are: One sink/BS node and N number of sensor nodes are distributed randomly in the cyber network. The sensor nodes in the network are uniform, and static, have a special ID, and are limited in terms of energy. The sink (BS) node is situated exterior of the network, free from resource limitations and full of security features. The use of symmetric and multi-hop communications among networked sensors. The network was divided using k-means clustering after deployment. The calculated distance utilizing RSSI (received signal strength indicator) between sensor nodes and/or BS nodes.

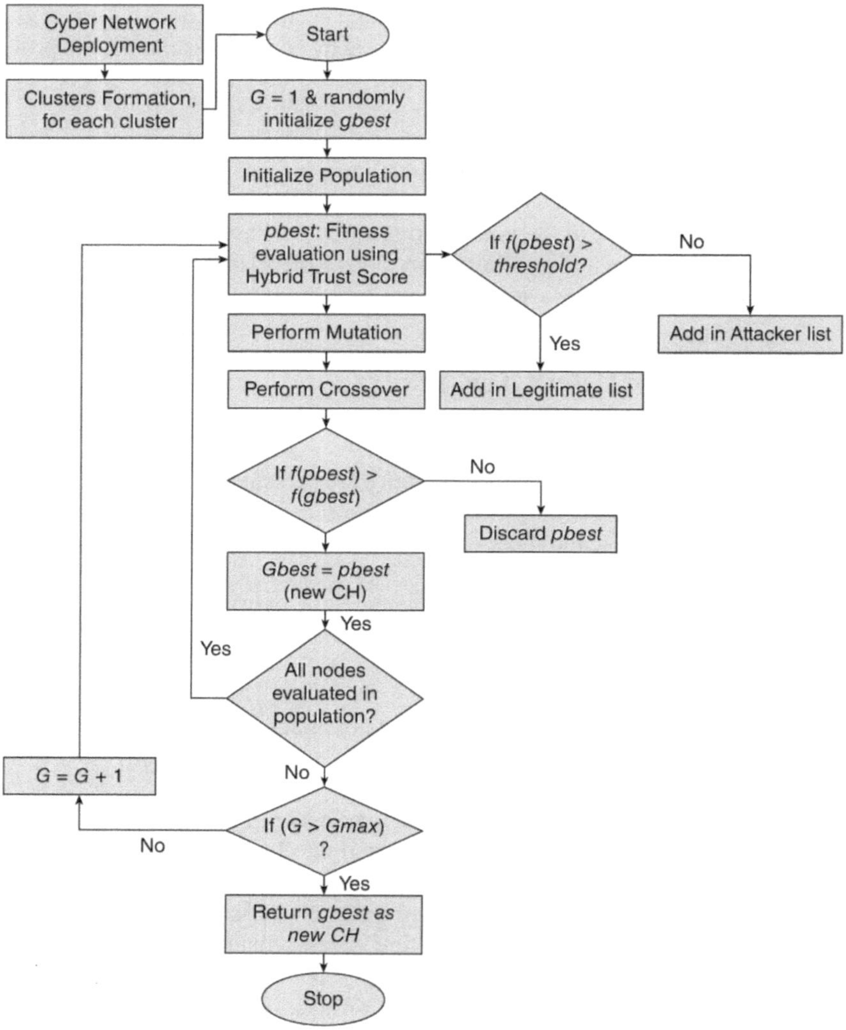

FIGURE 18.1 DE algorithm for secure clustering using hybrid trust score.

18.3.2 Implementation of Hybrid Trust Model

The trust model described in this section is utilized to determine each sensor node's fitness value or trustscore throughout the process of choosing the best CH. Because the trust value calculation is dependent on parameter selection, selecting the accurate parameters for the sensor nodes becomes crucial for network performance. The DE algorithm's fitness evaluation is based on a mix of different parameters that characterize the behavior of the sensor nodes and provide the secure DSS for the best CH choice. The suggested clustering system should be able to reliably choose the CH node while preventing any cyber threats thanks to a firm trust concept.

The implemented hybrid trust model utilizes four factors that encompass the various protocol layers in this study. To detect attacker nodes and select the optimal CH, the settings were chosen without losing generality. These four parameters are taken to calculate the overall trust score of the sensor node s^i in ith position, which is defined as the fitness value $f(s^i)$.

a. *Packet Transmission Behavior*: In the case of one-hop neighbors for sensor nodes, *ptb* states that s^j will compute the direct trust of s^i at the time t. Therefore, in each periodic trust evaluation round, the s^j node directly observes the s^i node.

$$ptb\left(s^i(t)\right) = \left(w1 \times pfr\left(s^{i,j}\right)\right) + \left(w2 \times af\left(s^{i,j}\right)\right) \tag{18.1}$$

where $w1$ and $w2$ are weight parameters whose sum is 1, $pfr\left(s^{i,j}\right)$ is the successful packet forwarding ratio of s^i by s^j and $af(s^i)$ is the availability factor of s^i by s^j. The availability factor $af(s^{i,j})$ and packet forwarding ratio $pfr\left(s^{i,j}\right)$ are calculated as follows:

i. *Packet Forwarding Ratio*: Node s^i forwarding data packets from a different sensor node involves sending four ACK packets. To calculate the number of sensing packets, s^j node gathers these ACK packets from the node s^i. Based on the calculation, the number of data packets being transmitted at time t is:

$$pfr\left(s^{i,j}(t)\right) = \frac{FP\left(s^{i,j}(t)\right) - FP\left(s^{i,j}(t-1)\right)}{FP\left(s^{i,j}(t)\right) + FP\left(s^{i,j}(t-1)\right)} \tag{18.2}$$

where $FP\left(s^{i,j}(t)\right)$ is the quantity of successfully forwarded packets sent through s^i at t time. By using this value, the network can be protected from online attacks and the hostile behavior of the node can be estimated.

ii. By using the Availability Factor parameter, the malicious behavior of sensor nodes in the network can be accurately defined. The HELLO packet that was sent by s^j to see if s^i was accessible to receive it. Once an ACK for a HELLO packet is sent through s^i to s^j is received, s^i's availability is defined. At time t, the availability factor $af\left(s^{i,j}(t)\right)$ is calculated as:

$$af\left(s^{i,j}(t)\right) = \frac{AH\left(s^{i,j}(t)\right)}{AH\left(s^{i,j}(t)\right) + NAH\left(s^{i,j}(t)\right)} \tag{18.3}$$

where $AH\left(s^{i,j}(t)\right)$ represents how much the acknowledged HELLO packets are and $NAH\left(s^{i,j}(t)\right)$ represents how much the non-acknowledged HELLO packets are.

In order to identify malicious nodes and choose the best CH, $ptb\left(s^i\right)$, once computed, will be used in conjunction with other trust factors, as shown in Figure 18.1. Higher $ptb\left(s^i\right)$ values increase the likelihood that s^i will be a valid node and, thus, the cluster's optimum CH node.

b. The Residual Energy Ratio is a parameter that has been utilized in many clustering algorithms for WSNs so far. Selecting nodes with higher residual energy values ensures less energy consumption and prolongs the network's lifetime. The $rer\left(s^i\right)$ at t time is calculated through Equation (18.4).

$$rer\left(s^i(t)\right) = \frac{E_{residual}\left(s^i(t)\right)}{E_{initial}\left(s^i\right)} \tag{18.4}$$

$E_{initial}\left(s^i\right)$ denotes the initial energy of the node, while $E_{residual}\left(s^i(t)\right)$ denotes the energy still present in the node at time t. Higher $rer\left(s^i(t)\right)$ nodes are ideal candidates for becoming CH.

c. *Degree of Connectivity*: This parameter is used to choose the node with the most nearby nodes, which assures the accurate clustering and data communication phases. The number of nodes in s^i's neighborhood determines its Degree of Connectivity (DoC), which is calculated as follows at time t:

$$D\left(s^i(t)\right) = \left[\frac{n_i}{dist\ (i,j)} < rssi\right] \tag{18.5}$$

where $i \neq jdist\ (i,j)$ denotes the separation in space between two nodes, and $rssi$ denotes the node i's communication range. This degree result leads to estimate the following DoC parameter:

$$doc\left(s^i(t)\right) = 1 - \left(\frac{1}{D\left(s^i(t)\right)}\right) \tag{18.6}$$

Given that there is only one hop between the cluster members and the CH node, this value allows for the best possible CH selection. This also reduces the possibility of malicious activity occurring during routine data exchanges within the cluster.

d. *Congestion Rate*: This parameter is used to detect rogue nodes in the network. Due to packet drops and a huge amount of ACK packets, the malicious or cyber threats that are constantly broadcasting bogus packets cause severe congestion around them. The node having a minimum congestion rate will be chosen as the most qualified node and ideal CH candidate. Node s^i's congestion rate is calculated as follows:

$$cr\left(s^i(t)\right) = 1 - \left(\frac{Q_{allocated}\left(s^i(t)\right)}{Q_{total}}\right) \tag{18.7}$$

where $Q_{allocated}\left(s^i(t)\right)$ denotes the number of packets currently being processed at the node s^i at time t and Q_{total} is the total capacity of the queue allotted to every sensor node.

Depending upon the node's s^i foregoing values at t time, the overall trust score taken into account, and represented as a fitness value, is calculated as follows:

$$f(s^i(t)) = ptb\left(s^i(t)\right)\times\lambda^1 + rer\left(s^i(t)\right)\times\lambda^2 + doc\left(s^i(t)\right)\times\lambda^3 + cr\left(s^i(t)\right)\times\lambda^4 \quad (18.8)$$

When weight parameters λ^1 through λ^4 are represented, and their sum is 1. In a specific TDMA schedule, the CH node that has the uppermost fitness value in DE is elected as the most secure and effective CH node. This method's versatility comes from the ability to change the weight factors to suit the needs of the network. The next section demonstrates how the aforesaid fitness function is used with a nature-inspired algorithm to accomplish optimal clustering and the detection of harmful nodes.

18.3.3 Secure and Optimal CH Selection

This section explains how to choose a CH in a safe and efficient manner using a hybrid trust computation approach. It is essential to choose the node as CH that has a higher trust score in order to protect the cyber threat or malicious node from being chosen. After comparison is taken into account between the current node and the previous best node score, the selection of optimal CH is carried out based on processes such as fitness evaluation, mutation, and crossover of DE. Prior to implementing the best CH selection, DSS has been implemented to identify malicious nodes based on their fitness value. If the fitness value of a node falls below the threshold, it is added to an attacker list; else is added to an authorized list. Each clustering round causes this list to refresh. The lists are used to support cluster-wide secure data transmission. At each TDMA schedule, as seen in method 1, an optimal CH selection process was carried through the function *getfitness* (). The fitness scour is being determined through Equation (18.8).

Algorithm 1: Secure and Optimal CH Selection

Inputs

$$M: Number\ of\ clusters$$
$$CM: set\ of\ nodes\ in\ cluster$$
$$Gmax: Number\ of\ maximum\ iterations$$
$$threshold = 0.4: to\ detect\ malicious\ node$$
$$lnodes: initialize\ list\ of\ legitimate\ nodes\ to\ null$$
$$mnodes: initialize\ list\ of\ malicious\ nodes\ to\ null$$

Output

$$CH: set\ of\ cluster\ heads\ selected\ during\ current\ round$$

For each cluster $j = 1 : M$
Parameter Initialized: $G = 1$
Initial population: $P = size\left(CM\left(j\right)\right)$
Random selection of gbest: $gbest = rand\left(P \rightarrow CM\left(j\right)\right)$
$i = 1 : P$ (For each population)

$$iteration = 1$$

The present population's fitness status can be evaluated using Equation (18.8).

$$f(CM\left(j, i\right)) = getfitness\ (CM\left(j, i\right))$$

If $f(CM\left(j, i\right)) > threshold$

$$lnodes = \left[lnodes; CM\left(j,i\right)\right]$$

Else

$$mnodes = \left[mnodes; CM\left(j,i\right)\right]$$

End If
Mutation: $f\left(gbest\right) = getfitness\left(gbest\right)$
Crossover: $pbest = CM\left(j,i\right)$
Perform selection
If $(f(CM\left(j, i\right)) > f\left(gbest\right)$
Update new finest solution: $gbest = pbest$

$$CH\left(j\right) = CM\left(j,\ i\right)$$

Else
Discard *pbest*

$$CH\left(j\right) = gbest$$

End If
If $(iteration > Gmax)$
return $CH\left(j\right)$
Else

$$iteration = iteration + 1$$

End If
End For
End For

18.3.4 Cluster Maintenance

The SCNIDSS method updates the CHs at each TDMA schedule in order to solve energy balance and QoS performance by reducing cyber threats.

18.3.5 Data Transmission

After cluster formation and the best CH selection, periodic data will be transmitted in both intra-cluster and inter-cluster modes. The routes are found using the traditional SPT method with a particular *mlist* check on each node. This guarantees safe data transfer throughout the network. At each CH node, duplicated data is checked and removed in order to reduce communication overhead and energy usage while carrying out the fusion.

18.4 RESULTS AND DISCUSSIONS

In NS2, the SCNIDSS method is implemented and assessed using cutting-edge protocols. The analyses were carried out utilizing several network circumstances, such as varying densities of senior nodes and cyber threats. To generate blackhole attackers in WSNs that carry out unwanted activity like packet dropping, the threat model was introduced. The SCNIDSS method's performance was future compared to that of three other protocols. The clustering algorithm is based on the standard LEACH [11], and trust management scheme (TMS) [23, 33]. The TMS method was chosen as a secure clustering against blackhole attackers. The eeTMFO/GA approach was recently implemented where nature-inspired algorithms and multiple sensor node trust factors are incorporated for optimal and secure solutions. SCNIDSS is judged on its efficiency and reliability performance by comparing it to secured (TMS & eeTMFO/GA) and unsecured LEACH. The comparison was made using performance parameters such as packet delivery ratio (PDR), average throughput, average energy consumption, communication overhead, and average delay in time. A wide range of performance measures were chosen to support the QoS, energy, and computational efficiencies. According to differences in density, the results are detailed in the following two sections.

18.4.1 Sensor Node Density Variations

The parameters used to set density variations for network simulation are presented in Table 18.1. Figures 18.2–18.6 show the outcomes of the implemented model. There are 10% blackhole attackers on each network.

TABLE 18.1
Parameters for Density Variations Used for Network Design

Number of sensor nodes	50–300
Threats ratio	10%
Pattern of traffic	CBR
Connection numbers	5
MAC	802.11
Location BS	1050 × 1050 m
Network size	1000 × 1000 m
Node deployment (topology)	Random
Energy (initial)	0.5 J
Energy consumption for transmitter	16.7 nJ
energy consumption at the receiver end	36.1 nJ
Time of simulation	100 seconds
Size of packet	512 kb

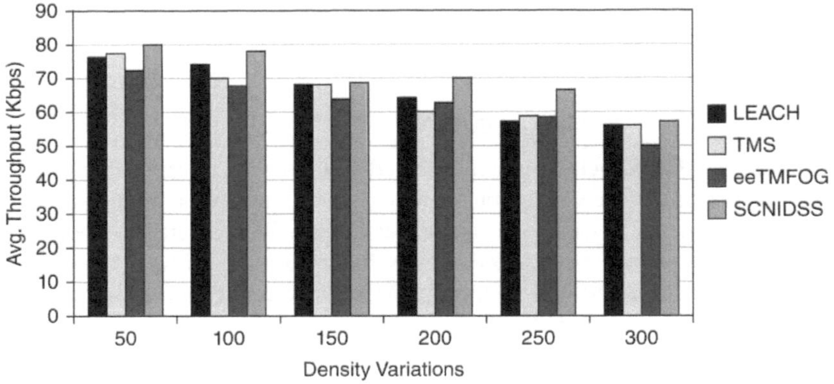

FIGURE 18.2 Result for average throughput for sensor nodes density variation.

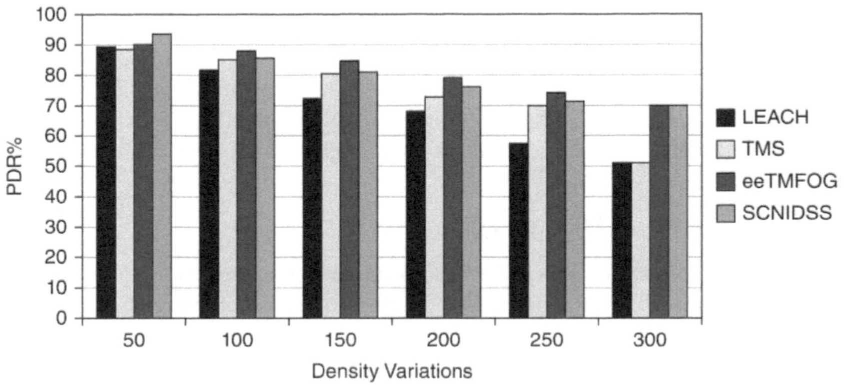

FIGURE 18.3 Performance of PDR for sensor nodes density variation.

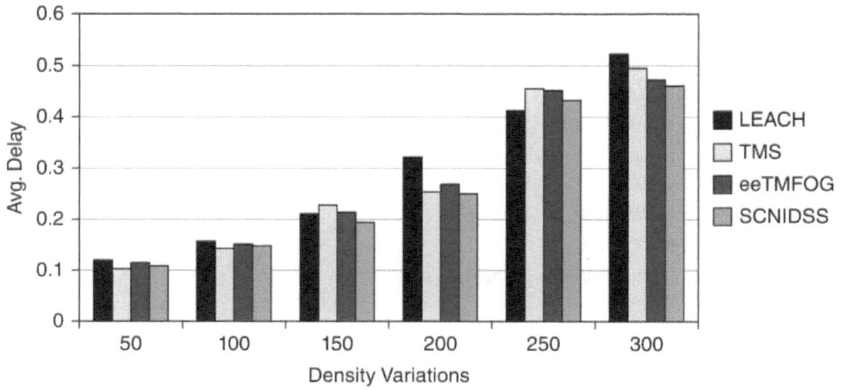

FIGURE 18.4 Sensor nodes with different sensor nodes have an average delay performance.

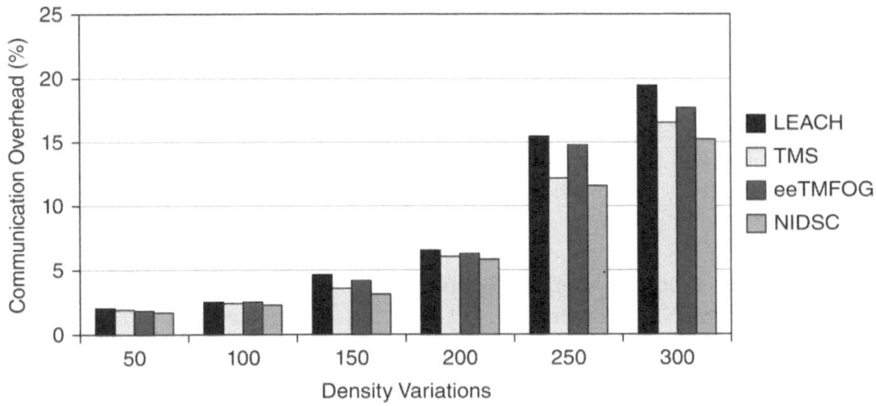

FIGURE 18.5 Performance of communication overhead for sensor nodes density variation.

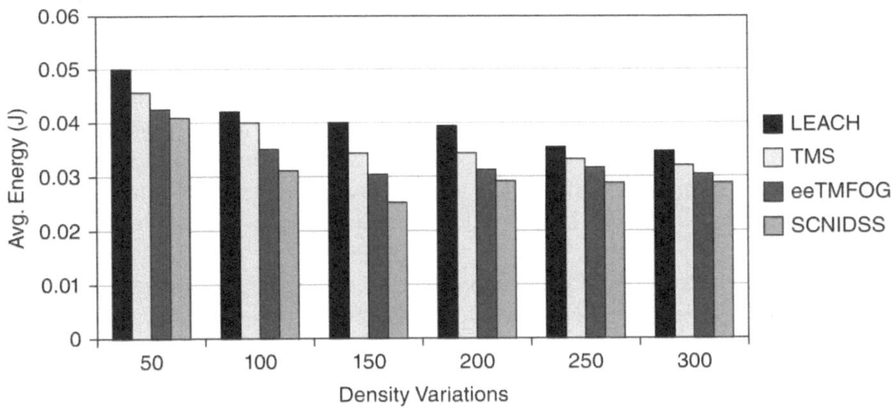

FIGURE 18.6 Performance of average energy consumption for sensor node density variations.

The average delay, average throughput, and PDR for the performance of clustering QoS are shown in Figures 18.2–18.4, respectively. These findings demonstrate that, in the network even if malicious nodes are present, the proposed SCNIDSS method enhances outcomes. The increase in population causes a decrease in throughput, PDR, and communication latency. This is caused by communication lines, and the quantity of clusters rises as the number of sensor nodes does as well.

There is no security defense mechanism implemented against the attackers, resulting in the nastiest performance for QoS in the traditional LEACH method.

In comparison to other secured methods, eeTMFO/GA performed better than TMS because they employed a mechanism to determine the trustworthiness of sensor nodes. The SCNIDSS further improves the QoS as compared to eeTMFO/GA because of the hybrid set of parameters and the provision of detecting attackers embedded in optimal CH selection. In eeTMFO/GA, the secured optimal CH selection was performed by using five trust parameters; however, the data communication was performed without bothering the presence of attackers in the path. The SCNIDSS method overcomes such problems which exist in TMS and eeTMFO/GA methods.

The effectiveness of communication overhead is seen in Figure 18.5. Comparing the SCNIDSS method to all other existing protocols demonstrates the reduction in communication overhead. The network's overabundance of routing packets produced during data transmission phases is what causes communication overhead. Increased network communication overhead may be caused by frequent cluster updating and route breakdowns. The SCNIDSS method minimized communication overhead by choosing more reliable CH and routes throughout the duration of the simulation. The DE algorithm's trust factors, which are used to calculate the fitness value, help to perform not only the best CH selection but also an accurate attacker identification that is not possible with the TMS and eeTMFO/GA methods.

Figure 18.6 presented average energy consumption, and when combined with other factors, SCNIDSS greatly reduced energy consumption when compared to other protocols. This shows that WSNs function better across their whole lifetimes when malicious nodes are present. When compared to other methods, the SCNIDSS method consumes less energy because of fewer rounds of clustering processes, frequent route breakdowns, and packet losses brought on by attackers.

The average results of all the existing and proposed approaches are presented in Table 18.2. From the table, it can be seen that the overall performance of the proposed work is better than the other existing approaches where Throughput is increased by around 3 kbps, PDR is increased by about 4%, latency is reduced by 0.01 seconds, the consumption of energy is decreased by approximately 0.003 J, and overhead in communication cut down by about 0.75%. Therefore, the overall performance of the proposed work is better than the other existing approaches.

TABLE 18.2
Average Result Analysis for Sensor Node Density Variations Scenario

Parameters	LEACH	TMS	eeTMFO/GA	SCNIDSS
Throughput (kbps)	66.326	65.308	62.845	**70.225**
PDR (%)	70.13	74.68	81.1	**79.675**
Delay (seconds)	0.2906	0.2788	0.2791	**0.2661**
Overhead (%)	8.458	7.133	7.905	**6.639**
Consumed Energy (J)	0.03885	0.0312	0.0284	**0.0259**

TABLE 18.3
Parameters of Attack Variation

Number of Sensor nodes	100
Traffic pattern	CBR
Cyber threats percentage	0–2–4–6–8–10%
MAC	802.11
Total connections	5
Node deployment (topology)	Random
Size of Network	1000 × 1000 m
Base Station location	1050 × 1050 m
Consumption Receiver energy	36.1 nJ
Initial level Energy	0.5 J
Consumption Transmitter energy	16.7 nJ
Packet size	512 kb
Simulation Time	100 seconds

18.4.2 Attacks Variations

In order to support its reliability and scalability, the second scenario to examine the performance of the SCNIDSS method is the attack scenario. Table 18.3 contains a list of all the simulation parameters. The networks simulate attackers who make up between 0% and 10% of the total 100 sensor nodes. The findings in Figures 18.7–18.11 demonstrate how attack variation strategies affect the effectiveness of each regimen.

Figures 18.7–18.9 illustrate the average delay, PDR, and average throughput for the QoS performances, respectively. These findings demonstrate how the network is being impacted by an increase in attackers. Even with the security measures in place, when the number of attackers rises, network performance for all parameters degrades due to packet loss and high congestion caused by the attackers. In comparison to other secured protocols, the QoS metrics for the LEACH unsecured method perform the

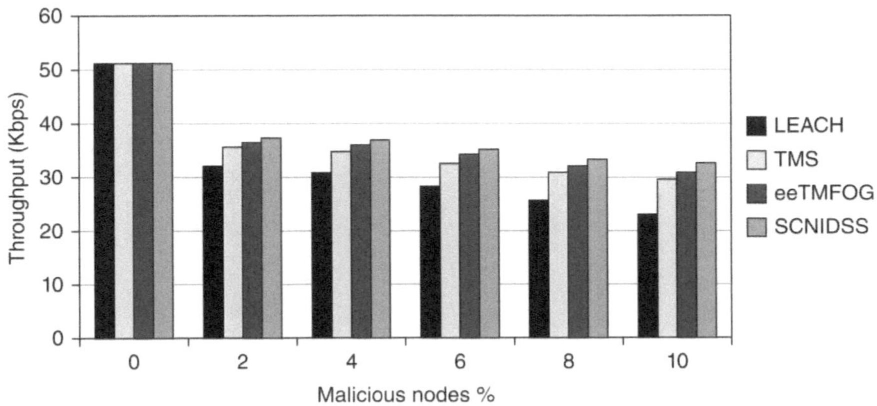

FIGURE 18.7 Performance of average throughput performances for attacker nodes variation.

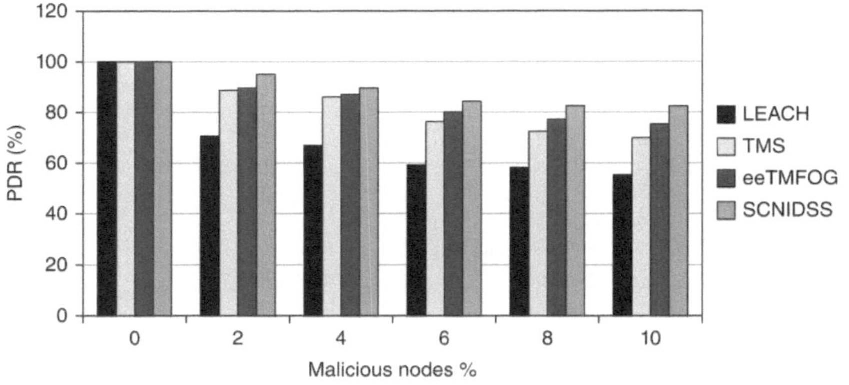

FIGURE 18.8 Performance of PDR for attacker nodes variation.

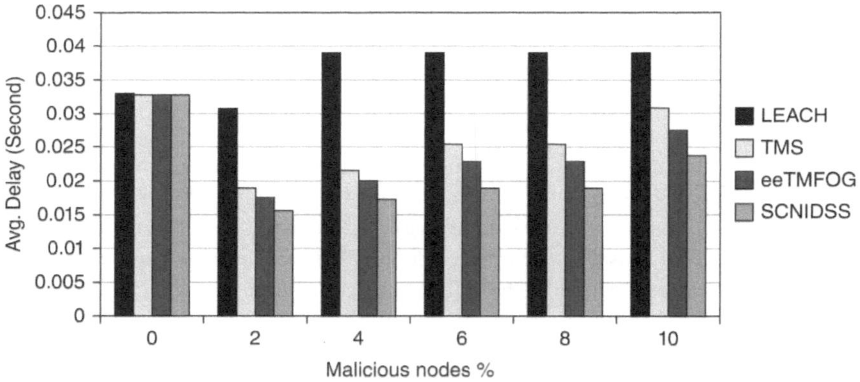

FIGURE 18.9 Performance of average delay for attacker nodes variation.

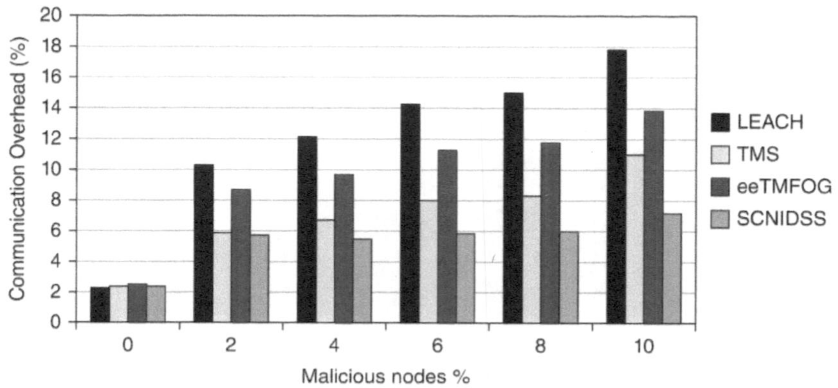

FIGURE 18.10 Performance of communication overhead for attacker nodes variation.

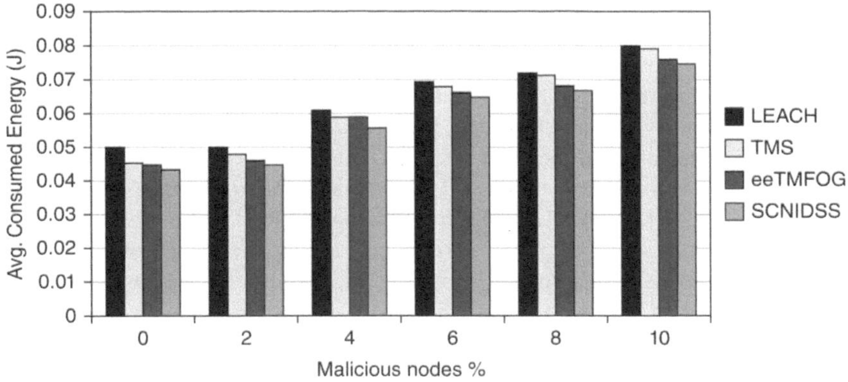

FIGURE 18.11 Performance of average energy consumption for attacker nodes variation.

poorest. Having the trustworthy trust metrics set and attackers' detection technique during clustering, the SCNIDSS method demonstrates that the attacker's impact is significantly decreased when compared to eeTMFO/GA and TMS methods.

The performance of the DE algorithm improved since it was able to deliver results more quickly than the methods employed in the TMS and eeTMFO/GA methods. In order to provide safe data transmissions in SCNIDSS, the parameters utilized in the computation of PTM aid in the accurate detection of intruders.

The performance of parameters communication overhead and average energy consumption are presented in Figures 18.10 and 18.11, respectively. When compared with the other methods used for this application, the SCNIDSS technique significantly reduces communication overhead and energy usage. In terms of communication overhead, the TMS protocol comes in second. Due to the intricate process of choosing the best CH using MFO and GA algorithms, the eeTMFO/GA performs the worst. The aforementioned section previously covered the causes of the SCNIDSS performance improvement.

Last but not least, the average findings are shown in Table 18.4, which illustrates that the SCNIDSS method enhanced the throughput by 37.76 (kbps), PDR values by 88.96 (%), and decreased average energy consumption by 0.0583 (J), Delay by 0.021238 (seconds), and communication overhead by 5.409 (%).

TABLE 18.4
Analysis of Average Results for Attack Variations Scenario

Parameters	LEACH	TMS	eeTMFOGA	SCNIDSS
Throughput (kbps)	31.84	35.77	36.78	**37.76**
PDR (%)	68.00	82.11	84.67	**88.96**
Delay (seconds)	0.03664167	0.025895	0.023985	**0.021298**
Communication Overhead (%)	11.9685	7.0245	9.66	**5.409**
Energy Consumption (J)	0.0636	0.0616	0.0599	**0.0583**

18.5 CONCLUSION

The implemented method aims to create innovative, secure, and effective clustered WSNs based on natural inspiration. To address the issues of modern clustering techniques and reliability trade-offs in the presence of attackers, the SCNIDSS was implemented. The fitness function's implementation allowed for precise attacker detection and steady, ideal selection of CH during each clustering round, which is the main contribution of this study. Each sensor node's Residual energy ratio, PDR, degree of connectivity, and conjunction ratio were used as a trust factor in the fitness function for better identification of attackers. The simulation findings employing various densities and attack situations support the efficacy of the proposed work. In comparison to contemporary and conventional approaches, the suggested SCNIDSS results provide a boost in throughput, PDR, and minimal delay, as well as lower energy usage and communication overhead.

There can be multiple directions to further investigate the performance of WSNs for secured data transmission based on nature-inspired algorithms. Following are a few of them such as using various new and evolving cyber threats, the performance of the presented algorithms can be compared, implementation on virtual machines followed by implementation on the actual network can be done to observe the security and efficiency parameters of proposed algorithms, as only simulation (on NS2) is conducted in the present work for performance analysis, For large data sets, a clustered routing protocol must be employed with data aggregation. This could be one of the immediate extensions of the proposed methods, and the presented algorithms can also be tested in unconventional WSNs like under water WSN.

REFERENCES

1. Mahajan, S.; Dhiman, P. K. Clustering in wireless sensor networks: A review. International Journal of Advanced Research in Computer Science. 2016 May 1; 7(3):198.
2. Mugheri, A. A.; Siddiqui, M. A.; Khoso, M. Analysis on security methods of wireless sensor network (WSN). Sukkur IBA Journal of Computing and Mathematical Sciences. 2018 Jun 26; 2(1):52–60.
3. Mallick, C.; Satpathy, S. Challenges and design goals of wireless sensor networks: A state-of-the-art review. International Journal of Computer Applications. 2018 Mar;179(28):42–47.
4. Almalkawi, I. T.; Zapata, M. G.; Al-Karaki, J. N.; Morillo-Pozo, J. Wireless multimedia sensor networks: Current trends and future directions. Sensors (Basel). 2010;10(7):6662–717.
5. Harjito, B.; Han, S. Wireless multimedia sensor networks applications and security challenges. In 2010 International Conference on Broadband, Wireless Computing, Communication and Applications 2010 Nov 4 (pp. 842–46). IEEE.
6. He, T.; Krogh, B.; Krishnamurthy, S.; Stankovic, J. A.; Abdelzaher, T.; Luo, L.; Hui, J. Energy-efficient surveillance system using wireless sensor networks. In Proceedings of the 2nd International Conference on Mobile Systems, Applications, and Services – MobiSYS '04, 2004. doi:10.1145/990064.990096.
7. Hussain, M. A.; kyung Sup, K. WSN research activities for military application. In 2009 11th International Conference on Advanced Communication Technology, 2009 Feb 15 (Vol. 1, pp. 271–74). IEEE.

8. Sharma, R.; Vashisht, V.; Singh, A. V.; Kumar, S. Analysis of existing clustering algorithms for wireless sensor networks. In System performance and management analytics, 2010 (pp. 259–77).
9. Cho, J. H.; Swami, A.; Chen, R. A survey on trust management for mobile ad hoc networks. IEEE Communications Surveys & Tutorials. 2010 Oct 14;13(4):562–83.
10. Willig, A.; Karl, H. Data transport reliability in wireless sensor networks. A survey of issues and solutions. PIK – Praxis Der Informationsverarbeitung Und Kommunikation. 2005, 28(2), 86–92. doi:10.1515/piko.2005.86.
11. Heinzelman, W. R.; Chandrakasan, A.; Balakrishnan, H. Energy-efficient communication protocol for wireless microsensor networks. In System Sciences, 2000 proceedings of the 33rd annual Hawaii International Conference, p. 10.
12. Hoang, D. C.; Yadav, P.; Kumar, R.; Panda, S. K.A robust harmony search algorithm based clustering protocol for wireless sensor networks. In 2010 IEEE International Conference on Communications Workshops, 2010 May (pp. 1–5).
13. Song, M. A.; Zhao, C. L. Unequal clustering algorithm for WSN based on fuzzy logic and improved ACO. The Journal of China Universities of Posts and Telecommunications. 2011 Dec 1; 18(6):89–97.
14. Enami, N.; Moghadam, R. A.; Ahmadi, K. D. A new neural network based energy efficient clustering protocol for wireless sensor networks. In 5th International Conference on Computer Sciences and Convergence Information Technology, 2010 Nov 30 (pp. 40–45). IEEE.
15. Bagci, H.; Yazici, A. An energy aware fuzzy unequal clustering algorithm for wireless sensor networks. In International Conference on Fuzzy Systems, 2010 Jul 18 (pp. 1–8). IEEE.
16. Geetha, V.; Chandrasekaran, K. A distributed trust based secure communication framework for wireless sensor network. Wireless Sensor Network. 2014 Sep 18; 6(09):173.
17. Guo, W. W.; Looi, M. A framework of trust-energy balanced procedure for cluster head selection in wireless sensor networks. Journal of Networks. 2012 Oct 1; 7(10):1592.
18. Tolba, F. D.; Ajib, W.; Obaid, A. Distributed clustering algorithm for mobile wireless sensors networks. In SENSORS, 2013 IEEE 2013 Nov 3 (pp. 1–4). IEEE.
19. Sahoo, R. R.; Singh, M.; Sardar, A. R.; Mohapatra, S.; Sarkar, S. K. TREE-CR: Trust based secure and energy efficient clustering in WSN. In 2013 IEEE International Conference ON Emerging Trends in Computing, Communication and Nanotechnology (ICECCN), 2013, Mar. 25 (pp. 532–38). IEEE.
20. Sahoo, R. R.; Singh, M.; Sahoo, B. M.; Majumder, K.; Ray, S.; Sarkar, S. K. A light weight trust based secure and energy efficient clustering in wireless sensor network: Honey bee mating intelligence approach. Procedia Technology. 2013 Jan 1; 10:515–23.
21. Nimbalkar, N. B.; Das, S. S.; Wagh, S. J.Trust based energy efficient clustering using genetic algorithm in wireless sensor networks (teecga). International Journal of Computer Applications. 2015 Jan 1; 112(9):30-33.
22. Dahane, A.; Berrached, N. E.; Loukil, A. Balanced and safe weighted clustering algorithm for mobile wireless sensor networks. In Computer Science and Its Applications: 5th IFIP TC 5 International Conference, CIIA 2015, Saida, Algeria, May 20–21, 2015, Proceedings 5 2015 (pp. 429–41). Springer International Publishing.
23. Sharma, R.; Vashisht, V.; Singh, U. eeTMFO/GA: A secure and energy efficient cluster head selection in wireless sensor networks. Telecommunication Systems. 2020 Jul; 74(3):253–68.
24. Wang, T.; Zhang, G.; Yang, X.; Vajdi, A. A trusted and energy efficient approach for cluster-based wireless sensor networks. International Journal of Distributed Sensor Networks. 2016 Apr 10; 12(4):3815834.
25. Juliana, R.; Maheswari, P. U. An energy efficient cluster head selection technique using network trust and swarm intelligence. Wireless Personal Communications. 2016, 89(2): 351–64.

26. Sharawi, M.; Emary, E. Clustering optimization for WSN based on nature-inspired algorithms. Nature-Inspired Computation in Engineering. 2016, 637:111–32.
27. Rehman, E.; Sher, M.; Naqvi, S. H.; Badar Khan, K.; Ullah, K. Energy efficient secure trust based clustering algorithm for mobile wireless sensor network. Journal of Computer Networks and Communications. 2017, Jan 1; 1:1630673
28. Mittal, N. Moth flame optimization based energy efficient stable clustered routing approach for wireless sensor networks. Wireless Personal Communications. 2019 Jan 30; 104:677–94.
29. Sharma, R.; Vashisht, V.; Singh, U. Nature inspired algorithms for energy efficient clustering in wireless sensor networks. In 2019 9th International Conference on Cloud Computing, Data Science & Engineering (Confluence), 2019, Jan. 10 (pp. 365–70). IEEE.
30. Sharma, R.; Vashisht, V.; Singh, U.EEFCM-DE: Energy-efficient clustering based on fuzzy C means and differential evolution algorithm in WSNs. IET Communications. 2019 May; 13(8):996–1007.
31. Pavani, M.; Trinatha Rao, P. Adaptive PSO with optimised firefly algorithms for secure cluster-based routing in wireless sensor networks. IET Wireless Sensor Systems. 2019 Oct;9(5):274–83.
32. Gilbert, E. P.; Baskaran, K.; Rajsingh, E. B.; Lydia, M.; Selvakumar, A. I. Trust aware nature inspired optimised routing in clustered wireless sensor networks. International Journal of Bio-Inspired Computation. 2019; 14(2):103–13.
33. Ramesh, S.; Yaashuwanth, C. Enhanced approach using trust based decision making for secured wireless streaming video sensor networks. Multimedia Tools and Applications. 2020 Apr; 79(15–16):10157–76.
34. Mahajan, H. B.; Badarla, A.; Junnarkar, A. A. CL-IoT: Cross-layer internet of things protocol for intelligent manufacturing of smart farming. Journal of Ambient Intelligence and Humanized Computing. 2021 Jul;12(7):7777–91.
35. Qureshi, S. G.; Shandilya, S. K. Advances in cyber security paradigm: A review. In Hybrid Intelligent Systems: 19th International Conference on Hybrid Intelligent Systems (HIS 2019) held in Bhopal, India, December 10–12, 20192021 (pp. 268–76). Springer International Publishing.
36. Qureshi, S. G.; Shandilya, S. K. Novel fuzzy based crow search optimization algorithm for secure node-to-node data transmission in WSN. Wireless Personal Communications. 2021 Mar 9;127:577–97.
37. Qureshi, S. G.; Shandilya, S. K. Novel hybridized crow whale optimization and QoS based bipartite coverage routing for secure data transmission in wireless sensor networks. Journal of Intelligent & Fuzzy Systems. 2021 Jan 1;41(1):2085–99.
38. Sharma, D. M.; Shandilya, S. K.; Satapathy, S. C. Maximizing blockchain security: Merkle tree hash values generated through advanced vectorized elliptic curve cryptography mechanisms. Concurrency and Computation: Practice and Experience. 2023 Oct 25;35(23):e7829.
39. Sharma, D. M.; Shandilya, S. K. An efficient cyber-physical system using hybridized enhanced support-vector machine with Ada-Boost classification algorithm. Concurrency and Computation: Practice and Experience. 2022 Sep 25;34(21):e7134.
40. Sharma, A. K.; Sharma, D. M.; Purohit, N.; Sharma, S. A.; Khan, A. Blockchain technology: Myths, realities and future. In Blockchain technology 2022 Apr 19 (pp. 163–80). CRC Press.
41. Sharma, D. M.; Shandilya, S.; Sharma, A. K. A comprehensive review on cyber physical system and its applications in robotic process automation. In Innovations in Bio-Inspired Computing and Applications: Proceedings of the 11th International Conference on Innovations in Bio-Inspired Computing and Applications (IBICA 2020), December 16-18, 2020, pp. 311–22. Springer International Publishing.

19 Transforming Network Protection via Open Innovation Strategies

Mohd. Noman Mansoori, Azra Siddiqui, and Priyanka Tripathi

19.1 INTRODUCTION

Open innovation is a corporate strategy that involves enabling internal ideas to flow outward and integrating external ideas, resources, and technologies into a company's innovation process. Henry Chesbrough coined the term "open innovation," which differs from more conventional "closed innovation" methods in which businesses would exclusively rely on internal research and development (R&D) to progress their technology and goods [1]. Open innovation can take many different shapes, including partnerships with third parties, academic institutions, startups, and even rival businesses. Partnerships, joint ventures, licensing deals, and acquisitions can all fall under this category. To expedite internal innovation and increase the markets for external applications, the key is to utilize external ideas and resources [2]. Different shapes of resources of open innovation are shown in Figure 19.1.

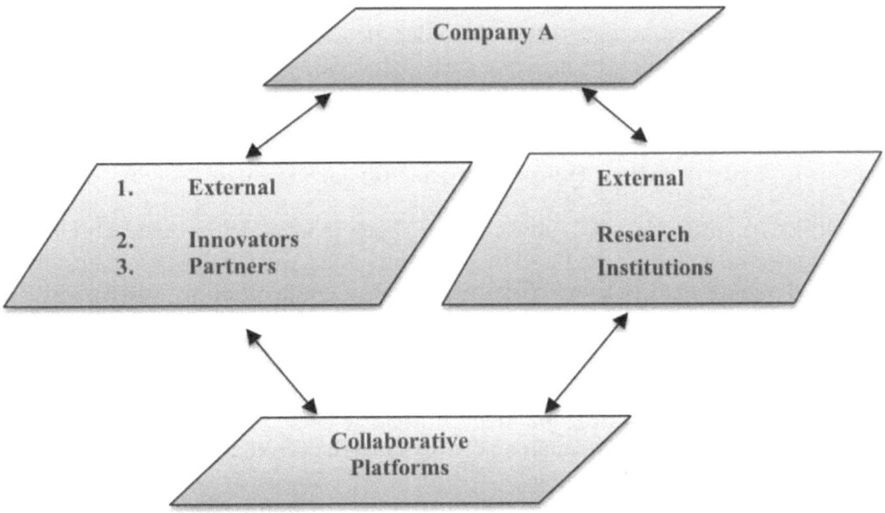

FIGURE 19.1 Different shapes of resources of open innovation.

DOI: 10.1201/9781003409571-19

Open innovation has become a crucial strategy for many businesses, especially in the technology and pharmaceutical sectors, to maintain their competitiveness and promote ongoing growth. Companies can access a worldwide network of talent and ideas through open innovation, promoting a culture of cooperation and accelerating invention at a faster rate than would be possible with old closed models [3]. Agile and adaptable security solutions have also emerged as a result of this collaborative environment, improving the capacity to identify and thwart sophisticated cyberattacks. Open innovation has promoted information exchange and skill growth, fostering the creation of a workforce capable of handling challenging security issues in the quickly changing technology environment [2].

19.2 NETWORK SECURITY

When trying to guarantee the confidentiality, integrity, and usefulness of computer networks, data, and assets, network security is essential. In order to stop unauthorized access, misuse, and alteration of network resources, it is important to deploy technologies such as firewalls, intrusion detection systems (IDS), and virtual private networks (VPNs). To fix known vulnerabilities, patch management and frequent upgrades are essential. By detecting and responding to unusual network behavior, continuous logging and monitoring help create a strong defense against online dangers like malware, hacking, and data breaches.

Putting policies in place to guard computer networks and data from abuse, disruption, alteration, and illegal access is known as network security. This includes using firewalls, IDS, encryption methods, and other protocols to protect sensitive data, hardware, and software. The goal of network security is to thwart cyberattacks that could jeopardize the confidentiality, integrity, and accessibility of network resources. Examples of these attacks include malware infections, phishing scams, and denial-of-service attacks. Maintaining a strong network security posture requires regular security assessments, updates, and user education. This ensures ongoing monitoring and threat mitigation to preserve the dependability and credibility of network operations.

19.3 NEED OF OPEN INNOVATION IN NETWORK

As network security becomes more complex, open innovation becomes more important for several reasons. First of all, it gives companies the flexibility to adjust to the always-changing field of cyber risks by fostering cooperation in recognizing and addressing new issues. Second, utilizing a larger reservoir of knowledge outside of organizational bounds lessens the impact of resource and talent constraints. Third, by encouraging partnerships that capitalize on a range of viewpoints and insights, open innovation enables organizations to keep ahead of technological advancements. Network security safeguards networks and data from cyber threats, ensuring reliability and integrity. It employs diverse tools and strategies to defend against evolving attack methods, emphasizing shared responsibility across all users [4]. It also encourages group learning, making it easier for many stakeholders to share knowledge. In conclusion, open innovation promotes creativity and conformity, stimulating the creation of innovative solutions while guaranteeing conformity to industry norms and guidelines. The network security must embrace open innovation to handle dynamic threats, overcome resource constraints, remain technologically relevant, promote collaborative learning, foster innovation, and maintain compliance.

19.4 RECOGNIZING NETWORK SECURITY ISSUES

Many different types of network security issues might occur. For a network to be adequately protected, these problems must be understood. This procedure entails ongoing observation, network data analysis, and system configuration assessment to identify any issues such as obsolete software, unwanted access attempts, or data breaches. For network administrators and cybersecurity specialists to take the appropriate action, safeguard sensitive data, and guarantee the stability and accessibility of the network, they must identify these problems as soon as possible [5]. The flow of problems identification regarding network security is shown in Figure 19.2.

Organizations experience a range of network security obstacles, such as viruses and ransomware that infiltrate via hacked websites or downloaded files. Distributed denial-of-service (DDoS) attacks cause network overload, disrupting activities, whereas phishing assaults result in data breaches by use of fraudulent means of collecting critical information. Employee carelessness and other internal problems are the root cause of insider threats, and weaknesses in access restrictions can lead to illegal access to data. Safeguarding sensitive data and guaranteeing confidentiality, integrity, and accessibility need the implementation of strong security measures like frequent evaluations, thorough training, and stringent protocols. When network security issues from the past and now are compared, it becomes clear that historical vulnerabilities continue to exist because effective solutions are adopted slowly, which leads to continuous risks and breaches. Network security is still a constant battle since advances in value generation and connectivity naturally invite adversary exploitation, necessitating a constant balancing act between innovation and defense [6].

FIGURE 19.2 Identifying problems regarding network security [2].

19.5 OPEN INNOVATION TO IMPROVE NETWORK SECURITY

In the constantly evolving field of cybersecurity, the use of open innovation as a strategic framework to improve network security has taken on increasing importance. A joint effort is undertaken to contribute creative concepts and answers targeted at enhancing network security within this dynamic environment. This is achieved by creating avenues for the efficient exchange of information and innovations, such as the dissemination of threat intelligence, participation in forums, and the use of collaborative platforms [7]. Open innovation is transforming network security by fostering collaborative, knowledge-driven approaches to combat cyber threats. Setting clear goals ensures accurate detection and addressing of malware and breaches [8]. Collaboration with external partners, through bug bounties and research, enhances security. This cooperative approach ensures more reliable and effective network security measures. The evolution and major topics in the literature on open innovation are examined in this study, which also identifies nine subject areas and makes recommendations for further research. It uses recent paper evaluations and text mining in a novel way to give thorough insights into the ever-changing open innovation scene [9].

Network security innovation thrives through crowdsourcing via hackathons and collaborations with suppliers and academia. Open (application programming interfaces) (APIs) and industry standards enhance security and interoperability. Key strategies include leveraging external expertise, promoting data sharing, and continuous improvement while addressing ethics, training staff, and maintaining communication and transparency. By utilizing the combined knowledge and ideas of a varied group of participants, open innovation can greatly strengthen network security. With this strategy, you may take proactive steps to address changing threats and implement cutting-edge solutions that might not be possible with a closed, in-house strategy [10].

19.6 INNOVATIVE NETWORK SECURITY THROUGH COLLABORATIVE APPROACHES IN CYBERSECURITY

Innovative network security techniques that emphasize collaboration to strengthen the digital fortifications of both enterprises and individuals are emerging in today's dynamic cybersecurity world. Utilizing the combined knowledge and insight of a wide range of stakeholders, such as cybersecurity professionals, researchers, companies, and the larger security community, is a key component of this approach. This strategy seeks to increase network security by encouraging collaboration and knowledge-sharing through a variety of channels, including forums, collaborative platforms, and information-sharing networks [11].

The fundamental concepts of this cutting-edge network security strategy include utilizing group skills to access a varied knowledge base for thorough cybersecurity awareness. By placing a strong emphasis on cooperative information exchange, the community can better comprehend security issues and keep one step ahead of cyber threats. Businesses may quickly modify procedures in reaction to new threats thanks to the adaptive security feature, which enables security measures to keep up with the changing threat landscape. This strategy, which is based on an innovation ecosystem, motivates participants to actively participate by offering bug bounty programs, open-source technology, and cooperative research, all of which improve network security [5]. The efficiency of each security layer is increased by utilizing many levels of defense, such as firewalls, IDS, encryption, and user awareness training. Rewards from this collaborative, intelligence-driven strategy include improved threat detection and prevention, quick reactions to new threats, and strengthened network security via information sharing. By using this approach, organizations may strengthen their defenses, react quickly to assaults, and maintain their leadership position in network security innovation.

19.7 OPEN-SOURCE NETWORK SECURITY SOLUTIONS

The goal of open innovation in network security is to increase the security of computer networks by collaborating with external parties such as people, companies, and the larger cybersecurity community. This strategy recognizes the need for a more transparent and collaborative approach to combat the continuously changing and more intricate digital threats. Some essential components of open innovation in network security are shown in Figure 19.3.

Collaboration and information sharing in open innovation promotes the sharing of knowledge, threat intelligence, and best practices among diverse businesses, governmental bodies, and the cybersecurity industry. By collaborating to keep ahead of new dangers, this collective exchange of knowledge contributes to improving overall security. Numerous businesses, including tech behemoths, have put in place bug bounty schemes. Through the provision of financial incentives for legitimate discoveries, these programs invite outside researchers to find and report vulnerabilities in their systems. Through the use of crowd-sourced security testing, this method enables the detection and correction of vulnerabilities that could otherwise go undetected [12]. Tools and software from the open-source community are important for network security.

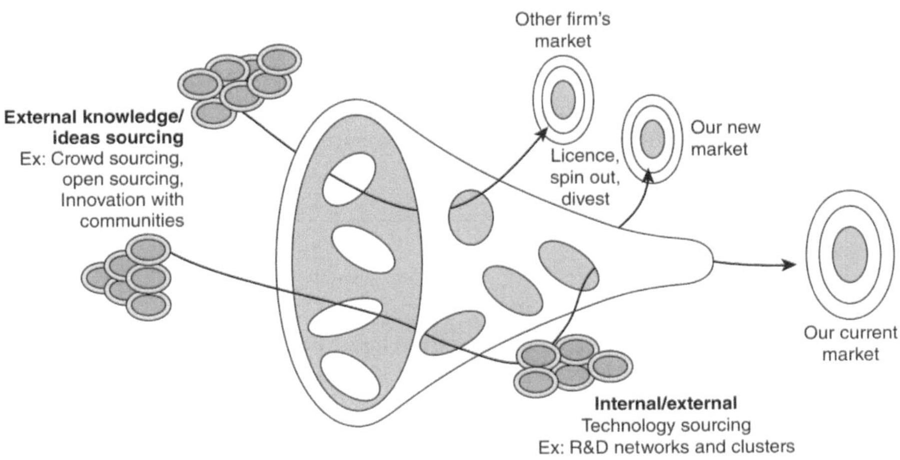

FIGURE 19.3 Network security with open-source solutions.

TABLE 19.1
Cybersecurity Innovations and Solutions

Problem/Challenge	Potential Solution	Contributing Innovators
Data breaches	AI-based threat detection	Company X, Research Group Y, Company Z
Network vulnerabilities	Blockchain-based security	Research Institute A, Company B
Insider threats	Behavior analysis tools	Company M, University N
Zero-day attacks	Community-based threat intelligence	Company C, Company D, Open-Source Group P
Regulatory compliance	Automated compliance monitoring	Company E, Regulatory Body Q

The open-source community creates and supports a large number of security tools and frameworks. Anyone can audit, enhance, and personalize these tools, enhancing their general security. Hacking competitions and Capture the Flag (CTF) competitions bring security enthusiasts together to compete in identifying vulnerabilities and resolving security issues. These contests offer a setting for developing and honing cybersecurity abilities. The innovations and solutions of cyber-security is given in Table 19.1.

The major problems caused by data breaches are handled by artificial intelligence (AI)-based threat detection, which is being spearheaded by Company X, Research Group Y, and Company Z. Blockchain-based security, supported by Research Institute A and Company B, addresses network vulnerabilities. University N and Company M's behavior analysis technologies help to reduce insider risks. Community-based threat intelligence from Company C, Company D, and Open-Source Group P is used to thwart zero-day attacks. Regulatory Body Q and Company E oversee computerized compliance monitoring, which assures regulatory compliance.

19.8 ORIGINS OF OPEN INNOVATION

Open invention has a long history, with examples going as far back as early human trade and cultural interactions. By deliberately acquiring businesses to mix hardware and software for new technology products that matched customer expectations in the late 1990s, Cisco emerged as a leading force in open innovation. Cisco was able to outperform industry heavyweights Bell Labs and Lucent because of this strategy, which successfully converted substantial technological investments into company growth. The origin of open innovation is given in Table 19.2.

TABLE 19.2
Origins of Open Innovation

Year	Origins of Open Innovation
2003	Henry Chesbrough was the first to establish the idea of open innovation.
2006	Chesbrough wrote "Open Business Models: How to Succeed in the New Era of Innovation."
2011	Chesbrough released "Open Services Innovation: Rethinking Business Strategies for a Changing World."
2019	Open innovation saw the start of garnering widespread acceptance across numerous industries.

Eli Lilly, a pharmaceutical corporation that aimed to overcome the customarily low success rates in pharmaceutical innovation, is another significant forerunner in the field of open innovation. Around 2000, Eli Lilly unveiled an open innovation platform that connects to world scientific knowledge. InnoCentive, an Eli Lilly spin-off created to support open innovation for other pharmaceutical and consumer goods firms, was founded on this platform. When Procter & Gamble (P&G) introduced their "Connect & Develop" process in 2000, it also made considerable progress in open innovation. P&G sought to maintain constant R&D spending while achieving a $2 billion annual growth target. P&G increased its R&D productivity by 60% by embracing external innovations and sourcing up to 50% of its fresh ideas from outside collaborators.

19.9 OPEN INNOVATION IN NETWORK SECURITY: RISKS AND REWARDS

Henry Chesbrough developed the idea of "open innovation," which refers to companies and organizations that invite outside input into their innovation processes. The target of this experiment is to analyze open innovation in network security, identify emerging trends, and suggest future lines of inquiry. Delineating nine critical domains, such as organizational dimensions, collaborative frameworks, and contextual dependencies, provides fresh perspectives for improving network security procedures [9]. Through this approach, businesses can leverage the knowledge, ideas, and resources of others to further their innovation initiatives. Although open innovation has shown benefits in many fields, there are dangers and benefits when it comes to network security.

19.9.1 Advantages of Open Innovation's Network Security

Network security open innovation has several benefits. Companies have access to a wide range of experts and professionals, each of whom brings distinct knowledge and insights to the table to tackle security problems in novel ways. Stronger security measures and shorter reaction times are made possible by this cooperative approach, which facilitates the quicker identification and resolution of security issues [13]. Working with outside organizations also makes it easier for innovative security methods and technologies to be adopted, which encourages the creation of more sophisticated security solutions to counteract changing cyber threats. Open innovation fosters flexibility and adaptability, which enables firms to quickly respond to evolving threat landscapes and technology breakthroughs.

19.9.2 Obstacles in Open Innovation's Network Security

Open innovation in network security comes with dangers and difficulties. When exchanging sensitive security-related data with outside partners, the collaboration may make data breaches and leaks more likely. Intellectual property theft is a problem since it can cause a company to lose market share if proprietary security technology is shared. It is risky to rely too much on outside parties for security innovation, and ending a partnership could affect the organization's security protocols. Conflicting interests among open innovation participants can also result in disagreements on security priorities, which could undermine collaborative efforts and jeopardize security goals.

Organizations should prioritize protecting sensitive data and intellectual property, create strong frameworks for collaboration, and establish clear criteria for information sharing in order to effectively leverage the benefits of open innovation in network security. To reduce possible dangers related to open innovation in network security, regular risk assessments and the application of strict security rules are essential.

19.10 CASE STUDIES: EFFECTIVE OPEN INNOVATION INTEGRATION IN NETWORK SECURITY

Collaborating with outside interests, such as cybersecurity specialists, startups, academic institutions, and industry peers, is necessary to integrate open innovation into network security and improve the efficacy and flexibility of security measures. This cooperative strategy encourages the sharing of knowledge, skills, and assets, resulting in a more thorough and proactive protection against cyberattacks and analyzing three case studies from the East of England to determine how publically supported infrastructure influences open innovation in particular areas. It highlights the difficulties in managing the varied stakeholder interests in promoting innovation networks and illustrates how disparate approaches to open innovation infrastructure impact regional innovation dynamics [14]. Real-world examples include Cisco Systems' improvement of security capabilities through the acquisition of creative security firms, Google's use of bug bounty programs and projects like Google Project Zero, and the employment of outside security experts to reinforce network security. Initiatives like IBM's X-Force Exchange and Palo Alto Networks' collaborations demonstrate how open innovation fosters information exchange and cooperative problem-solving, which in turn improves network security. These initiatives are essential given the linked and dynamic digital environment of today.

Here are some examples of how open innovation has successfully been incorporated into network security (Table 19.3).

TABLE 19.3
Enhancing Network Security through Collaborative Initiatives

Case Study	Company	Description
1	Cisco Systems	Collaboration with experts and startups to enhance threat detection and acquisitions strengthens security product portfolio.
2	Bug bounty programs	Successful use of bug bounty programs to engage external security researchers and improve network security.
3	IBM X-Force Exchange	A platform for sharing threat intelligence and security information, strengthening network security through open collaboration.
4	Palo Alto Networks	Collaboration with industry groups, government organizations, peers, and Google (including Project Zero) to share threat intelligence and develop security solutions.
5	Symantec (NortonLifeLock)	Partnership with academic institutions and independent researchers to stay at the forefront of network security research and product development.

19.10.1 Bug Bounty Programs

Prominent corporations such as Google, Microsoft, and Facebook have implemented bug bounty initiatives, enlisting the help of ethical hackers worldwide to identify and disclose security flaws. These companies use a wide range of experts to improve their security through open innovation.

19.10.2 Cisco's Security Technology Alliances (CSTA) Programmed

Through partnerships with outside IT firms, CSTA Program integrates its security solutions with Cisco's infrastructure. Cisco's security services are strengthened by this external collaboration, which offers strong solutions against contemporary cyber threats. In today's changing organization, the chapter highlights the benefits of open innovation for small and medium-sized businesses. It also shows how this approach can offer up new opportunities for growth and development, especially for smaller businesses that are limited by different constraints. It proposes that businesses can better create value in their innovation activities by utilizing open innovation platforms to help them seize important benefits and opportunities [15].

19.10.3 IBM'X-Force Exchange

Practitioners and researchers may more easily share security insights, threat data, and best practices thanks to IBM's open threat intelligence platform, X-Force Exchange. By enabling proactive defense against cyber threats, this cooperative ecosystem advances network security as a whole.

19.10.4 Mozilla Foundation's Security Initiatives

Through programs like Mozilla Open-Source Support and the Mozilla Security Bug Bounty Program, Mozilla actively participates in open innovation. Mozilla can consistently improve the security features of its products, such as the Firefox web browser, by collaborating with the broader security community.

19.10.5 Google's Project Zero

The goal of Google Project Zero is to find and fix zero-day vulnerabilities in a variety of software applications. These vulnerabilities are found by a group of knowledgeable researchers, who work with software developers to create fixes. If a patch is not supplied, strict disclosure standards that include a 90-day reporting window guarantee public disclosure.

19.11 THE FUNCTION OF OPEN INNOVATION IN NETWORK SECURITY

Network security benefits greatly from open innovation since it fosters collaboration and makes use of a wide spectrum of expertise to tackle complex problems. It encourages a culture of constant learning and modification, enabling the investigation of fresh concepts and cutting-edge security solutions. Cybercrime in technology companies is impacted by open innovation, underscoring the difficulty of striking a balance between innovation and cybersecurity. In order to reduce risks and encourage open innovation, it proposes coopetition. It also provides insights on how to manage cyber dangers in technologically advanced environments [16]. In the end, this strategy strengthens the security framework by encouraging the sharing of insightful knowledge and cooperative resources, which results in the creation of more resilient and creative security solutions. Open innovation has a wide and important role in network security, greatly influencing the state of cybersecurity today. It performs several vital tasks, as shown in Figure 19.4.

Open innovation in network security encourages cooperation between various stakeholders, such as organizations, professionals, and researchers, enabling group efforts to tackle difficult problems. It makes use of outside viewpoints and specialist knowledge to acquire a thorough grasp of novel risks and weaknesses. This strategy fosters a dynamic environment and keeps firms up to date with new developments in network security trends and technology by encouraging continual learning and adaptability. Subject matter expert SMEs can improve their sustainability performance even with low resources by using open innovation to leverage external resources, as demonstrated by this sustainability-focused open innovation strategy. The research identifies important organizational and knowledge factors that support sustainable innovation in SMEs and highlights the positive relationships between innovativeness, inbound open innovation, and sustainability performance through partial least squares structural equation modeling on 199 Batik Tulis-producing

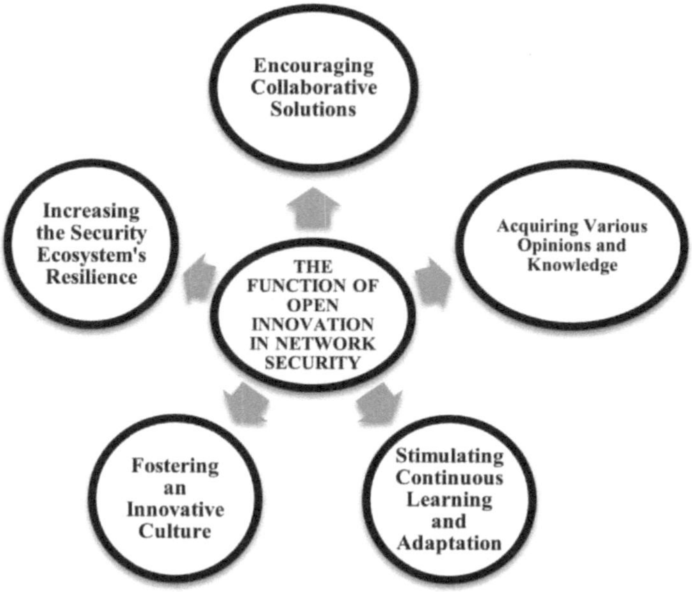

FIGURE 19.4 Open innovation's purpose in network security.

Small and Medium Enterprises (SMEs) [17]. Open innovation fosters a culture of creativity to address security-related concerns with a persistently inventive mindset by encouraging the investigation of novel approaches and instruments. Moreover, it strengthens the security ecosystem's resilience by facilitating cooperative information exchange, offering strong protection against cyberattacks, and prompting fixes for security flaws. All in all, open innovation fosters a dynamic and cooperative atmosphere that propels the creation of adaptable and successful security frameworks to traverse the always-changing digital terrain.

19.12 CONCLUSION

Open innovation has the potential to significantly impact the future of network security by offering a plethora of opportunities and ground-breaking developments. This change is anticipated to encourage cooperative threat intelligence sharing among enterprises, enabling prompt responses to complex cyber threats and the relationship, using structural equation modeling, between openness, open innovation capacities, and business performance. Results emphasize the role that desorptive capacity which is reinforced by knowledge management capacity plays in effective outbound open innovation processes. These findings provide guidance for open innovation networking strategies [18]. The use of blockchain technology to improve data integrity, the growth of multi-stakeholder cybersecurity initiatives, and the incorporation of sophisticated automation and AI into security systems are some notable advancements in the relationship between open innovation and strategic corporate social responsibility Corporate Social Responsibility (CSR) initiatives, with a particular focus on the energy sector's Ente Nazionale per l'Energia (National Entity for Electricity) ENEL case study. It also looks at how open innovation might act as a catalyst for sustainability [19]. This cooperative strategy of public, corporate, governmental, and academic groups aims to build global standards and guidelines with a focus on privacy, data protection, and the creation of all-encompassing frameworks for cyber resilience. Network security in the linked digital age will be defined by a dedication to proactive threat prevention, quick incident response, and ongoing monitoring, resulting in strong, flexible, and all-encompassing security ecosystems.

REFERENCES

1. Chesbrough, H. W. (2003). Open Innovation: The New Imperative for Creating and Profiting from Technology. Harvard Business School Press, Boston, MA.
2. Brant, J. (2014). The open innovation model. *Innovation and Intellectual Property Series*, 1080 (2), 28. Retrieved from https://www.iccwbo.be/wp-content/uploads/2012/03/20140325-The-Open-Innovation-Model.pdf
3. Busarovs, A. (2013). Open innovation: Current trends and future perspectives. *Humanities and Social Sciences: Latvia*, 21 (2), 103–119.
4. Nick, B. (2022, October). Network Security. Retrieved from TechTarget: https://www.techtarget.com/searchnetworking/definition/network-security
5. Ennomotive. (2019, Oct 23). Open Innovation: Accelerating Your Innovation Results. Retrieved from Medium: https://ennomotive-33439.medium.com/open-innovation-accelerating-your-innovation-results-196f3110b67e

6. Goeringer, S. (2016, May 24). The Future of Network Security. Retrieved from CableLabs Logo: https://www.cablelabs.com/blog/the-future-of-network-security
7. Alberti, F. G. and Pizzurno, E. (2017). Oops, I did it again! Knowledge leaks in open innovation networks with start-ups. *European Journal of Innovation Management*, 20 (1), 50–79.
8. Amponsah, C. T. and Adams, S. (2017). Open innovation: Systematisation of knowledge exploration and exploitation for commercialisation. *International Journal of Innovation Management*, 21 (3), 1750027.
9. Bigliardi, B. (2020). The past, present, and future of open innovation. *European Journal of Innovation Management*, 24 (4), 1130–1161.
10. Badawy, M. K. (2011). Is open innovation a field of study or a communication barrier to theory development? A perspective. *Technovation*, 31 (1), 65–67.
11. Reed, R., Storrud-Barnes, S. and Jessup, L. (2012). How open innovation affects the drivers of competitive advantage: Trading the benefits of IP creation and ownership for free invention. *Management Decision*, 50 (1), 58–73.
12. Allen, T. J. (1977). Managing the Flow of Technology: Technology Transfer and the Dissemination of Technological Information within the R&D Organization. MIT Press, Cambridge, MA.
13. Hartmann, D. T. (2009, December). Why 'open innovation' is old wine in new bottles. *International Journal of Innovation Management*, 13 (4), 715–736.
14. Minshall, T. (2014). Developing infrastructure to support open innovation: Case studies from the East of England. *International Journal of Innovation and Technology Management*, 11 (1), 1440006-1 – 1440006-21.
15. Ndou, V. (2010). Open innovation networks: The role of innovative marketplaces for SME value creation. *International Journal of Innovation and Technology Management, 8(3),* 437 – 453.
16. Ratten, V. (2019). The effect of cybercrime on open innovation policies in technology firms. *Information Technology & People*, 32 (5), 1301–1317.
17. Kurniawati, A. (2022). Sustainability-oriented open innovation: A small and medium-sized enterprises perspective. *Journal of Open Innovation: Technology, Market, and Complexity*, 8 (2), 69.
18. Ahn, J. M., Ju, Y., Moon, T. H., Minshall, T., Probert, D., Sohn, S. Y. and Mortara, L. (2016). "Beyond absorptive capacity in open innovation process: The relationships between openness, capacities and firm performance. *Technology Analysis and Strategic Management*, 28 (9), 1009–1028.
19. Lippolis, S. (2023, November). Open innovation for sustainable transition: The case of ENEL "open power. *Business Strategy and the Environment*, 32 (7), 4202–4216.

20 Emerging Role of Digital Library in Evaluating Copyright Regulations

Priyanka Tripathi, Preetika Tripathi, and Mohd. Noman Mansoori

20.1 INTRODUCTION

The landscape of copyright regulation has undergone a dramatic change as a result of the rise of digital libraries. The influence of digital libraries on copyright regulation has increased more than ever in the current digital era, as knowledge and information are more readily available online. Digital libraries are now a potent force in the fight for a fair handling of copyright. They act as knowledge repositories, providing a substantial selection of digitized books, research papers, old documents, and numerous multimedia materials. This abundance of digital content has spurred crucial discussions about copyright reform and the necessity for intellectual property rights [1]. Digital libraries have to maintain a balance between the protection of intellectual property rights and achieving universal access to knowledge. Copyright laws are very important to writers and might impede the free flow of ideas. More easygoing copyright regulations are supported by digital libraries because they facilitate the digitization and distribution of works, which benefits both the general public and writers. Additionally, they support fair use, open access, and Creative Commons licenses to guarantee that research and instructional materials are freely accessible to a larger audience. This is in line with India's mission to promote creativity, learning, and research to increase information accessibility and inclusivity. Additionally, digital libraries contribute to the cause of cultural heritage archiving and preservation. They endeavor to digitize and make accessible historical records and cultural objects, and they frequently advocate for exceptions to copyright laws that permit the preservation of priceless but deteriorating works [2].

20.2 DIGITAL LIBRARY

A digital library is an online collection of organized, easily available digital resources, such as books, documents, films, and photos. It functions as a virtual repository, giving users access to a huge selection of data and media from their desktops or portable devices. Digital libraries reduce the need for physical visits to traditional libraries and increase accessibility and convenience. They frequently have searchable categories, interactive elements, and other features. With the effective storage, retrieval, and dissemination of knowledge made possible by the digital transformation of libraries, more individuals around the world are now able to

DOI: 10.1201/9781003409571-20

access information in easier and more convenient ways. Digital libraries have a brief history, but it is one that is rapidly changing. These libraries act as online storage facilities for numerous types of digital content, such as publications, multimedia, and other types of knowledge. Here is a brief summary of significant turning points in the development of digital libraries.

20.2.1 Early Digitalization Efforts, 1940s–1970s

Digital libraries originated in the 1940s as individuals and organizations began converting documents into digital versions. Ted Nelson experimented with electronic information retrieval in the 1960s with Project Xanadu. The first electronic databases were developed in the 1970s, which paved the way for the creation of digital library systems.

20.2.2 Emergence of Digital Library Projects, 1980s–1990s

Digital libraries gained popularity in the 1980s, particularly in educational and research settings. Digital content was organized with the creation of standard formats such as SGML (**Standard Generalized Markup Language**). The National Center for Supercomputing Applications (NCSA) released Mosaic, the first web browser, in 1993, and it was a major factor in increasing accessibility to digital libraries and the World Wide Web.

20.2.3 Expansion and Standardization in the Late 1990s and Early 2000s

In the late 1990s and early 2000s, there was a notable surge in the programs and collections offered by digital libraries. Interoperability and resource sharing were made easier by standardized efforts, such as the creation of metadata standards like Dublin Core and protocols like OAI-PMH (Open Archives Initiative Protocol for Metadata Harvesting). During this time, notable projects like the Digital Public Library of America (DPLA) were already progressing.

20.2.4 Improvements and Diverse Content from the Mid-2000s to the Present

Digital libraries are growing in the academic, cultural, and governmental domains in the twenty-first century. Digitizing historical documents, artwork, photographs, and audiovisual resources is highly prioritized in order to establish collections that are available to all people worldwide. Copyright and accessibility issues surfaced with the launch of Google Books in 2004. Projects such as the Internet Archive have been essential in advancing open access to a wide range of digital content.

20.2.5 Current Issues and Innovations

Digital libraries are still facing difficulties with copyright, sustainability, and preservation. Technological developments in artificial intelligence (AI) (AI) and machine learning, in particular, improve these libraries' search and discovery features. It is now more important than ever to guarantee the continued usage of digital materials, which has led to a greater focus on digital preservation strategies. The notion of digital longevity is becoming increasingly prevalent as institutions endeavor to effectively arrange and maintain vast digital archives. Global changes in data storage, access, and sharing patterns, as well as continuous technology advancements and the need for increased accessibility, all influence how digital libraries develop.

20.3 RESEARCH PUBLICATION AND RESEARCH ETHICS

20.3.1 Research Publication

Research publication is the process of disseminating research study findings and outcomes to the general scientific community and the general public. Writing a research paper or article outlining the research's methodologies, findings, analysis, and conclusions is usually required. The publishing may appear in academic journals, conference proceedings, books, or online platforms, among other formats. Dissemination of knowledge, adding to the body of knowledge, and enabling other researchers to examine, repeat, or expand upon the work are the main goals of research publication.

20.3.2 Research Ethics

The moral precepts, regulations, and standards that direct the conduct of research operations are referred to as research ethics. It entails making sure that research is conducted ethically and responsibly, taking into account the rights and welfare of every person involved in or impacted by the research. This covers the safeguarding of human and animal test subjects, the preservation of data integrity, the openness with which results are reported, and the avoidance of conflicts of interest. Respecting research ethics guarantees that research benefits society without doing any harm and helps to preserve the confidence and integrity of the scientific community.

20.4 COPYRIGHT RULES AND REGULATION

The laws governing copyright are meant to secure the artists' intellectual property rights and protect their original works, whether they take the shape of literature, music, artwork, or other forms of creative expression in copyright. In addition to giving artists the sole right to reproduce, distribute, exhibit, and perform their works, copyright also permits restricted uses that fall under the "fair use" concept. These laws, which might differ from nation to nation, seek to strike a balance between fostering innovation and creativity and upholding the rights of content creators. Understanding and following copyright regulations is crucial for creators, users, and organizations to successfully negotiate the complicated world of intellectual property, whether in the analog or digital sphere [3].

20.4.1 Digital Library Copyright Compliance

E-books, movies, and online articles are all subject to traditional copyright regulations, which digital libraries must abide by. These frameworks govern the usage and distribution of digital content. This entails integrating public domain and open access content, as well as following fair use guidelines and getting the necessary authorizations or licenses. Protecting copyrighted content requires the use of digital rights management (DRM) technologies. In order to stay up to speed with the constantly changing copyright laws, users need to stay informed and compliant. To this end, organizations like Digital Libraries Frequently (DLF) provide tools and information.

20.4.2 Important Facets of Copyright Policies for Digital Libraries

Acquiring licenses, navigating the complexities of fair use, and incorporating public domain or open access materials are all necessary to ensure respect for copyrighted works in the digital sphere. For protection and control, DRM technology is required. It is essential to follow international copyright treaties, give credit where credit is due for works protected by copyright, and update regularly to stay up to speed with new laws. One important function of DLF is to give users the knowledge and resources they need to comprehend copyright laws and follow them when using digital resources.

20.5 CHALLENGES IN COPYRIGHT SERVICES

Copyright services face a variety of challenges in the digital age. These include preventing digital piracy, navigating the complexities of fair use and Creative Commons, managing inconsistent global enforcement, dealing with DRM challenges, monitoring user-generated content, keeping up with technological advancements, maintaining a balance between access and conservation, dealing with privacy concerns, managing legal costs, and ensuring equitable distribution. In order to effectively address these concerns and protect the rights of content producers in the digital age, cooperation between copyright services, lawmakers, and the creative industry is required. The digital age and the constantly changing legal landscape have created numerous obstacles in the field of copyright services [4]. These difficulties can be divided into several important categories:

- **Digital Piracy:** Enforcing intellectual property rights has become increasingly difficult due to the widespread unauthorized copying and sharing of copyrighted materials, such as music, movies, and books, made possible by the digital age.
- **Fair Use Management and Creative Commons:** Managing "fair use" and Creative Commons licenses while maintaining the rights of writers and the public makes copyright protection more difficult to define in the present generation.
- **Global Implementation Challenges:** It is challenging to protect copyrighted works globally due to differences in copyright laws and enforcement procedures throughout states, underscoring the necessity of international cooperation.

- **Orphan Works:** A large number of copyrighted works have ambiguous ownership, which makes it difficult to handle licensing and restricts access to items of significant cultural value.
- **DRM Restrictions:** DRM technology can be unduly restrictive, which can cause compatibility problems and restrict user access even though its goal is content protection.
- **Difficult Licensing Processes**: Copyrighted content licensing entails complex negotiations, contracts, and royalties that present difficulties for smaller and startup content creators.
- **Changes in Revenue Patterns:** The digital era has disrupted traditional revenue patterns for content creators, raising questions about the best ways to commercialize creative works. In order to develop workable solutions that protect artists' rights, encourage innovation, and guarantee the ethical use of copyrighted information in the digital age, cooperation between copyright services, legislators, and the creative industry is imperative.

20.6 ROLE OF RESEARCH PAPERS IN DIGITAL LIBRARIES METHODOLOGY DETAILS

Research papers are essential elements of digital libraries that provide priceless knowledge resources to a global audience. These libraries house a wide variety of intellectual materials that are useful to academics, students, and the general public. The function of research papers in digital libraries is described in full below, along with the research methodology used and an illustration of a table or graphic that would appear in how e-service excellence and e-trust affect e-loyalty among digital library users in the digital economy, finding that e-service quality positively predicts e-trust, which in turn predicts e-loyalty, with e-trust mediating the relationship between e-service quality and e-loyalty [5].

- **Review of the Literature:** The research process should begin with a thorough analysis of the body of literature that has been written about digital libraries, taking into account their historical context, significance, and the function that research articles play in these repositories.
- **Accessibility and Storage:** Research publications are electronically stored in digital libraries, allowing for global access. Users can identify and obtain research articles without being restricted by physical libraries because of this accessibility that cuts across geographical boundaries [6].
- **Finding and Discovery:** To find research papers, users can use keywords, authors, titles, and metadata. Locating pertinent research papers is made easier with the help of search engines and categorization systems.
- **Reference and Citation:** Scholars locate sources for their work in digital libraries. The variety of research papers allows them to mention relevant literature, which is essential for supporting their study.
- **Cooperation and Sharing:** By allowing scholars to share their work, digital libraries foster collaboration. In order to share their work with colleagues and the larger academic community, researchers can upload their articles to these libraries.

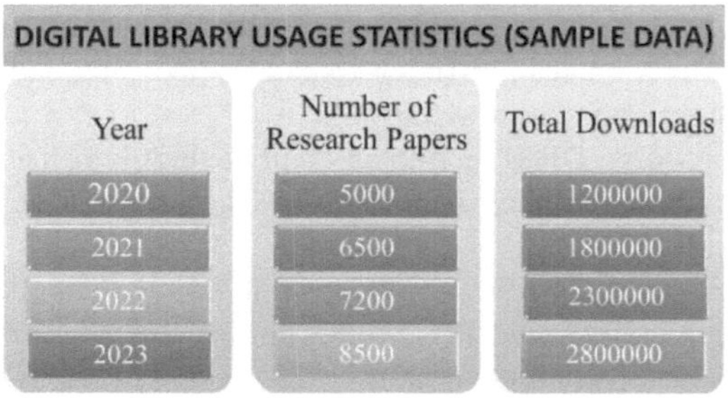
DIGITAL LIBRARY USAGE STATISTICS (SAMPLE DATA)

Year	Number of Research Papers	Total Downloads
2020	5000	1200000
2021	6500	1800000
2022	7200	2300000
2023	8500	2800000

FIGURE 20.1 Digital library usage statistics (sample data).

- **Storage and Preservation:** Academic journals, conference proceedings, and research articles are frequently stored in digital libraries. This protection guarantees that intellectual contributions will be preserved for future generations.
- **Cross-Disciplinary Study:** Digital libraries act as centers for the convergence of research from several academic fields. This encourages the discovery of novel ideas and facilitates multidisciplinary study.
- **Table or Diagram Illustration:** Here is an example table that might be used to illustrate a book chapter on how digital libraries affect the dissemination of research, as shown in Figure 20.1.

The table below illustrates a digital library's rise in research papers and downloads over a four-year period. These charts and info-graphics efficiently communicate the major patterns and ideas in your book chapter. The digital library research approach comprises multiple critical stages. It starts with an extensive examination of the literature, examining the historical background and significance of digital libraries while emphasizing the function of research articles. After that, data are gathered by combining information from digital libraries and other sources, such as user surveys, usage statistics, case studies, and the metadata of scholarly articles. Following collection, the data are examined for patterns and trends, including user preferences and the effect of digital libraries on the spread of information. Case studies on particular digital libraries offer insightful information on effective tactics and difficulties. Understanding user experiences and viewpoints is further enhanced via user surveys and interviews. With the use of both qualitative and quantitative methods, this approach guarantees a comprehensive understanding of the topic.

20.7 WHY COPYRIGHT ISSUE IS MEASURING CHALLENGES IN RESEARCH FIELD: PARAMETERS OF RESEARCH ETHICS

When using copyrighted resources for research and publication, copyright issues can present serious difficulties in the field of academic copyright challenges. Here is an illustration of how these difficulties might be assessed and displayed using a table or diagram (Table 20.1).

TABLE 20.1
Research Challenges Affecting Copyright

Difficulties	Overview	Data/Statistics
Utilization of copyrighted materials	Accessing books, papers, or other resources that are protected by copyright may be challenging for researchers.	Percentage of researchers who reported access problems.
Fair use restriction	When using copyrighted resources in their research, researchers are required to follow fair use principles.	Percent of researchers who change their studies as a result of fair use issues.
Restrictions on publication	When publishing research with copyrighted material, researchers could face limitations. A number of copyright infringement complaints and takedown requests.	Percentage of published research that is subject to copyright limitations.
Costs of licensing	Licensing fees are frequently required for researchers to use copyrighted content allocation of copyright licenses to research funding.	Costs of licensing on a yearly average for research institutions.
Plagiarism concerns	When citing copyrighted content, researchers must avoid accidental plagiarism, plagiarism that has been found in research publications.	The proportion of researchers that are concerned about accidental plagiarism.

Please take note that the information in the Table 20.1 and chart (Figure 20.2) is basically meant to be illustrative. The exact data would differ based on the geography, the type of research being done, and the particular research field. To evaluate copyright concerns in their particular context, researchers may carry out

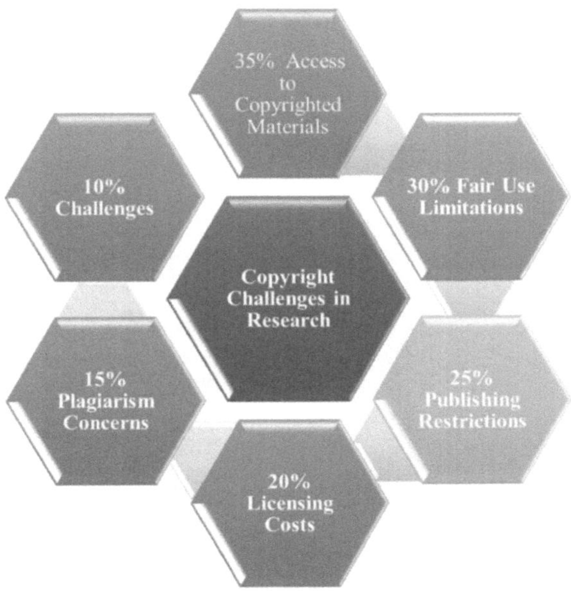

FIGURE 20.2 Research challenges affecting copyright.

surveys or gather pertinent data [6]. Research ethics are the fundamental rules and concepts that direct the conduct of studies involving people, animals, or private information. Before including participants in the study, researchers must seek their informed consent. Participants must voluntarily agree to participate in the study and be fully informed of its goal, methods, risks, and rewards. For instance, before taking part in a clinical study for a new treatment, participants receive comprehensive information on the medication's potential negative effects and are required to sign a consent form [7].

Regarding researchers to be morally accountable, safe, and objective, research ethics are essential. It is imperative to maintain security and secrecy, assuring participants that their identity and personal data will be kept secret. Researchers must refrain from inflicting injury, be it physical or psychological, and endeavor to mitigate any possible adverse consequences while optimizing favorable outcomes. In medical research, where new medicines should be more beneficial than current ones, goodwill is especially important. To be considered scientifically true, researchers must do their job with integrity and accurately explain their findings without fabrication or falsification. Research ethics are ensured by treating all volunteers fairly and equally, regardless of their age, gender, or color. Below is an easy-to-understand diagram outlining the essential elements of research ethics (Figure 20.3).

In actuality, research ethics make sure that researchers put the welfare and rights of their subjects first, uphold the caliber and integrity of their work, and maintain a high level of ethical behavior. These criteria are essential in all study areas, including the social sciences, psychology, and other medical and scientific disciplines. The unethical practice of using someone else's thoughts, words, or works without giving them due credit or securing their permission and presenting them as your own is known as plagiarism. This dishonest behavior manifests in a variety of ways, as mentioned below.

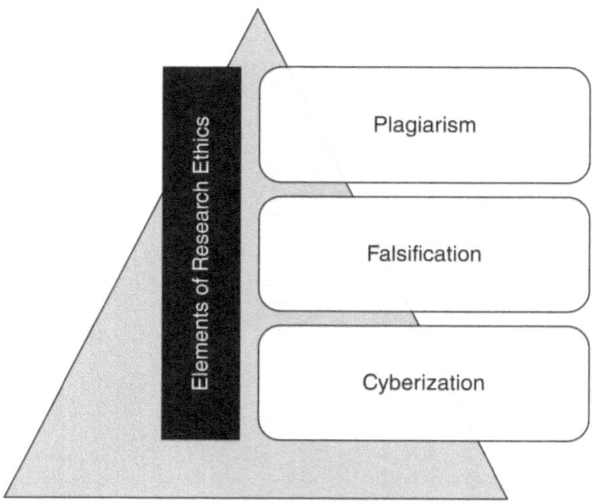

FIGURE 20.3 Elements of research.

20.7.1 Plagiarism

Plagiarism is the practice of using another person's thoughts, words, or creative expressions without giving them fair credit or asking their permission while passing it off as one's own. This behavior is generally considered to be unethical and to violate the standards of academic and professional honesty. Plagiarism is the act of using someone else's words, ideas, or artistic works without giving proper credit or gaining permission and then passing them off as one's own. In scholarly and artistic circles, this unethical action is well-acknowledged. There are many misunderstandings regarding the nature and gravity of plagiarism, making it a complex issue that has to be reevaluated. The authors propose that students' research skills and ethical awareness be improved by means of digital settings and explicit training [8]. It is essential to provide acknowledgment to the original author or source if you borrow their concepts, language, or works in order to avoid plagiarism. Your choice of citation and citing styles, such as APA, MLA, or Chicago, will be determined by the style manual that applies to your academic or professional field. It's important to note that companies, publishing houses, and academic institutions frequently impose rigorous regulations and consequences against plagiarism [9]. The components of plagiarism are shown in Figure 20.4.

When you recreate someone else's work in your own words without giving due credit to the original author even when that author's work significantly influences your content you are engaging in paraphrasing plagiarism (Table 20.2).

Plagiarism is strongly discouraged in academic and professional settings due to severe consequences such as academic sanctions, legal repercussions, damage to one's reputation, and a decline in support from mentors and peers. Properly crediting and referencing sources using the appropriate citation style is crucial to avoid plagiarism. Plagiarism and copyright infringement occur when individuals use copyrighted text without giving credit, with plagiarism involving the

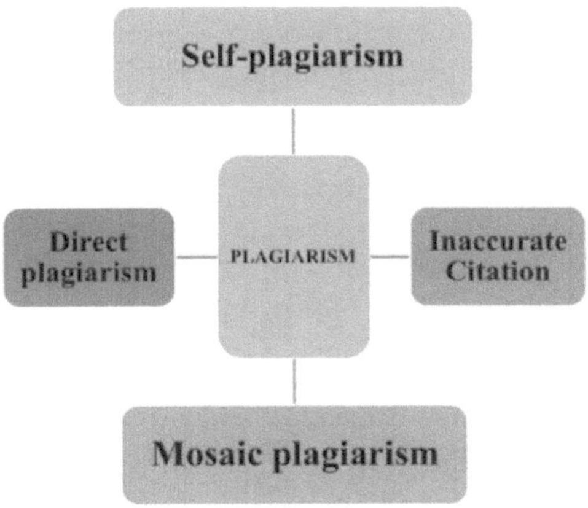

FIGURE 20.4 Components of plagiarism.

TABLE 20.2
Description of Plagiarism

Type of Plagiarism	Definition	Illustrative Example
Direct plagiarism	Replicating someone else's work word-for-word	Copying and pasting a paragraph from a source without proper citation
Self-plagiarism	Reusing your own prior work without appropriate citation	Submitting a paper previously turned in for another class without authorization
Paraphrasing plagiarism	Rewriting someone else's work with minor alterations	Restating the content from a source in your own words without adequate citation
Mosaic plagiarism	Integrating content from multiple sources without proper attribution	Blending information from several sources without giving credit where it's due
Incomplete citation	Neglecting to cite sources correctly	Mentioning a source but omitting essential citation details, such as page numbers
Collusion plagiarism	Collaborative copying with another's knowledge	Sharing a classmate's assignment with consent leads to both submitting similar work

additional ethical violation of not citing the source. These acts compromise academic and creative integrity, emphasizing the importance of ethical writing practices. When using another person's intellectual property, it is imperative to give due credit and, if required, get permission in order to prevent moral and legal issues. The copyright rules that are relevant to your particular location or jurisdiction must also be understood [10].

20.7.2 Falsification

Falsification is a key idea in the philosophy of science that was made popular by Karl Popper. It entails the thorough examination and potential refutation of scientific theories and hypotheses (Table 20.3).

According to Popper, a theory or hypothesis can only be considered scientific if it can be refuted by empirical experiments. A scientific theory should produce precise and substantiable predictions about the physical cosmos, according to the falsification principle. To examine these hypotheses, scientists do observations or experiments; if the outcomes differ from the predictions, the hypothesis is either insufficient or falsified. When a hypothesis regularly withstands attempts at falsification and is accepted by the scientific community, it gains credibility. A key idea in the philosophy of science is falsification, which is used to distinguish between concepts that are scientific and thoughts that are not. Non-scientific or pseudoscientific beliefs frequently cannot be thoroughly verified or refuted using empirical methods, in contrast to falsifiability. Discussions about the demarcation problem, which entails separating true science from non-scientific or pseudoscientific ideas,

TABLE 20.3
Description of Falsification

Type of Falsification	Description	Example
Data falsification	The manipulation of research data with the intent to achieve specific outcomes.	Altering data points or selectively excluding data to present study results in a more favorable light.
Credentials falsification	Providing untrue information about one's qualifications.	Falsely claiming to possess a degree or certification on a job application when one does not have it.
Resume falsification	Exaggerating or misrepresenting work experience.	Inflating job responsibilities or tenure on a resume to create a more impressive professional history.
Plagiarism falsification	Presenting someone else's work as your own.	Copying and submitting a colleague's project or using plagiarized content for academic assignments.
Witness falsification	Offering false or deceptive testimony.	Providing untruthful statements in court or misleading authorities through inaccurate information.
Document falsification	Forging or altering official documents.	Modifying dates, signatures, or content on documents such as contracts or academic transcripts to deceive or gain an advantage.

have been greatly affected by Popper's contributions to the idea of falsifiability [11]. Falsification largely focuses on the testing and potential refutation of scientific ideas and theories, whereas copyright law concentrates on protecting intellectual property and original creative works [12]. Understanding key terms like copyright infringement, which occurs when someone uses, reproduces, distributes, or profits from a copyrighted work without the proper authorization, and plagiarism, which is the unethical act of using someone else's work without giving them due credit, is essential in the field of copyright law. The philosophical idea of falsification is distinct from these legal and ethical ideas.

20.7.3 Cyberization

The term "cyberization" has several different meanings depending on the context in which it is used. In general, it has to do with how our lives are becoming more and more integrated with and dependent upon digital technologies, the internet, and cyberspace. The term "cyberization" refers to the widespread use of and reliance on digital technology, particularly the internet and computer systems. It includes the shift toward a society that is increasingly digitally focused, linked, and technologically advanced in all areas of business, personal life, and society. This phrase comes up regularly in discussions about the expanding digital transformation and reliance

TABLE 20.4
Description of Cyberization

Cyberization Type	Description	Example
Digital transformation	The integration of digital technologies into various aspects of an organization.	An established retailer shifting to e-commerce and online marketing to reach a wider audience.
Augmentation	Enhancing human abilities through cybernetic or digital enhancements.	Implanting a microchip in a person's hand to enable contactless payment and access control.
Cybersecurity	The protection of computer systems and networks from cyber threats.	Implementing firewalls, encryption, and security protocols to safeguard an organization's data.
Internet of Things (IoT)	The interconnection of everyday objects through the internet.	Using smart thermostats to remotely control and monitor home heating and cooling systems.
Data digitization	Converting analog data into digital formats.	Scanning and digitizing paper records for efficient storage and retrieval.
Virtual reality (VR)	Immersive experiences in computer-simulated environments.	Using VR headsets to explore virtual worlds and environments for entertainment or training.

on technology across various industries [13]. The detailed description of cyberization is given in Table 20.4.

The term "cyberization" has no fixed meaning or general recognition in the field of copyright law. Particularly in the digital age, copyright law is largely concerned with protecting the rights to original creative works, such as software, music, art, and literature. The creation, dissemination, and accessibility of copyrighted work have all been profoundly altered by the internet and digital technology, although the term "cyberization" has no exact legal or academic definition in this context. DRM, dealing with internet piracy, managing license agreements, controlling fair use, and prosecuting copyright infringement actions are some of the major issues in copyright law. In the digital age, these elements are essential to protecting intellectual property. It is advised to use the established legal structure and jargon that governs copyright law when addressing any specific queries or worries you may have regarding copyright and its ramifications in the digital sphere [14].

20.8 ROLE OF INFORMATION AND COMMUNICATION IN DIGITAL LIBRARIES

Digital libraries place a high value on information and communication, which has a significant impact on how users engage with and use information resources. They play important roles in digital libraries in a number of key ways, including the following:

- **Facilitating Information Retrieval:** Digital libraries act as access points to a sizable digital informational repository. Users make use of metadata, search engines, and information systems to quickly find and retrieve specified resources. Users' ability to discover the information they need quickly depends on processes for information access and retrieval that are effective.

- **Information Dissemination:** To make information available to their users, digital libraries use a variety of communication channels. These channels include social media sites, email alerts, and web interfaces. The main objective is to inform patrons about the library's offerings, services, and developments.
- **User Engagement:** Digital libraries have communication capabilities that enable users to interact with the system, librarians, and other users. These resources could include feedback methods, discussion forums, and chat services. In order to provide assistance, respond to inquiries, and improve the user experience overall, user participation is essential.
- **Organization and Metadata:** The organization and retrieval of content rely heavily on information about digital resources, often supplied in the form of metadata. Important information about resources is provided via metadata, which includes titles, authors, topics, and publication dates. The foundation for organizing and discovering information is effective metadata and cataloging.
- **Remote Access and Mobility:** Through the internet, digital libraries expand users' access to information. Users can access digital library materials from a variety of devices, including smartphones and tablets, thanks to their usage of mobile technologies and responsive design, ensuring smooth communication and access.

In the context of digital libraries, information and communication are crucial, especially when it comes to maintaining and enforcing copyright. Digital libraries are digital repositories for many kinds of content, such as publications, photographs, and multimedia. The following are some examples (Figure 20.5) of how information and communication play a role in this situation.

In conclusion, communication and information are crucial for managing copyright in digital libraries. Digital libraries can make sure that copyrighted content is accessible in a way that is legally compliant and respects intellectual property rights by managing information effectively, controlling access, interacting with rights holders, educating users about copyright, and putting DRM technologies to use [15].

FIGURE 20.5 Roles of information and communication in digital libraries.

20.9 PLAGIARISM SOFTWARE TO CHECK DUPLICATION

If you want to check for duplicate content in written work, there are many different types of plagiarism detection tools accessible. Students, teachers, writers, and professionals frequently utilize these tools to guarantee the originality of their work. The following is a list of popular solutions for plagiarism detection software (Figure 20.6).

- **Turnitin:** Frequently used in educational institutions, Turnitin compares academic papers and essays to a sizable database of scholarly sources and websites in order to detect plagiarism.
- **Grammarly:** Grammarly's plagiarism checker, which is part of its premium subscription, compares your content to a sizable database of online pages and academic publications.
- **Copyscape:** By inputting a URL or uploading a document, users can check for duplicate content online using the Copyscape tool. Making sure their content is original is extremely popular among website owners.
- **Plagscan:** Plagscan is a plagiarism detector that is employed by both corporations and schools. It offers thorough reports on stuff that might have been plagiarized.
- **Quetext:** Quetext is an easy-to-use online plagiarism detector with a free and commercial edition. It performs a plagiarism check on the material and assigns a similarity score.

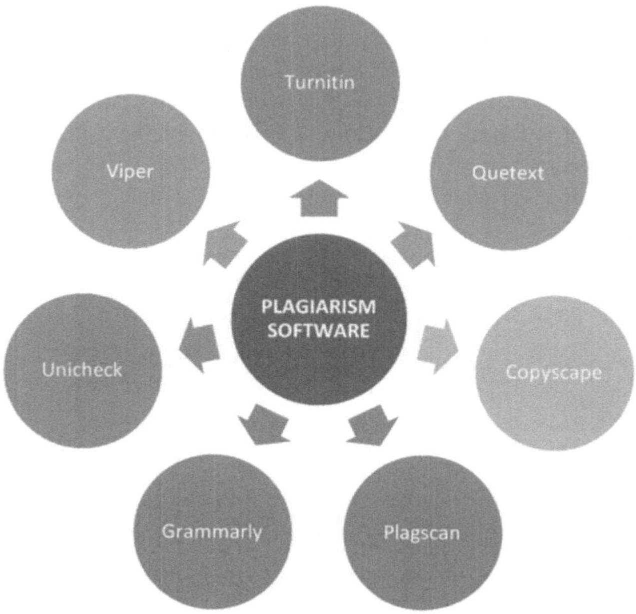

FIGURE 20.6 Plagiarism software to check duplication.

- **Unicheck:** Unicheck is a plagiarism detector made specifically for academic settings. Learning management systems like Moodle and Canvas are integrated with it.
- **Viper:** Viper is a free plagiarism detector that enables you to examine your writing for resemblance and receive a report.

It's critical to comprehend the unique capabilities, restrictions, and costs of each program before employing plagiarism detection software. Other programs could have usage restrictions or additional expenses, while some offer more thorough checks and platform integration. Furthermore, it's crucial to carefully interpret the findings because common phrases or references could be marked as matches even when they don't actually constitute plagiarism. Avoid depending only on these tools to determine whether or not something is plagiarized and only use them as tools to help you uphold your academic or professional integrity [16].

20.10 DATA-BASED STORED IN PLAGIARISM CHECK SOFTWARE

Databases are used by plagiarism detection tools to compare submitted text to published sources. These databases include a vast selection of literature with both academic and non-academic roots. Different software developers have different database structures. An overview of the many databases used by plagiarism detection software is provided below:

- **Academic Databases:** These databases, which contain academic papers, research articles, theses, and dissertations, are a common component of plagiarism detection software. To find overlaps with already published scholarly work, the content is drawn from academic journals and organizations.
- **Internet and Web Content:** Web pages and online content from the internet are indexed and stored by some software. This gives the software the ability to compare the given information with sources that are freely accessible online.
- **Documents Previously Submitted:** Educational institutions frequently employ plagiarism detection software that saves copies of papers and assignments that students have previously turned in. This is primarily done within the university to catch self-plagiarism.
- **Books and Publications:** A few plagiarism detection tools keep databases of text taken from books, periodicals, and other published works. This is crucial for identifying similarities between published and copyrighted works.
- **Institution-Specific Databases:** Institution-specific databases may be created by educational institutions and other businesses that use plagiarism detection technologies. These databases may contain student work as well as exclusive materials.
- **User-Uploaded Papers:** Some software platforms enable users to submit their own papers for inclusion in the database for upcoming comparisons. User consent is often required for this process, which frequently has institutional or educational goals.

The size and content of these databases should be taken into consideration by organizations and individuals employing plagiarism detection software since they can affect how well the program detects similarities within particular types of content. In order to avoid any potential copyright violations, users must also make sure they have the required authorizations and rights to utilize the content that is kept in these databases [17]. Turnitin, Copyscape, Grammarly, and Unicheck are little-known software for detecting plagiarism; each has its own special features and cost structures. It's crucial to assess these tools in light of your unique needs and spending capacity. To obtain the most accurate findings, combining computerized plagiarism detection with manual assessment is also advised. It's crucial to keep in mind that no technology is perfect, and human judgment is frequently required to comprehend the context and intention of potentially copied content [18].

20.11 IMPLICATION OF THE ROLE OF INFORMATION AND COMMUNICATION TECHNOLOGY TO REDUCE PLAGIARISM, FALSIFICATION, AND CYBERIZATION IN THE COPYRIGHT DOMAIN IN THE PRESENT AND FUTURE SUSTAINABLE APPROVAL

Information and communication technology (ICT) is essential to the battle against copyright infringement, including plagiarism, falsification, and cyberization because it offers both short-term remedies for the here and now and long-term strategies for the future. ICT can help with these issues in the following ways.

20.11.1 Detecting Plagiarism

Plagiarism detection and content verification rely heavily on a variety of ICT tools. Programs such as Turnitin, Copyscape, and Grammarly use large databases to find possible instances of plagiarism in content that is submitted. Sophisticated techniques are used by text analysis algorithms to examine texts for similarities and produce reports that point out possible plagiarism. Utilizing large databases and online resources as a comparison, plagiarism detection software provides an effective way to spot copied text in student papers and manuscripts. It helps educators and editors uphold academic integrity and deal with accidental mistakes or deliberate plagiarism. However, subtleties like text recycling should be taken into account when interpreting similarity rates, which makes it necessary to make an educated decision when evaluating plagiarism cases [19]. Through learning from fresh examples, AI and machine learning systems are able to adapt to evolving plagiarism strategies, hence continuously improving detection capabilities. ICT, in particular blockchain technology, also guarantees digital ledgers for protected content that are safe, unchangeable, authentic, and traceable. Authorship and originality verification in the digital sphere are further supported by methods such as digital watermarks, signatures, steganography, metadata, and time stamping.

- **Verifying Content and Preventing False (Falsification) Information**: ICT offers a number of instruments to strengthen copyrighted content protection. One important component is blockchain technology, which creates an unalterable, safe digital ledger for copyrighted content, guaranteeing authenticity and traceability. Another method uses steganography to create digital watermarks and signatures that allow data to be hidden inside digital files while still authenticating the source of the content. ICT also uses time stamping and metadata to embed pertinent information, proving authorship and preserving the integrity of original content. These developments strengthen the foundation for protecting intellectual property in the digital sphere.

- **Enforcement and Protection of Copyright:** ICT is used in the DRM approach to monitor and protect copyrighted content. These mechanisms exercise control over usage and access rights for copyrighted content through the use of ICT-enabled DRM systems. On many ICT sites, automated takedown techniques are also very important. Copyright in India protects original works of authorship, including literary, artistic, musical, and dramatic creations, granting exclusive rights such as reproduction and adaptation to the owner, with registration serving as prima facie evidence of ownership [20]. These systems make use of technology to automatically identify and remove content from websites and social media platforms that violate copyright laws. In the digital sphere, this automatic procedure protects the rights of content creators and maintains the integrity of copyright.
- **Education and Information:** Online tools and courses can be developed to teach people about copyright laws, fair use, and appropriate citation techniques by utilizing ICT. The purpose of these digital teaching resources is to decrease accidental plagiarism and increase awareness. Furthermore, ICT-based resources play a critical role in enabling students to recognize plagiarism and steer clear of it in their academic work by offering workable alternatives for upholding academic integrity. The foundation of copyright law is derived from both positive and natural law conceptions, giving rise to two different kinds of copyrights: common law and statutory. Natural law-based fair use was developed to restrict the unrestricted use of other people's creations and strengthen the rights of copyright owners. The development of copyright from its inception to contemporary interpretations shows how judicial principles and legislative actions interact [21].
- **Frameworks for the Law and Reporting:** With the use of online reporting tools for copyright violations, ICT is essential to streamlining copyright enforcement. The process of reporting and resolving copyright issues is streamlined by these user-friendly websites. ICT also makes it easier to establish digital copyright registers, which are centralized databases that are managed by organizations and governments. By effectively capturing and preserving copyright data, these registers support the general administration and protection of intellectual property in the digital sphere. Continuous ICT research and development is necessary to remain ahead of ever-evolving challenges and assure future sustainable solutions. A standardized framework for copyright protection in the digital age must be established, which calls for international cooperation and standardization efforts. Campaigns for education and awareness will also be essential in advancing moral conduct and the observance of intellectual property rights [22].

20.12 CONCLUSION

This chapter concludes by highlighting how digital libraries have a revolutionary effect on copyright dynamics. Digital libraries have come a long way, from the first attempts at digitalization to the problems and innovations of today, which emphasizes how important they are to strike a balance between the protection of intellectual property and public access to knowledge. Several individuals can access the library digitally at once in the future, which raises concerns about plagiarism. Even while copyright may appear difficult, its removal could result in even less user-friendly settings since it offers limitations and preserves that are advantageous to both providers and users. If copyright rules were to vanish, librarians might not be able to make use of their benefits [23]. Legislators, copyright services, and the creative sector must work together to resolve copyright issues. With its instruments for content verification, enforcement, education, and plagiarism detection, ICT is vital. Sustainable copyright solutions require worldwide cooperation, established frameworks, and continuous ICT research as we negotiate the complicated digital terrain.

REFERENCES

1. Sheat, K. (2004). Libraries, copyright and the global digital environment. *The Electronic Library*, 22(6), 487–491. https://doi.org/10.1108/02640470410570785
2. Handa, S., & Bhatt, K. (2015). Intellectual property rights (IPR) in digital environment: An overview in Indian digital environment. *International Journal of Digital Library Services*, 5(2), 117–123.
3. Adav, R. P. (2015). Digital libraries and intellectual property rights. In *International Conference on Digital Libraries 2010* (Vol. 7637, pp. 1177–1187).
4. Hombal, S. G., & Prasad, K. N. (2012). Digital copyright protection: Issues in the digital library environment. *DESIDOC Journal of Library & Information Technology*, 32(3), 233–239. https://www.proquest.com/openview/2348c926442a7d7fb4cc029cd695aad0/1?pq-origsite=gscholar&cbl=2028807
5. Anser, M. K. (2020). Toward the e-loyalty of digital library users: Investigating the role of e-service quality and e-trust in digital economy. *Library Hi Tech*. Advance online publication. DOI: 10.1108/LHT-07-2020-0165
6. Muir, A. (2003). Copyright and licensing issues for digital preservation and possible solutions. In *Proceedings of the 7th ICCC/IFIP International Conference on Electronic Publishing* (pp. 89–94).
7. Tomaszewski, Z. (2000). Copyright and its effect on digital libraries, 1–8. Retrieved from http://zach.tomaszewski.name/uh/copyrightpaper.html
8. Evering, L. C., & Moorman, G. (2012). Rethinking plagiarism in the digital age. *Journal of Adolescent & Adult Literacy*, 56(1), 35–44.
9. Liu, G.-Z., Lo, H.-Y., & Wang, H.-C. (2013). Design and usability testing of a learning and plagiarism avoidance tutorial system for paraphrasing and citing in English: A case study. *Computers & Education*, 69, 1–14. https://doi.org/10.1016/j.compedu.2013.06.011
10. George, T. (2023, November 21). How to avoid plagiarism | Tips on citing sources. *Scribbr.* https://www.scribbr.com/proofreading-editing/
11. Al-Adawi, S., Ali, B. H., & Al-Zakwani, I. (2016). Research misconduct: The peril of publish or perish. *Oman Medical Journal*, 31(1), 5–11. https://doi.org/10.5001/omj.2016.02. PMCID: PMC4720946. PMID: 26816563.
12. Morgan, M., Brickell, G., & Harper, B. (2008). Applying distributed cognition theory to the redesign of the 'Copy and Paste' function in order to promote appropriate learning outcomes. *Computers & Education*, 50(1), 125–147. https://doi.org/10.1016/j.compedu.2006.04.006

13. Yin, M. (2023, May). Cyberbodies in spectacular landscapes: Exploring the intersection of aesthetics, technology, and culture in cyberpunk films. *Art and Design Review*, 11(2). USC-SJTU Institute of Cultural and Creative Industry, Shanghai Jiao Tong University, Shanghai, China. https://doi.org/10.4236/adr.2023.112014
14. Lupovici, A. (2022, October 18). Ontological security, cyber technology, and states' responses. *European Journal of International Relations*, 29(1). https://doi.org/10.1177/13540661221130958
15. Abbasi, F., & Zardary, S. (2012). Digital libraries and their role in supporting E-Learning. *AWER Procedia Information Technology & Computer Science, 1*, 809–813. In *Proceedings of the 2nd World Conference on Information Technology (WCIT-2011)*
16. Bigby, G. (2023, September 11). Top 15 plagiarism tools for finding duplicate content. In *Search Engine Optimization*. Dyno Mapper. Retrieved from https://dynomapper.com/blog/21-sitemaps-and-seo/445-top-15-plagiarism-tools-for-finding-duplicate-content
17. Anson, C.. (2023). Plagiarism Detection and Intertextuality Software. In O. Kruse (Ed.), Title of the Book (pp. 231–243). Publisher. DOI: 10.1007/978-3-031-36033-6_15. Retrieved from https://www.researchgate.net/publication/373941060_Plagiarism_Detection_and_Intertextuality_Software
18. Harris, L., Harrison, D., McNally, D., & Ford, C. (2020). Academic integrity in an online culture: Do McCabe's findings hold true for online, adult learners? *Journal of Academic Ethics*, 18, 419–434.
19. Von Isenburg, M. (2019). Plagiarism detection software and its appropriate use. *Nurse Author & Editor*, 29(1), 1–9.
20. Online Copyright. (1957). Copyright Law in India. Legalserviceindia.com. Link: https://www.legalserviceindia.com/article/l195-Copyright-Law-in-India.html
21. Patterson, L. R. (1992). Understanding fair use. *Law and Contemporary Problems*, 55(2), 249–266.
22. Curtis, G. J., & Clare, J. (2017). How prevalent is contract cheating and to what extent are students repeat offenders? *Academic Ethics*, 15, 115–124. https://link.springer.com/article/10.1007/s10805-017-9278-x
23. Ginsburg, J. C. (1993). Copyright without walls. *Representations*, 42(Spring), 53.

Index

S

T

V

W

Z